Probability and Mathematical Statistics (Continued)

PURI and SEN • Nonparametric Methods in General Linear Models
PURI, VILAPLANA, and WERTZ • New Perspectives in Theoretical and Applied Statistics
RAO • Asymptotic Theory of Statistical Inference
RAO • Linear Statistical Inference and Its Applications, *Second Edition*
ROBERTSON, WRIGHT, and DYKSTRA • Order Restricted Statistical Inference
ROGERS and WILLIAMS • Diffusions, Markov Processes, and Martingales, Volume II: Îto Calculus
ROHATGI • A Introduction to Probability Theory and Mathematical Statistics
ROSS • Stochastic Processes
RUBINSTEIN • Simulation and the Monte Carlo Method
RUZSA and SZEKELY • Algebraic Probability Theory
SCHEFFE • The Analysis of Variance
SEBER • Linear Regression Analysis
SEBER • Multivariate Observations
SEBER and WILD • Nonlinear Regression
SERFLING • Approximation Theorems of Mathematical Statistics
SHORACK and WELLNER • Empirical Processes with Applications to Statistics
STAUDTE and SHEATHER • Robust Estimation and Testing
STOYANOV • Counterexamples in Probability
STYAN • The Collected Papers of T. W. Anderson: 1943–1985
WHITTAKER • Graphical Models in Applied Multivariate Statistics
YANG • The Construction Theory of Denumerable Markov Processes

Applied Probability and Statistics

ABRAHAM and LEDOLTER • Statistical Methods for Forecasting
AGRESTI • Analysis of Ordinal Categorical Data
AGRESTI • Categorical Data Analysis
ANDERSON and LOYNES • The Teaching of Practical Statistics
ANDERSON, AUQUIER, HAUCK, OAKES, VANDAELE, and WEISBERG • Statistical Methods for Comparative Studies
ASMUSSEN • Applied Probability and Queues
*BAILEY • The Elements of Stochastic Processes with Applications to the Natural Sciences
BARNETT • Interpreting Multivariate Data
BARNETT and LEWIS • Outliers in Statistical Data, *Second Edition*
BARTHOLOMEW, FORBES, and McLEAN • Statistical Techniques for Manpower Planning, *Second Edition*
BATES and WATTS • Nonlinear Regression Analysis and Its Applications
BELSLEY • Conditioning Diagnostics: Collinearity and Weak Data in Regression
BELSLEY, KUH, and WELSCH • Regression Diagnostics: Identifying Influential Data and Sources of Collinearity
BHAT • Elements of Applied Stochastic Processes, *Second Edition*
BHATTACHARYA and WAYMIRE • Stochastic Processes with Applications
BIEMER, GROVES, LYBERG, MATHIOWETZ, and SUDMAN • Measurement Errors in Surveys
BIRKES and DODGE • Alternative Methods of Regression
BLOOMFIELD • Fourier Analysis of Time Series: An Introduction
BOLLEN • Structural Equations with Latent Variables
BOULEAU • Numerical Methods for Stochastic Processes
BOX • R. A. Fisher, the Life of a Scientist
BOX and DRAPER • Empirical Model-Building and Response Surfaces
BOX and DRAPER • Evolutionary Operation: A Statistical Method for Process Improvement

Continued on back end papers

*Now available in a lower priced paperback edition in the Wiley Classics Library.

MATH
STAT.
LIBRARY

Statistical Applications Using Fuzzy Sets

Statistical Applications Using Fuzzy Sets

KENNETH G. MANTON
MAX A. WOODBURY

Center for Demographic Studies
Duke University
Durham, North Carolina

H. DENNIS TOLLEY

Department of Statistics
Brigham Young University
Provo, Utah

A Wiley-Interscience Publication
JOHN WILEY & SONS, INC.
New York • Chichester • Brisbane • Toronto • Singapore

This text is printed on acid-free paper.

Library of Congress Cataloging in Publication Data:

Manton, Kenneth G.
Statistical applications using fuzzy sets/by Kenneth G. Manton, Max A. Woodbury, H. Dennis Tolley.
p. cm. -- (Wiley series in probability and mathematical statistics. Probability and mathematical statistics)
"A Wiley-Interscience publication."
Includes bibliographical references and index.
ISBN 0-471-54561-9 (acid-free)
1. Fuzzy sets. 2. Mathematical statistics. I. Woodbury, Max A. II. Tolley, H. Dennis, 1946– . III. Title. IV. Series.
QA248.M285 1994
519.5--dc20 93-2324

Printed in the United States of America

10 9 8 7 6 5 4 3 2 1

Contents

Preface

The cross-sectional and longitudinal analysis of high dimensional, sparse, categorical data is a topic of wide interest for statisticians, biostatisticians, actuaries, and others involved in quantitative analyses. This type of data is frequently encountered in demographic, sociological, epidemiological, biostatistical, clinical, biological, and other applications.

Though a statistical topic of general interest, usual methods for dealing with high dimensional sparse, categorical data have limitations. One approach relies on distributional approximations to allow categorical data to be treated as if generated from an intrinsically continuous latent multivariate distribution with the discreteness of the measures being an artifact of the "coarseness" of instrumentation. A second approach considers discretely coded variables as representing "events" so the statistical models used deal directly with discrete probability distributions and their added complexity (i.e., the need to deal with higher order and correlated moments).

The first approach (e.g., factor analysis of tetrachloric correlation) is often used in the analysis of high dimensional problems with large samples. However, even in very large samples, the phenomena may be both discrete and relatively rare (conditions generally increasing the importance of high order moments) so that approximating distribution assumptions to restrict or "smooth" the number of moments to be analyzed may not well approximate the data. In the case of factor analysis, solutions can be produced that do not fulfill even basic mathematical assumptions of the model; e.g., the tetrachloric correlation matrix input to the factor analysis may not be Grammian, leading to improper, mathematically undefined solutions. The second approach (i.e., using exact methods for discrete distributions) can become computationally refractory except for problems of limited dimensionality, i.e., with few cases and variables.

In this book we present a third approach using a different strategy to resolve the measure theoretic issues involved in analyzing sparse, high dimensional discrete data. These approaches, in effect, assume that *all* data, no matter how finely measured, are *finite*. Thus, the data realized for any example finitely (and necessarily sparsely) populate a continuous general, theoretical response space, even though the values associated with individual data points might be partitioned by further data collection to any desired, but finite, degree of precision to increase density strategically in portions of the latent continuous response space. Thus, both "continuous" and "discrete" response *data* are, in fact, necessarily finite in dimension and never continuous; i.e., they can never completely "cover" the theoretical response space with its infinite number of potential data points.

To take advantage of the intrinsic finiteness of data, statistical models based on convex sets are used here to generate statistical models using "fuzzy set" principles. In fuzzy set models, the continuous nature of data is generated by the fact that statistical partitions generated by finite data only approximate real, continuous phenomena. Thus, it is assumed that the "fuzziness" in the real world can be explicitly represented by specific model parameters; i.e., "grade of membership" is represented in the use of finite data to describe an infinite dimensional, continuously varying world. Such fuzzy set or fuzzy partition models, developed initially in electrical engineering, are increasingly being used in other physical science (e.g., computer science) areas. They have mathematical and statistical properties that are less restrictive than assuming that true state "fuzziness" is due to the projection of truly discrete phenomena onto a probabilistic measure, i.e., assuming fuzziness in uncertainty in repeated trials instead of true event variations. Thus, a fundamental conceptual distinction has to be made among probability measures that suggest that a proportion p of trials will produce a specific identical outcome, and that each outcome is unique and can only be related using a grade of membership to other cases which are partially similar in a theoretical, infinite dimensional space.

One problem with fuzzy state models is that their theoretical nature has not often been translated appropriately into the necessary statistical tools so that concepts were represented by parameters that are inappropriately estimated. Thus, there is a mismatch of fundamental model concepts and the estimation strategy, which means that the analytic power of fuzzy sets or partitions is not fully realized.

In this book we present statistical techniques to bridge this gap; i.e., statistical procedures that directly reflect fuzzy set principles in the estimation of parameters of a model are presented. To make the translation, we use well known theorems regarding convex sets, polytopes, and the duality of spaces (e.g., Weyl, 1949; Karlin and Shapley, 1952). What we have done is to embed those principles in the formulation of likelihood estimation principles.

The ability to exploit such convex set models has been restricted, until recently, by computational power. However, recent advances in computational power, especially that available in scientific workstations that can be dedicated to resolution of complex problems for relatively long periods of time, now make the direct implementation of fuzzy set statistical procedures to even quite large problems practical.

To demonstrate this, we perform three tasks in this book. One is to carefully define notation and concepts for fuzzy set statistical models to avoid confusion on fundamental principles. For example, there is a general tendency to try to interpret the fuzzy set grade of membership parameters as a type of probability measure when it, in its convex space representation, is *not* a probabilistic entity but a mathematical distance measure in a specific type of topological space. After having carefully defined notation, we present the geometric and mathematical basis for the fuzzy set models and their corresponding estimation procedures. Certain standard statistical concerns, such as identifiability, consistency, sufficiency, and efficiency, must be treated carefully for, though applicable to fuzzy set models, they involve unique arguments and proofs where the dual nature of data representation (i.e., cases and variables) are fully preserved and exploited in the bilinear, convex forms of the fuzzy set models. Finally, we illustrate these modeling and estimation methods with empirical applications from a number of different substantive areas showing the behavior of the models and procedures in problems with quite different characteristics and with data generated from a wide range of observational plans. We not only apply these procedures to statistical analyses but also to the representation of complex stochastic processes in order to deal with discrete time series and longitudinal data.

Statistical Applications Using Fuzzy Sets

CHAPTER 1

Crisp and Fuzzy Sets in Statistics

1.1. PURPOSE

In this book we present methods for modeling high dimensional discrete response data using the theory of "fuzzy" sets and partitions. Most statistical methods are based on the assumption that sets of persons or objects are organized into "crisp" sets. That is, an observation is either from a set identified with specific characteristics or parameter values, or it is not. Estimation of the characteristics of the set, in the face of uncertainty in the observation process, is the task of the statistician or data analyst. When we do not make the assumption that objects have "crisp" (i.e., "all" or "none") membership in a set, but rather permit "fuzzy" set membership (i.e., a person may have "degrees" of membership in multiple sets), it is possible to use more general families of probability models to model observations. Because classical crisp set theoretic assumptions can be viewed as special cases of assumptions used in fuzzy set theory, many standard statistical methods currently in use can also be shown to be special cases of statistical methods based on fuzzy sets—or, conversely, that fuzzy set methods generalize standard statistical models based on crisp set logic in order to better deal with additional and more complex sources of heterogeneity in the data.

The basic fuzzy set theory and assumptions that we use in generalizing statistical models is outlined in Chapter 2. However, for the most part, this book describes and illustrates the use of fuzzy "partition" statistical methods, based on fuzzy set mathematical principles, as a means of analyzing high dimensional data with categorical responses. Statistical procedures such as estimation, testing of hypotheses, and goodness of fit evaluation will be based on maximum likelihood principles adapted to fuzzy set mathematics and logic (e.g., Kaufmann and Gupta, 1985). These methods are more computationally demanding than those for most standard statistical applications because of the need to deal with parameters for individuals. Other possible estimation techniques such as least squares, minimum information, and minimum distance methods are not considered in this book.

1.2. DATA CHARACTERISTICS

We assume that the data are provided as J discrete response variables with only *one* of finite number (say L_j) of response categories for the jth variable occurring or that continuous variables can be recoded into discretely defined intervals to generate a categorical variable without loss of significant information. For variables of an intrinsically discrete nature (e.g., sex, current marital status) this coding is direct. In this case we may view the data as consisting of either J multinomial variables x_{ij} with L_j response levels for the jth variable or, equivalently, L_j binary variables y_{ijl} defined for the jth variable (i.e., a total of $\Sigma_{j=1}^{J} L_j$ binary variables).

To use continuous variables the analytic and computational effort will be greater because *cutpoints* that form interval categories representing all "meaningful" information in the continuous distribution of outcomes on the original variable have to be selected. Fortunately, it can be shown that, for most standard types of continuous random variables, the number of categories L_j needed to represent a given variable j seldom needs to be larger than 15 (Terrell and Scott, 1985). In many cases 5 to 10 interval categories will satisfactorily approximate the information contained in a continuous variable. In practice this additional recoding effort has the potential advantages that (a) often continuous variable data is actually reported "clumped" (e.g., blood pressure readings rounded to 0 and 5; income reported in categories) so that the discrete categories "filter" or "smooth" extraneous variation from measurement processes from the distribution and (b) without the necessity of using nonlinear scale transformations, the effects of outlying observations, or distributional anomalies (e.g., bimodal distributions) are reduced.

1.3. CONCEPTS OF FUZZY SETS

1.3.1. Discrete versus Fuzzy Sets

The concept of a "set" is familiar in mathematics. Elementary courses in mathematics define a *set* as a collection of objects; with each object in the collection defined to be an *element* of the set. Much of mathematics, and consequently much of the theoretical underpinnings of statistics, is determined by the additional constraints, assumptions, and characteristics we define for crisp sets in models developed to analyze data generated from specific observational plans.

During the last 25 or more years, the concept of a fuzzy set has been developed and implemented in solving many engineering problems (e.g., Zadeh, 1965). In contrast to fuzzy sets, the classical type of set is referred to in this book as a *crisp set*. The major difference between fuzzy and crisp sets is in the essential nature of the membership of individual elements in sets. For example, membership for crisp sets requires that individual elements either be a member of a set—or not. Two sets are disjoint if they have no elements in common. In fuzzy sets, an element can be a *partial* member of multiple sets. As a consequence, the membership of any particular element relative to a collection of fuzzy set is not crisp but "fuzzy." The concept of disjoint or crisply defined set boundaries or partitions is not directly applicable in fuzzy set methods except as a special extreme case. More formally,

Definition 1.1. For each element in a fuzzy set there is a *Grade of Membership* score (denoted as g_{ik}) that represents the degree to which the element i belongs to the kth set. These scores are assumed to vary between 0 and 1 inclusive. A 0 indicates that the element has no membership in the set, and a 1 indicates that the element has complete, or crisp, membership in the set.

If the Grade of Membership score for every element in a set is either 1 or 0, the the set is crisp. Otherwise the set is fuzzy. Thus, the presence of Grade of Membership scores that vary continuously between 0 and 1 implicitly defines a fuzzy set. The degree to which elements are members of fuzzy sets is represented by the Grade of Membership scores. The nature of fuzzy set statistical methods are determined by the definition of the properties of these scores. Grade of Membership scores are essential for fuzzy set methods since they represent state parameters for elements and must be *explicitly* defined and manipulated. In crisp set calculations, where the Grade of Membership scores can only be 0 or 1, they are generally handled implicitly by using probability models where set membership is unknown or measured with error, and probabilities are used to make statements about uncertainty of crisp set membership. Because set membership is often implicit in crisp set statistical models there is often confusion about the mathematical structure of the two types of models because of the coincidence that the g_{ik} (i.e., grade of membership scores or intensities) vary between 0 and 1.0 over k. Thus, at first glance the g_{ik} scores resemble probabilities. As shown below, this interpretation is not correct.

At first it may seem that there is little need for fuzzy sets in areas where rules of behavior or organization can be clearly and "crisply" defined. However, in the context of engineering and computer science, where mathematical representation is the norm, it is becoming increasingly clear that fuzzy sets are necessary to describe the behavior of complex, naturally occurring physical systems. For example, devices that are controlled by electronic computers are often given "crisp," digital feedback for system control. In the case of the computer-controlled "bullet trains" of Japan, the digital feedback used to exert braking pressure resulted in perceptibly "jerky" stops. By using fuzzy, rather than discrete, digital feedback, where the digital signal is given a component of "fuzz," the jerkiness in digitally controlled stops could be eliminated. Similarly, fuzzy methods have been used in computer parsing of words and phrases, in robotic pattern recognition, in representing constraints of a limited linguistic system, in designing electrical circuits and microprocessors, and as illustrated above, in modeling uncertainty in automated systems. For example, in language recognition models, the same words uttered by different individuals are intended to carry a clear "command" even though the words are spoken in a variety of ways (i.e., the message content is "fuzzy", it varies in loudness, dialect, etc.).

Besides being useful as a mathematical model in engineering applications, fuzzy set methods are proving useful in areas where the individual objects are *intrinsically* fuzzy members of sets. For example, we have all used words with indefinite, "fuzzy" meanings such as "old," "handsome," "happy" to describe the "core" or essential characteristics of states of individuals which vary widely on other attributes. Though any individual for which these adjectives might apply is a crisp, well-defined element, his or her membership in the set of old people, or handsome people, or happy people may be fuzzy. With the exception of gender and possibly a few isolated measurements such as weight and height taken at fixed times (even an individual's weight and height can vary over the course of a single day), most descriptions of humans entail fuzzy statements since each individual can be characterized by so many variables that each person is different. The differences go beyond simple explanatory characteristics such as genealogy, education, and age. At the level of molecular biology even sex and age, which are generally viewed as defining crisp sets, become fuzzy concepts (e.g., "sex" is related to a complex mix of hormonal factors that vary in level and effect, even in different types of tissues within an individual assigned to a specific gender category). The high dimensional, nonstochastic differences in individuals comprising a population we will refer to as *heterogeneity* implying differences in multiple statistical moments, so as not to be confused by use of the term "variation" which could be taken to imply only differences in second-order moments (i.e., variances).

Intuitively, any mathematical or statistical method based on crisp sets assumes that there is some set of which everyone is a crisp member. This is tantamount to assuming that there exists some crisp partitioning of the collection of individuals for which there is *no* heterogeneity. This is often not a practical assumption in working with either human or animal populations. For example, an area where fuzzy set logic has been applied in medicine is in formal diagnostic problems where the expression of a disease (especially chronic diseases) is a function of (a) the other diseases present, and (b) the state of the host physiology. Thus, though a physician can often attach a label to a disease process, say, such as breast cancer, the exact characteristics of the disease process (e.g., aneuploidy, estrogen and progesterone receptor status, loss of specific growth control functions related to p53 and CDC genes) and of the host (e.g., age, menopausal status, comorbidity) makes each such disease manifestation in an individual effectively unique. Thus, physicians both work towards labelling disease states and, in the ideal case, individualizing therapeutic responses to the individual as the patient responds (or not) to treatment (Woodbury and Clive, 1974; Woodbury et al., 1978). This has even been formalized in concepts of clinical trials of individuals, i.e., examining the change over time of the patient's response to interventions so as to optimize (i.e., individualize) treatment processes as more individual characteristics are assessed. This obviously is conceptually different than standard statistical logic where "certainty" or "likelihood" increases with the number of "identical" replicates, rather than increasing with the number of variables to uniquely characterize and differentiate individuals. The fuzzy set models have to quantitatively bridge these two views of the world, i.e., that of the *typical* and that of the *unique*.

When heterogeneity is present in the population being studied, it is necessary either to approximate this heterogeneity with a crisp set model (say, by using a variety of explanatory variables linked to the phenomena by regression coefficients to "condition out" heterogeneity by generating a marginal distribution averaged over heterogeneity) or to use a fuzzy set model where the heterogeneity of individuals is explicitly parameterized in the model. When a fuzzy set model is used, the fuzziness does not simply represent limitations of our classification system, but rather is a model of individuals whose membership *is* fuzzy in any defined set. As sample size increases, the common features of individuals increase the certainty of the defining profiles for each fuzzy set, which increases with the number of variables, which, in turn, helps us more confidently describe the individual or fuzzy features of the model. Thus the fuzzy set model uses both types of "increase" in information (i.e., increases in the dimensionality of the "case" and "variables" space of observations).

1.3.2. A Paradigm for the Fuzzy Set or Fuzzy Class Model

Throughout this book we will explain concepts by providing illustrations of the type of data we are trying to model and the potential sources of individual heterogeneity. A generic example begins with the assumption that a sample of I individuals is drawn from a population. For each individual there are J different questions or response measures illicited. For question j, the individual has a choice of L_j different responses. Each individual gives one, and *only* one, response for each question.

This paradigm introduces the following notation used in the remainder of the book.

Notation. The variable I represents the *number of individuals* in the sample.

Notation. The variable J represents the *number of variables* measured on each individual.

1.3.3. Crisp Sets in Statistical Methods

The use of crisp sets in the development of statistical models is currently the predominant practice. Those models assume that a person is a member of one, and only one, of the sets that span the state space for the phenomena of interest. By *state space* we mean the set of conditions defining the state of each individual in the population. Thus, the state space is spanned by all variables of relevance in explaining the study phenomena, where explanation means that past and future (or, alternatively, spatially contiguous) events are independently distributed after conditioning on state variables. The constraints implicit in a crisp state space makes it necessary to assume that the set to which a person belongs is homogeneous. Each set to which a person may be assigned is assumed to occupy a *distinct* set of coordinates or a *point* in the J dimensional state space defined by values on the J measured variables. Thus, all members in a given set occupy identically the same physical point (i.e., have the same J coordinates) in the state space and there is no individual heterogeneity within any of the crisp sets. Any variation among individuals in the same set is attributed to stochastic error generated by the measuring instrument or sampling experiment.

The implications of crisp set concepts are sometimes ambiguous in specific statistical models because the geometric interpretation of the discrete sets is left implicit (i.e., is an unstated assumption) in the mathematical and statistical statement of the model. For example, there may be no formal specification of a crisp set model's deterministic, parametric, or stochastic structures. Of particular interest is the distinction between models where the membership in each set is assumed *known* a priori, as opposed to models where much membership is *not* assumed known and is identified probabilistically.

An example of the situation where membership is assumed known (and fixed) a priori is linear discriminant analysis (Eisenbeis and Avery, 1972). In these models, membership in each crisp set or group for each individual used in estimating discriminant coefficients is a priori defined and assumed to be determined without error. The group membership (i.e., the state variables g_{ik}) are used to a priori collect or assemble cases into crisp sets for which the mean and covariance structure of all cases in each set can be calculated for all of the observed variables. Inference is based on the determinantial ratios of between- to total-group generalized variances as expressed by determinants of the relevant variance–covariance matrices (e.g., Van de Geer, 1971). Though the discriminant function calculations clearly require exclusive assignment of each case to a *single* homogeneous set, a parameterization (e.g., the g_{ik}) is often not devised to explicitly represent set membership ($g_{ik} = 1.0$). Nonetheless, it is mathematically appropriate in such models to conceive of a *set assignment function* where all cases are given a value of 1.0 in the group to which they are assigned and a zero for all other groups. Though there is variation in the J variable measurement space, there can be no variation in the set assignment function in the crisp set model; i.e., the set assignment function is a *non*stochastic 0–1 quantity. Each individual's set membership is exactly determined by the set assignment function. Since this is unknown in new data (i.e., data not used to estimate the discriminant function coefficients), we need to estimate a probabilistic set assignment function for new cases from the observed data.

Because an estimated probabilistic assignment function is used to classify *new* members into crisp sets, such discriminant or classification methods are sometimes confused with fuzzy set methods. Using the estimated probabilistic assignment function, the probability of membership of new cases in one of the crisp sets [i.e., $\Pr(g_{ik} = 1.0)$] of each value of K may be estimated *after* the discriminant function is fit and coefficients estimated (and assumed fixed). The estimated probability of membership of any individual in a particular crisp set is the a priori probability of the individual being in one of the crisp sets. A priori, this is a random assignment. However, a postiori, the group membership is *fixed*, but *unknown*. In both cases, the individual is, by definition, a member of only one crisp set. The uncertainty lies in our ability to identify which crisp set this is. The probabilistic set assignment function is based on an estimate of the likelihood that any particular individual is a member of each crisp set, usually assigning membership to the set with the largest likelihood. However, the estimated likelihood is the probability that the individual is a full member of each crisp set ($g_{ik} = 1.0$) and *not* the degree of membership of a case in all sets.

Though the existence and characteristics of the set assignment function may be reasonably clear when classes are assumed to be known a priori (i.e., the difference between $\Pr(g_{ik} = 1.0)$ versus g_{ik}), things can become, conceptually, less clear when the class assignment is *not* known. In this case the crisp membership of each element in the "training" data is unobserved (latent).

One set of models, where the sets are crisp but membership is "latent," is the latent class model (LCM) for analysis of multiple categorical variables (Lazarsfeld and Henry, 1968). In these models the crisp set assignment functions for the training set are not known but must be estimated. What causes confusion is that developing direction vectors and derivatives for estimating the crisp set assignment function parameters using, say, maximum likelihood procedures is numerically difficult (i.e., with $g_{ik} = 1.0$ or 0.0 the second-order derivatives in a Newton–Raphson search are discontinuous) and possibly violates the Neyman–Pearson lemma identifying conditions for consistent estimation of parameters. As a consequence, most empirical applications of latent class models do *not* attempt to generate estimates of the set assignment function parameters (i.e., g_{ik}) which, for the crisp set model must be 0 or 1.0. Instead, these methods attempt to estimate only selected features of the frequency distribution of cases over the crisp sets (Kiefer and Wolfowitz, 1956). In this way, information is averaged over multiple individuals so that information on a parameter increases with increases in the size of I, making it possible for mathematically identifiable parameters to be consistently estimated. This averaging may assume the form of any one of a number of different smoothing operators or functions. This estimation procedure provides an empirical description of those latent crisp sets to which observations either do or do not belong.

For example, the EM algorithm (e.g., Dempster et al., 1977) has been used to estimate the location parameters of the K crisp sets in the state space (i.e., for a J discrete variable space, these may be written as λ_{kjl}, i.e., the probability that a person crisply in set k has the jth response to the lth variable) and the posterior frequency of case classification in the K crisp sets (P_{ik}). Everitt (1984), for example, presents a modified EM algorithm for estimating the set location parameters (denoted λ_{kjl}) and the posterior probability that a case is in one of the K crisp sets (denoted P_K). Everitt refers to work by Madansky (1959) where, using a hill-climbing algorithm, an attempt *is* made to identify the membership of an element in a discrete set.

Thus, what causes confusion is a tendency to interpret or view the posterior probabilities of classification as analogous to the set assignment function parameters. These are *not* analogous. They have different mathematical properties, parametric functions, and substantive interpretations. In particular, the set assignment function parameters are not stochastic, but are mathematical measures of the physical properties of the elements as described by partial membership in fuzzy sets or full membership in crisp sets. Substantively, by partial membership we mean that the observed attributes of an individual may be described as a continuous mixture of each extreme type of profile of attributes. Classically, such a continuous mixture arises naturally in chemical reactions where two or more liquids can be mixed in continuously varying proportions. In human data, for example, a similar graded mixture of profile types may be defined by many attributes. The larger the number of attributes, the closer the mixture comes to approximating a truly continuous mixture. However, even with fluids, the mixture becomes discrete at the molecular level. Likewise, as J increases, the state space becomes more finely partitioned (and, with appropriately increasing sample size, more densely populated). The g_{ik} scores thus may be reasonably viewed as the intensity of expression of each profile where the space is bound by each purely expressed profile. A number of analytic procedures attempt to generate such scores. In factor analysis, scores are generated on latent dimensions, but there is no set of boundary conditions to score responses on an absolute basis. In the Older American Resource

Study (Maddox and Clark, 1992), scores are assigned on multiple dimensions according to a subjectively generated algorithm. Fuzzy set mathematics allow this to be done on a more rigorous basis.

In both standard regression and analysis of variance methods, we also implicitly assume a crisp set classification. In both methods one usually explicitly deals with crisp sets or classes of individuals. A crisp set is implicitly assumed if, conditional upon regression parameters, the responses over different individuals are assumed to be drawn from the same probability distribution. In other words, individuals with the same values of significant independent variables are assumed to be members of the same crisp set.

1.3.4. Fuzzy Partitions

In this book we will discuss a class of statistical models that are based on generalizations of discrete set concepts to those for fuzzy sets (Bellman and Zadeh, 1970; Zadeh, 1965). In the fuzzy set model, the set assignment function attributes a *degree* or *grade* of membership for each element for each and every defined set. These grade of membership scores g_{ik} can take any value between 0 and 1 for one or more sets or classes of objects or persons. Since a person can be a partial member of multiple sets or classes, we refer to this as a *fuzzy set model*. By allowing grade of membership scores for each individual that vary in magnitude to represent the degree of membership in K fuzzy sets, we represent individual variation or heterogeneity of individuals within each set. In this situation, classification is generalized from assigning cases to discrete sets to identifying features of the structure of the multivariate density function of the grade of membership scores of the individuals up to the resolution which is statistically possible with the available data. The modeling of heterogeneity of the individuals is thus analogous to identifying various characteristics of the multivariate density function which describes the distribution of the grade of membership scores across the populations of individuals of interest. This multivariate density function is crucial to the analysis of heterogeneous data. As we will see later, in some fuzzy set models we may specify distributions (e.g., the Dirichelet) from which the individual grade or membership values are drawn. This unconditional form of a fuzzy set model imposes a prior distribution in order to smooth parameter estimates in small samples or with small numbers of variables.

A statistical methodology based on fuzzy sets requires us to identify and estimate a grade of membership score for each individual relative to any set. This is accomplished by constructing a *fuzzy partition*.

Definition 1.2. A *fuzzy partition* of a collection of objects or persons consists of a number of crisp, single element sets called *extreme profiles* and a set of Grade of Membership scores for each element in the collection *relative* to these extreme profiles.

A fuzzy partition is simple to construct since grade of membership scores (i.e., positions in the state space) are constructed relative to the extreme profiles. Extreme profiles have a role similar to basis vectors in linear spaces, which allow representation of points in the space as linear combinations of the basis vectors. In particular, all elements of a fuzzy partition can be expressed as a convex combination of extreme profiles. In the early literature of fuzzy partitions, extremes profiles were referred to as *pure types* (e.g., Woodbury and Clive, 1974; Woodbury et al., 1978). As will be seen subsequently, however, the justification for the phrase "extreme profiles" is that all *potential* outcomes or profiles for individuals are "interior" to a convex set generated by the collection of extreme profiles. Since the extreme profiles are single-element crisp sets, they are unambiguously defined. As a practical matter, the extreme profiles are chosen so that a wide variety of characteristics of the collection are represented by the profile. For example, in representing the health of a human population of elderly males we might choose "frail, age 100" and "active, age 50" to be two extreme profile sets. An individual's grade of membership score would represent how physically active and how old he is relative to these extreme profiles. As we will see in subsequent chapters, the definitions of extreme profiles can entail characteristics in very large numbers of dimensions. The performance of fuzzy set models generally improves as the dimensionality of the measurement space is increased.

An important problem with regard to a fuzzy partition is determining how many extreme profiles are necessary to represent individuals in a given finite data set. Continuing with the analogy to basis vectors, the extreme profiles span the space of individuals in the measurement space. The partition based on these extreme profiles should include the minimum number of extreme profiles necessary to construct this span. Unlike vector analysis, however, it may not be easy to determine when we fully span the space. This issue is taken up in a subsequent chapter when we attempt to represent the distribution of the grade of membership scores using extreme profiles.

In the preceding discussion, we have referred to several key points and notation used in fuzzy partitions. For completeness, we will now formally state these.

Notation. The variable g_{ik} is used to represent the Grade of Membership score for the ith individual for the kth extreme profile in a fuzzy partition.

A g_{ik} value for an individual that is 0 for some k indicates that the individual has no membership in the set represented by extreme profile k. If the value of g_{ik} is 1.0, the individual is said to be a complete member of the fuzzy set represented by the extreme profile k. A value between 0 and 1 indicates the individual is a partial member of the extreme profile set. In a sense the g_{ik} value represents the proportion or intensity of membership in each extreme profile. This is *not* the same as the probability that the individual is from extreme profile k because the proportion or intensity is a mathematical measure of quantity (e.g., how many of J attributes a person of a given extreme profile manifests) and not a statement about the frequency of a specific, crisp outcome in repeated trials. The individual is not from any single extreme profile unless he has a value of unity for one of the g_{ik}. In the limit, as $J \rightarrow \infty$, the likelihood of $g_{ik} = 1.0$ should be 1.0. In all other cases he is a convexly weighted mixture of the attributes associated with the extreme profiles.

Notation. The integer variable K is used to represent the number of extreme profiles used in the fuzzy partition.

The manner in which we have described a fuzzy partition imposes some constraints on the g_{ik}. Explicitly, they must all be nonnegative and sum to unity over values of k. We will refer to these constraints as Condition I.

Condition I (Basic Constraints on g_{ik})

$$g_{ik} \geq 0 \qquad \textit{for each } i \textit{ and } k\,,$$

$$\sum_{k=1}^{K} g_{ik} = 1 \qquad \textit{for each } i\,.$$

As a result of these constraints, values of g_{ik} are contained in a unit simplex of dimension $K - 1$.

In summary, discrete set models are special cases of fuzzy set models where the g_{ik} must be equal to 0 or 1. The set assignment functions defined in the state space for fuzzy sets or partitions may adopt any point interior to (or on the boundary of) a $K - 1$ dimensional simplex projected down from a J dimensional measurement space. In a discrete or crisp set model (e.g., LCM) the J to $K - 1$ dimensional projection is far more restricted, mapping to $K - 1$ points or simplex vertices, rather than to any interior point of the $K - 1$ dimensional simplex. The additional components of heterogeneity that can be represented by allowing the g_{ik} to take any value *within* the $K - 1$ dimensional simplex is what generalizes the discrete set classification model.

1.4. PARAMETERIZATION OF FUZZY PARTITION MODELS

1.4.1. Model Assumptions

Though well accepted in engineering and computer science and employed in electronic chip design, fuzzy set principles have not been extensively used in statistical models to be applied in data analysis. One shortcoming of past attempts to implement fuzzy set statistical models has been the difficulty in merging standard computational and statistical machinery based on probability theory and discrete set logic with the different conceptual basis of fuzzy set logic and the calculation of possibilities.

We will show that fuzzy sets can be employed in statistical analyses if the underlying principles of the statistical models and numerical algorithms are made consistent with the mathematical principles of the fuzzy set models. One of the difficulties possibly encountered in efforts to make the translation is that, even by those applying fuzzy sets, the heterogeneity of measurement still tended to be viewed as an artifact of flaws in the ability to represent (e.g., incomplete linguistic system) or measure heterogeneity (i.e., it was stochastic) rather than being an actual physical (deterministic) property of highly complex, multidimensional systems. This difference in view leads to different geometric representations that generate parameter spaces with different mathematical and statistical properties.

To begin this effort we start by defining a basic form of fuzzy set model (see Woodbury and Clive, 1974; Woodbury et al., 1978). This statistical implementation of a fuzzy set model has been referred to as the Grade of Membership (GoM) model, where as defined previously, Grades of membership are parameters (scores) that represent the degree to which a given case or individual i is a member of the kth extreme profile. These are estimated from observations.

Notation. For each question or measure j for individual i let the response represented by a set of L_j binary random variables be Y_{ijl}. We denote y_{ijl} as empirical realizations of the random variables Y_{ijl}.

To formulate the model we need the following assumptions.

Assumption 1. *The random variables Y_{ijl} are independent for different values of i.*

Assumption 2. *The g_{ik}, $k = 1, 2, \ldots, K$, values are realizations of the components of the random vector $\boldsymbol{\xi}_i = (\xi_{il}, \ldots, \xi_{ik})$ with distribution function $H(\mathbf{x}) = \Pr(\boldsymbol{\xi}_i \leq \mathbf{x})$.*

Assumption 3. *Conditional on the realized values g_{ik} for individual i, the Y_{ijl} are independent for different values of j.*

Assumption 1 states that the responses for different individuals are independent. Initially we assume that the Y_{ijl} are randomly sampled from the population. As we will see subsequently, however, different sampling models can be used by appropriately modifying the form of the likelihood.

Assumption 2 states that the Grade of Membership scores (g_{ik}) are random variables that are realized when an individual is selected from the population. The sample distribution of the realizations (the g_{ik} or "scores" in the sample) provides estimates of the distribution function $H(\mathbf{x})$. Although much of the discussion is based on the g_{ik}, it is important to realize that a different sample of individuals will yield a different set of g_{ik} values for the fuzzy partitions.

Assumption 3 is the conditional independence assumption common in latent class models and in discrete factor analysis (see, e.g., Bollen, 1989). This assumption, for fuzzy set models, states that if the grade of membership "scores" g_{ik} are known, the responses of individual i to the various questions (i.e., Y_{ijl}) are independent over values of j.

The role of the fuzzy partition in the model is defined by Assumptions 4 and 5.

Assumption 4. *The probability of a response l for the jth question by the individual with the kth extreme profile is λ_{kjl}.*

Recall that though many individuals may be a partial member of the kth type, by assumption there is at least one, possibly theoretical, individual that is a crisp member of the kth type. Assumption 4 gives the probabilities of response for this individual to the various levels of each question. Clearly, the probabilities λ_{kjl}, satisfy the constraints required of discrete probabilities. We state this formally as Condition II.

Condition II (Constraints on the λ)

$$\lambda_{kjl} \geq 0 \qquad \textit{for each } k,\ j, \text{ and } l\,,$$

$$\sum_{l=1}^{L_j} \lambda_{kjl} = 1 \qquad \textit{for each } k \textit{ and } j\,.$$

Assumption 5. *The probability of a response of level l to the jth question by individual i, conditional on the g_{ik} scores, is given by the bilinear form*

$$Pr(Y_{ijl} = 1.0) = \sum_{k=1}^{K} g_{ik} \cdot \lambda_{kjl} \tag{1.1}$$

Assumption 5 bridges the definition of the fuzzy partition with a parametric expression for the probability of a specific discrete response. Implicit in this assumption is that the value of K is fixed. Initially we assume that it is known. Subsequently we will present statistical methods of testing for the correct value of K. It is important to note that a fuzzy partition will necessarily entail a set of g_{ik} values for each element of the collection. However, it is not necessary that such a partition requires the probability of response to be given by (1.1). In Section 6, (1.1) is relaxed. A test of

Assumption 5 is formulated there. Although the model in equation (1.1) does remarkably well in fitting data from diverse situations and in producing forecasts, there is no explicit test for the validity of Assumption 5. There is, however, a geometric argument for the naturalness of this representation by simply assuming J variables have been selected, which can represent the behavior of interest, and each are categorical in form.

Assumptions 1–5 provide a basis for forming a probability model for constructing a likelihood estimation procedure. Explicitly, the probability model for a random sample is the product multinomial model with cell probabilities given by

$$E(Y_{ijl}) = \sum_k g_{ik} \cdot \lambda_{kjl}$$

or, equivalently,

$$p_{ijl} = \sum_k g_{ik} \cdot \lambda_{kjl}$$

where g_{ik} are assumed to be known and satisfy Condition I. Different probability models (e.g., for certain types of contaminated or mixed distributions, such as the hypergeometric) can be formed for other sampling designs (e.g., spatially clustered sampling) using Assumptions 1–5.

1.4.2. Probability Distribution of the Basic Model

From Assumption 1 and the fact that each of the J questions contains a finite (L_j) number of levels, the observed counts for each question will follow the multinomial probability model. From Assumption 3 we know that the overall model for the data, across questions, will be a product multinomial model, given the g_{ik} values. Assumptions 4 and 5 are used to determine the cell probabilities for each question. Therefore, from these assumptions, the likelihood of the model can be written,

$$L(\mathbf{y}) = \prod_{i=1}^{I} \prod_{j=1}^{J} \prod_{l=1}^{L_j} \left(\sum_{k=1}^{K} g_{ik} \lambda_{kjl} \right)^{y_{ijl}} . \tag{1.2}$$

For pedagogical reasons, in subsequent chapters, this form of the product multinomial likelihood is used in lieu of the more common form involving factorials. Note that, for all combinations of i and j, the y_{ijl} will be zero for all values of l except one, for which it will be unity.

1.5. EXAMPLES OF PHENOMENA AMENABLE TO FUZZY SET INTERPRETATION

The original motivation for the GoM model derived from problems of formal diagnosis of chronic diseases—especially Bayesian diagnosis of chronic diseases. Bayesian procedures, which assumed small numbers of discrete disease processes, often proved ineffective in the formal diagnosis problem for chronic diseases—especially those with lengthy natural histories in elderly populations (Woodbury, 1963). The problem was that the typical or textbook case was an exceedingly rare, if ever observed, phenomenon in such a high dimensional space. Thus, relying on data to identify diseases from classical or textbook clinical presentations often proved unsuccessful because diseases, interacting with other diseases, and the host's physiology (and changes in the diseases and host physiology with time), produced symptomatic manifestations that almost always had unique elements for each individual—at least in any reasonably sized (i.e., finite) sample. Such problems were frequently found in a number of clinical specialty areas such as psychiatric disorders where disease entities (e.g., those whose underlying mechanisms may be complex physiochemical changes in neurophysiological processes) had to be inferred from large numbers of symptomatic behaviors or in medical oncology, where the disease strongly interacted with the host physiology at multiple levels of organization (e.g., Firth, 1991; Lipsitz and Goldberger, 1992).

Strauss et al. (1979), for example, found difficulty in using multivariate procedures to assess psychiatric disorder manifestations at different levels of the health service delivery system (e.g., outpatient, clinic, inpatient) because, despite being generated by the same disease process, the "early" manifestation of a given disease process tended to manifest greater symptomatic heterogeneity and diffusiveness (i.e., the signal to noise ratio was less, a standard statistical problem, *and* the signal itself was more heterogeneous—the uniquely "fuzzy" aspect of the problem).

In oncology, though a tumor may begin in a particular type of tissue, it undergoes (a) genetic mutations as it replicates with mutations occurring at greater than normal rates due to the accumulating damage of DNA material controlling cell growth (including mechanisms for DNA replication which loses fidelity at increasing rates) and function (i.e., the essential nature of *neo*plastic disease) (b) the process of metastization that involves multiple physiological subprocesses governing dissemination, tissue invasion, and clonal growth, and (c) the hormonal activity of the tumor cells which allow it to influence the behavior of normal tissue even beyond direct physical contact (e.g., the release of angiogenesis factor allowing tumors to vascularize, cachexia, various autocrine and paracrine growth factors, such as insulin growth factor 1 and the angiogensis factor).

Making issues more complex is the fact that heterogeneity in disease expression might also be induced by a complex set of *exogenous* factors. For example, schizophrenia in a WHO sponsored study (Manton et al., 1993) was manifest with very distinct symptomatic patterns depending upon the culture in which the disease was expressed (e.g., in less developed societies the manifestation of visual hallucinations was more prevalent).

Heterogeneity can also emerge from the measurement process (e.g., the interaction of the nature of the phenomena with measurement procedures, a situation well known in the physical sciences). For example, anthropologists will ask questions of subjects by selecting from an extremely large repertoire of potential questions after cognitively analyzing results of earlier observations (some possibly of a noninterview type, i.e., inferences made from the organization of holes or gaps in the data that are nonrandomly distributed).

A similar type of positively and negatively directed information gathering and organization in an extremely high dimensional variable space occurs when physicians go through a large repertoire of potential tests and asking questions to eliminate (rule out) less probable diagnostic options sequentially until the patient is, in essence, *uniquely* characterized (i.e., the disease and the patient's manifestation of the disease) (Manton et al., 1991).

Even in the design of surveys, we can find fuzziness. For example, the concept of a response rate is based on the notion that a sample can be drawn at a set time. In longitudinally followed populations, transitions that affect which response set a person is in change continuously over time. Thus, in such longitudinal scenarios, the concept of response sets becomes fuzzy sets for a survey; it is a discrete set only at an arbitrary and fixed time t—it is fuzzy when one considers the entire data collection period for a survey (e.g., Manton and Suzman, 1992).

The need to uniquely characterize a patient is a practical one in designing a treatment regimen with the greatest likelihood of success by the physician. This approach becomes particularly important in geriatric medicine where the patient may have a multiplicity of conditions and comorbidities which change over time (e.g., Rubenstein and Josephson, 1989), or in intensive care management where there is an intensive, multidimensional clinical effort to maintain multiple physiological variables in homeostasis over real time (Thompson and Woodbury, 1970). These principles are antithetical to a statistical analysis of patients where increasing numbers of replications of identical outcomes are necessary—those for which standard principles of model estimation and inference were designed.

To illustrate the issue further, in an emergent area of statistical analysis, meta-analyses have become common in the clinical literature because seldom can a single clinical trial resolve issues about treatment of a disease. However, even for diseases as common as female breast cancer or heart disease, the universe of complete trials is inadequate to answer anything but the most basic questions. For example, the Early Breast Cancer Trialists Collaborative Group (1988; 1992a,b) analyzed 133 trials done over 30 years involving 75,000 patients. From this large set of data the conclusions that could be made with confidence were relatively limited (e.g., polychemotherapy is better than monochemotherapy; chemotherapy improves survival in women under 50; hormonal therapy improves survival in women over 50). However, there was inadequate experience to judge the newest chemotherapeutic drugs, or to evaluate all but the most fundamental aspects of dosage levels and timing. As a consequence, increased effort is being expended on mathematical modeling of cell kill kinetics for drugs administered multiply, in different combinations, at different times. The issues are made more complex, when the goal is not simply the level of cell kill but includes, say, the use of antihistamines or nonsteroidal anti-inflammatory agents to restrict the ability to metastasize.

In Antman et al. (1992; Lau et al., 1992) the problem of using trials accumulated over time was addressed for treatment of heart disease. Often the success or failure of a therapy was ascertained at a particular desired confidence level much earlier in the experience base (i.e., the set of all relevant trials). However, there were many situations where a single contradictory trial caused the continuation of studies—even though the amount of statistical information in the trial was small. Detsky et al. (1992) tried to promulgate dimensions of quality in order to weight trials and found that it was difficult to establish consensus on these dimensions. This was amplified in Ravnskov (1992) where the deletion of trials on qualitative grounds often could be interpreted as a bias towards a particular outcome or result. The fundamental problem is the lack of statistically rigorous models to combine trial information (Thompson and Pocock, 1991). The issue becomes extremely difficult when we realize that greater heterogeneity of patients in a trial increases its generalizability (a type of validity) but at the cost of less statistical power regarding acceptance or rejection of a hypothesis. Thus, there is an operating characteristic curve that describes the functional relation of within trial heterogeneity to statistical power. However, the difficulty is that such functions are not formally used in the decisions about conducting further trials (and of what type). They are used, to a degree, in deciding when to terminate a trial manifesting high efficacy.

The fuzzy set model allows us to examine such problems (i.e., of combining evidence from studies of heterogeneous populations) by retaining the uniqueness of individuals (so that, to a degree, trial identity is noninformative given the individual state descriptions) yet still abstracting the fundamental dimensions of their *partial* commonality. If each patient were wholly (instead of partly) unique, the world would effectively be in a maximum entropy state and nothing could be done.

1.6. OUTLINE OF THE BOOK

We have presented a series of concepts to make fundamental distinctions between classical multivariate statistical models based on discrete set logic and multivariate statistical models based on fuzzy set logic. We also have shown that the fuzzy set models are generalizations of discrete set models but, if implemented using statistical and numerical procedures specifically appropriate to the fuzzy set mathematical formulation, that model estimation, inference, description, and forecasting may have better operating characteristics than the simpler discrete set models (especially in complex, high dimensional substantive applications).

In Chapters 2 and 3 we deal with basic issues in estimation and inference. Chapter 4 covers specialization of the GoM model presented in Chapter 2 to deal with aggregate data. In Chapter 5 we extend the model to deal with time and multivariate event history processes. In Chapter 6 we discuss a generalization of the GoM model to deal with empirical Bayesian conditions where heterogeneity is allowed in the λ_{kjl} structural parameters as well as in the g_{ik} (e.g., $\bar{p}_{ijl}$ is estimated rather than p_{ijl}). Chapter 7 describes how the parameters estimated from the model may be used to make various types of health forecasts. In Chapter 8 we discuss how GoM may be used to combine data from multiple sources. Finally, in Chapter 9 we summarize the use of fuzzy set models, review the progress made in their development to date, and identify additional needed areas of statistical and mathematical research.

In each of these chapters, we apply the the GoM model to multiple data sets with widely varying characteristics. In these comparisons we evaluate the performance of GoM and other multivariate procedures in common data sets, and we examine the performance of different specifications of GoM type models in the same data sets. The data sets examined are drawn from a wide variety of sources. They generally have large numbers of discretely measured variables and a reasonable range of sample sizes. Many of the data sets are longitudinal in nature—though with highly variable temporal sampling plans. Some are derived from multiple sites selected without being representative of larger populations. Others require direct comparisons of interventions. Thus, multiple versions of fuzzy set models are extensively tested and illustrated under a wide range of empirical conditions and substantive applications. We believe that it is the demonstrated operational characteristics of a statistical model under widely varying empirical conditions that is the ultimate test of the validity of any statistical model.

REFERENCES

Antman, E. M., Lau, J., Kupelnick, B., Mosteller, F., and Chalmers, T. C. (1992). A comparison of results of metaanalyses of randomized control trials and recommendations of clinical experts: Treatments for myocardial infarction. *Journal of the American Medical Association* **268**(2): 240–248.

Bellman, R. E., and Zadeh, L. A. (1970). Decision-making in a fuzzy environment. *Management Science* **17**(4): B-141.

Bollen, K. A. (1989). *Structural Equations with Latent Variables*. Wiley, New York.

Dempster, A. P., Laird, N. M., and Rubin, D. B. (1977). Maximum likelihood from incomplete data via the EM algorithm (with discussion). *Journal of the Royal Statistical Society, Series B* **39**(1): 1–38.

Detsky, A. S., Naylor, C. D., O'Rourke, K., McGeer, A. J., and L'Abbe, K. A. L. (1992) Incorporating variations in the quality of individual randomized trials into meta-analysis. *Journal of Clinical Epidemiology* **45**(3): 255–265.

Early Breast Cancer Trialists' Collaborative Group (1988): Effects of adjuvant tamoxifen and of cytotoxic therapy on mortality in early breast cancer: An overview of 61 randomized trials among 28,896 women. *New England Journal of Medicine* **319**: 1681–1691.

Early Breast Cancer Trialists' Collaborative Group (1992a). Systematic treatment of early breast cancer by hormonal, cytotoxic, or immune therapy—Part I. *The Lancet* **339**: 1–15.

Early Breast Cancer Trialists' Collaborative Group (1992b). Systematic treatment of early breast cancer by hormonal, cytotoxic, or immune therapy—Part II. *The Lancet* **339**: 7185.

Eisenbeis, R. A., and Avery, R. B. (1972). *Discriminant Analysis and Classification Procedures: Theory and Application*. D. C. Heath, Lexington, Massachusetts.

Everitt, B. S. (1984) *An Introduction to Latent Variable Models*. Chapman and Hall, London and New York.

Firth, W. J. (1991). Chaos—predicting the unpredictable. *British Medical Journal* **303**: 1565–1568.

Kaufmann, A., and Gupta, M. M. (1985) *Introduction to Fuzzy Arithmetic: Theory and Applications*. Van Nostrand Reinhold, New York.

Kiefer, J., and Wolfowitz, J. (1956). Consistency of the maximum likelihood estimator in the presence of infinitely many parameters. *Annals of Mathematical Statistics* **27**: 887–906.

Lau, J., Antman, E. M., Jimenez-Silva, J., Kupelnick, B., Mosteller, F., and Chalmers, T. C. (1992). Cumulative metaanalysis of therapeutic trials for myocardial infarction. *New England Journal of Medicine* **327**(4): 248–254.

Lipsitz, L. A., and Goldberger, A. L. (1992). Loss of "complexity" and aging. *Journal of the American Medical Association* **267**(13): 1806–1809.

Lazarsfeld, P. F., and Henry, N. W. (1968). *Latent Structure Analysis*. Houghton Mifflin, Boston.

Madansky, A. (1959). Partitioning methods in latent class analysis. Report P-1644, RAND Corporation.

Maddox, G. L., and Clark, D. O. (1992). Trajectories of functional impairment in later life. *Journal of Health and Social Behavior* **33**: 114–125.

Manton, K. G., and Suzman, R. M. (1992) Conceptual issues in the design and analysis of longitudinal surveys of the health and functioning of the oldest old. Chapter 5 in *The Oldest-Old* (R. M. Suzman, D. P. Willis and K. G. Manton, Eds.). Oxford University Press, New York, pp. 89–122.

Manton, K. G., Dowd, J. E., and Woodbury, M. A. (1991) Methods to identify geographic and social clustering of disability and disease burden. In *Measure-*

ment of Health Transition Concepts (A. G. Hill and J. Caldwell, Eds.). Australian National University Press, Canberra, pp. 61–84.

Manton, K. G., Korten, A., Woodbury, M. A., Anker, M., and Jablansky, A. (1993). Symptom profiles of psychiatric disorders based on graded disease classes: An illustration using data from the WHO International Study of Schizophrenia. *Psychological Medicine*, in press.

Ravnskov, U. (1992). Cholesterol lowering trials on coronary heart disease: Frequency of citation and outcome. *British Medical Journal* **305**: 15–19.

Rubenstein, L. Z., and Josephson, K. E. (1989) Hospital based geriatric assessment in the United States: The Sepulveda VA Geriatric Evaluation Unit. Danish Medical Bulletin: *Gerontology*, Special Supplement Series No. 7, pp. 74–79.

Strauss, J. S., Gabriel, K. R., Kokes, R. I., Ritzler, B. A., Van Ord, A., and Tarana, E. (1979). Do psychiatric patients fit their diagnoses? Patterns of symptomatology as described with the biplot. *Journal of Nervous and Mental Disease* **167**: 105–112.

Terrell, G. R., and Scott, D. W. (1985). Oversmoothed nonparametric density estimates. *Journal of the American Statistical Association* **80**: 209–214.

Thompson, H. K., and Woodbury, M. A. (1970). Clinical data representation in multidimensional space. *Computers and Biomedical Research* **3**: 58–73.

Thompson, S. G., and Pocock, S. J. (1991). Can meta-analysis be trusted? *The Lancet* **338**: 1127–1130.

Van de Geer, J. P. (1971). *Introduction to Multivariate Analysis for the Social Sciences*. W. H. Freeman, San Francisco.

Woodbury, M. A. (1963). Inapplicability of Bayes' Theorem to diagnosis. *Proceedings of Fifth International Conference on Medical Electronics*, sponsored by the International Institute for Medical Electronics and Biological Engineering, Liege, Belgium, July 16–22, 1963.

Woodbury, M. A., and Clive, J. (1974). Clinical pure types as a fuzzy partition. *Journal of Cybernetics* **4**: 111–121.

Woodbury, M. A., Clive, J., and Garson, A. (1965). Mathematical typology: A Grade of Membership technique for obtaining disease definition. *Computers and Biomedical Research* **11**: 277–298.

Zadeh, L. A. (1965). Fuzzy sets. *Information Control* **8**: 338–353.

CHAPTER 2

The Likelihood Formulation of the Fuzzy Set Partition

INTRODUCTION

The fundamental strategy used to estimate statistically the parameters of a fuzzy or GoM set model is the method of maximum likelihood (Cox and Hinkley, 1974). This methodology entails first specifying the probability distribution of the realized data, parameterizing the processes believed to generate the data, and then constructing the appropriate likelihood functions. Then it is necessary to examine the statistical properties of the numerical procedures used to estimate the parameters of the function which maximize its value. In this chapter we examine the structure of the likelihood function for the GoM model. In Chapter 3 we examine the statistical properties of parameter estimates produced by maximizing the likelihood.

As noted in Chapter 1, the probability model we usually assume to generate the data is the multinomial. The parameters of this model are the cell proportions (i.e., the λ_{kjl} for an individual completely in the kth extreme profile set). The model specifies outcome probabilities for individuals (i.e., p_{ijl}). This is a very general model of observational heterogeneity. The g_{ik}, or grade of membership scores, parameterize such heterogeneity by modeling variability across individuals. Thus, data can be combined across individuals to estimate the structural or location parameters (the λ_{kjl}) of the model. Given the mathematical structure of the fuzzy set portion of this model, there are several aspects of the statistical estimation theory to be considered. These are the basic concepts of identifiability (sufficient and necessary conditions), consistency, efficiency, and the distribution of test functions which are reviewed in Chapter 3.

The Grade of Membership (GoM) model specifies grade of membership parameters g_{ik} for each individual, location or structural parameters λ_{kjl}, the binary structure of the discrete response random variables Y_{ijl}, and how the discrete data is linked to the probability of outcomes (p_{ijl}) using the fuzzy set parameters. In this chapter we consider these model elements in a formal sense to construct a likelihood function for a GoM model. We compare this likelihood with those of other multivariate procedures commonly used for categorical variables. The nature of the fuzzy set model differs fundamentally from the models typically used in multivariate statistical inference. GoM represents a strategy for estimating parameters of semimetric maximum likelihood functions where expending computational effort is preferred to making strong distributional assumptions. With recent increases in computational power, models that can *directly* utilize the moment structure of observations in even large, complex data sets (e.g., $I > 100{,}000$; $K > 10$; $J > 100$) are feasible.

2.1. LIKELIHOOD EQUATION METHODS FOR FUZZY SETS

2.1.2. The Likelihood Equation

As described in Chapter 1, the random variable Y_{ijl} is binary with outcomes 0 or 1. For fixed values of i and j there is only one nonzero realization. If l is an index for the cells of a multinomial, Y_{ijl} are indicators for individual outcomes in the cells of a multinomial random variable under Assumptions 1 and 3 of Chapter 1. Thus the probability that Y_{ijl} is equal to 1, denoted p_{ijl}, for $l = 1, \ldots, L_j$, sums to unity over all values of l.

We can specify a general model by allowing the multinomial random variable Y_{ijl} to have different probabilities associated with each value of i and j. The paradigm used regarding these multinomials in Chapter 1 was to consider i an index for individuals, j an index for questions or characteristics to be observed on each individual, and l an index for the response of the ith individual to the jth question. The GoM model specifies that the p_{ijl} associated with these outcomes be parameterized to capture heterogeneity *across* individuals. This is done by assuming that p_{ijl} (corresponding to each $p(Y_{ijl} = 1)$ may be represented as a bilinear form where one set of coefficients are the grade of membership scores for an individual, denoted g_{ik}, and the other are probabilities of response for question j for the kth extreme profile, denoted λ_{kjl}. Explicitly, as given in Assumption 5 of Chapter 1,

$$\Pr(Y_{ijl} = 1.0) = \sum_k g_{ik} \cdot \lambda_{kjl} .$$

From Assumptions 1, 3, 4, and 5 the form of the likelihood is that of a product multinomial model,

$$L = \prod_i \prod_j \prod_l \left(\sum_k g_{ik} \cdot \lambda_{kjl} \right)^{y_{ijl}} . \tag{2.1}$$

The problem in implementation is how to estimate the parameters g_{ik} and λ_{kjl} and to determine the statistical properties of those parameter estimates once generated from the data. The likelihood given in (2.1) is the basis for implementing the model.

The assumptions of Chapter 1 dictate the form of the likelihood in (2.1). Of importance is Assumption 3, which produces a likelihood that is the product of independent multinomials over index j. As a result of these assumptions the likelihood function factors into components for each value of i, j, and l (Suppes and Zanotti, 1981). The ability to factor the likelihood into components containing only single values of i, j, and l is important in estimation. Throughout the book we will refer to the fact that "the likelihood can be factored" to refer this condition. It does not refer to factorization of the solution space into structural and individual parameter spaces which is often assumed in standard maximum likelihood estimation (e.g., that functions of individuals, the g_{ik}, can be factored from structural parameters, the λ_{kjl}; p. 3, Cox and Hinkley, 1974).

2.1.2. The Unconditional Likelihood

Though the likelihood appears simple, estimation of the parameters using the likelihood seems difficult (i.e., computationally burdensome) because there are as many sets of g_{ik} values as there are individuals. Thus, increasing the number of individuals in a data set also increases the number of unknown parameters. This was anticipated in part by Assumption 2 of Chapter 1, i.e., that the g_{ik} values for individual i are realizations of a random vector $\boldsymbol{\xi}_i$ with distribution function $H(\mathbf{x}) = \Pr(\boldsymbol{\xi}_i \leq \mathbf{x})$. The likelihood in (2.1) is "conditional" on outcomes g_{ik} of the random variable ξ_{ik}. Rather than estimate the g_{ik} for each individual, the usual strategy is to form the likelihood unconditionally on the realizations g_{ik}. This entails multiplying the conditional likelihood in (2.1) by the marginal density of ξ_{ik} and integrating over the range of $\boldsymbol{\xi}$. For GoM, this produces the unconditional likelihood function,

$$L^* = \int \prod_i \prod_j \prod_l \left(\sum_k \xi_k \lambda_{kjl} \right)^{y_{ijl}} dH(\boldsymbol{\xi}) . \tag{2.2}$$

In equation (2.2) we suppressed the index i on $\boldsymbol{\xi}$ since the $\boldsymbol{\xi}$ are assumed to be independent and identically distributed over values of i.

Equation (2.2) contains the structural parameters, λ_{kjl}, and the parameters of $H(\mathbf{x})$—the multivariate distribution function for g_{ik}. If we know the form of H, L^* can be used to estimate parameters. This method does *not* provide for estimation of the realizations g_{ik}, but only of the parameters of H, the moments of the distribution of ξ_{ik}. In general, however, we will not specify $H(\mathbf{x})$, but approximate $H(\mathbf{x})$ to the degree identifiable by the data, as determined by a fixed number of moments. Maximizing the conditional likelihood L in (2.1) with respect to the g_{ik} and the λ_{kjl} asymptotically maximizes L^* when a set number of moments of $H(\mathbf{x})$ are estimated in the likelihood maximization process (see Tolley and Manton, 1994). As a consequence, though the unconditional likelihood L^* is the appropriate function for an application with (a) a limited number of variables and (b) a large sample, we will refer to and use L as the *likelihood*. In addition, we will formulate the problem and the estimation procedure with L because it has useful numerical properties.

2.1.3. Identifiability

A second question about the likelihood function in (2.1) is identifiability. To illustrate, note that, in principle, by judiciously choosing a set of scalars with which to multiply and divide the individual components of the summand, it may be possible to transform the g_{ik} and λ_{kjl} values, without changing the likelihood, but still satisfy Conditions I and II of Chapter 1. If this is possible, there will be more than one parameterization of the likelihood that gives the identical likelihood function value for the random variables. If such a situation exists, the parameter estimates based on maximizing (2.1) are not identified, and their particular numerical values have no meaning.

To see the problem symbolically, suppose that the set of multinomial probabilities is written in matrix form as

$$\mathbf{P} = \begin{pmatrix} p_{111} & p_{112} & \cdots & p_{1JL_J} \\ p_{211} & p_{212} & \cdots & p_{2JL_J} \\ \vdots & \vdots & & \vdots \\ p_{I11} & p_{I12} & \cdots & p_{IJL_J} \end{pmatrix} \tag{2.3}$$

We can write (1.1), for a set of I observations, in matrix form as

$$\mathbf{P} = \mathbf{\Lambda G} \tag{2.4}$$

In this expression $\mathbf{\Lambda}$ is an $I \times K$ matrix of λ_{kjl} values and $\mathbf{G}$ is a $K \times M = \Sigma_{j=1}^{J} L_j$ matrix of g_{ik} values. If $\mathbf{A}$ is any nonsingular $K \times K$ matrix, then $\mathbf{P}$ may be rewritten,

$$\mathbf{P} = \mathbf{\Lambda A}^{-1}\mathbf{AG} \tag{2.5}$$

As long as entries of $\mathbf{\Lambda A}^{-1}$ satisfy Condition II, and the entries of $\mathbf{AG}$ satisfy Condition I of Chapter 1, the entries of these two matrices can be used to parameterize $\mathbf{P}$. Thus, the parameters of (2.1) really are one point in a span of possible parameter values generated by a series of nonsingular matrices, $\{\mathbf{A}_n\}$ that satisfy Conditions I and II. Any parameterization in this span will yield the same likelihood value. Clearly the result holds for the unconditional likelihood of the previous section where $\mathbf{A}$, in effect, changes (rotates) the parameters (or moments) of $H(\cdot)$.

One approach to insuring identifiability of the parameters might be to require the parameterization to be the closest element in the span to, say, the singular value decomposition of **P**. Since the singular value decomposition is unique, it may be possible to determine a unique point in the span that would be closest to this point. This could be used as the parameterization of the span. Why choose the point based on the singular value decomposition? Though such a decomposition is unique, it is unclear that such an approach has any special justification for constructions of fuzzy partitions.

To be consistent in notation with previously published material this section presents the development of published work with a minor change in notation from x_{ij} to x_j^i, y_{ijl} to y_{jl}^i, p_{ijl} to p_{jl}^i and λ_{kjl} to λ_{jl}^k. This change was made to bring the notation closer to compatibility with tensor usage.

An alternative approach employs theorems about the properties of convex polytopes (Weyl, 1949). Let $\mathbf{Y}^i$ be a vector of binary random variables representing J independent multinomial variables observed for N persons, i.e., $\mathbf{Y}^i$ is the vector $\mathbf{Y}^i = ((Y_{jl}^i, l = 1, 2, \ldots, L_j), j = 1, 2, \ldots, J)$, where Y_{jl}^i has the value of 0 or 1, with only one nonzero entry for each J. We avoid trivial cases by assuming $L_j > 1$ for all j. Denote the expectation $E(Y_{jl}^i)$ (i.e., the probability of an event) as p_{jl}^i for all i, j, and l. The vector of expectations is $p^i = E(Y^i) = (p_{jl}^i, l = 1, \ldots, L_j), j = 1, \ldots, J)$ for an individual.

To generalize from discrete mixtures to convex sets, we define, for each j $(j = 1, 2, \ldots, J)$, $\mathfrak{M}_j$, the set of all $\mathbf{p}_j = (p_{jl}, l = 1, 2, \ldots, L_j)$, where $p_{jl} \geqq 0$ and $\Sigma_l\, p_{jl} = 1$. $\mathfrak{M}_j$ is a simplex with L_j extreme points *and* L_j facets. This implies that every $\mathbf{p}_j$ is the convex combination (unit weighted average) of L_j basis vectors $\mathbf{u}_{j1} = (1, 0, 0, \ldots, 0)$, $\mathbf{u}_{j2} = (0, 1, 0, \ldots, 0), \ldots,$ $\mathbf{u}_{jL_j} = (0, 0, 0, \ldots, 1)$, i.e.,

$$\mathbf{p}_j = p_{j1}\mathbf{u}_{j1} + p_{j2}\mathbf{u}_{j2} + \cdots + p_{jL_j}\mathbf{u}_{jL_j} \tag{2.6}$$

where $\mathbf{p}_j$ is the vector of probabilities for $\mathbf{Y}_j$ with L_j dimensions; one element of $\mathbf{Y}_j$ is unity, the rest are zero, i.e., $p_{jl} = \Pr[Y_{jl} = 1 \mid \mathbf{p}_j]$. The direct sum,

$$\mathfrak{M} = \mathfrak{M}_1 \oplus \mathfrak{M}_2 \oplus \cdots \oplus \mathfrak{M}_J$$

is the set of all "profiles" $\mathbf{p}^{\mathrm{T}}(\mathbf{p}_1^{\mathrm{T}}, \mathbf{p}_2^{\mathrm{T}}, \ldots, \mathbf{p}_J^{\mathrm{T}})$. Thus, $\mathfrak{M}$ is the space of all discrete joint probability distributions for J variables.

Lemma 1. *$\mathfrak{M}$ is a polytope.*

Proof. Since each $\mathfrak{M}_j$ is a polytope with L_j extreme points, $\mathfrak{M}$ has at most $L = \Pi\, L_j$ extreme points. Let P be the convex hull of extreme points. Then P is a polytope defined by the intersection of its supporting half spaces at the extreme points. Thus, $\mathfrak{M} \subset P$. If $p \in P$, then $p = \Sigma_{k=1}^{L} \alpha_k \xi_k$ where ξ_k are extreme points of P (and $\mathfrak{M}$). Define p_j as the vector of elements $\Sigma_{n=1}^{j-1} L_n + 1$ through $\Sigma_{n=1}^{j} L_n$ of p. Clearly, $\mathbf{p}_j = \Sigma_{k=1}^{L_j} \alpha_k^j \mu_k^{(j)}$ where $\alpha_k^{(j)} = \Sigma\, \alpha_k$ and $\mu_k^{(j)}$, $k = 1, \ldots, L_j$ are basis vectors (as in (2.6)). The sum is over k where the $(\Sigma_{n=1}^{j-1} L_n + k)$th element of ξ_k is unity. Thus, $\Sigma_{k=1}^{L_j} \alpha_k^j = 1$ and $\alpha_k^{(j)} \geq 0$ for all K and j. As a result, $p_j \in \mathfrak{M}_j$ and $p \in \mathfrak{M}$. Therefore, $\mathfrak{M} \subseteq P$. Hence, the result.

Define a probability space over $\mathfrak{M}$ by $(\mathfrak{M}, F, \Pr)$, where F is a σ-field of measurable subsets of $\mathfrak{M}$ and Pr is a σ-complete probability measure defined for each set in F. Each profile p in $\mathfrak{M}$ is an element of the probability space and the Nicodyn–Radon derivative (Hahn and Rosenthal, 1948), $f(\mathbf{p} \mid \theta)$ of Pr is assumed to exist. To obtain the density of $\mathbf{Y}$ as a function of θ it is combined with $h(\mathbf{y} \mid \mathbf{p})$.

$$\phi(y \mid \theta) = \Pr[y \mid \theta] = \int_{\mathfrak{M}} h(\mathbf{y} \mid \mathbf{p}) f(\mathbf{p} \mid \theta)\, d \Pr(\mathbf{p}) .$$

The solution is identified if the mapping $\phi : \theta \rightarrow \phi(\cdot \mid \theta)$ is one to one with probability one. Let S be the set of measurable functions mapping $\mathfrak{M}$ to $\mathbb{R}^1$. Define $L_{\mathfrak{B}}$ as the linear space generated by $\mathbf{p} \in \mathfrak{M}$ using

$$L_{\mathfrak{B}} = \left\{ \mathbf{p}^x = \int X(\mathbf{p}) \mathbf{p}\, d \Pr(\mathbf{p}), \text{ for } X \in S \right\} .$$

$L_{\mathfrak{B}}$ is a linear space equivalent to $\mathbb{R}^n$ for $n < M$ where $M = L_1 + L_2 + \cdots + L_J$. Of interest is $n \ll M$. Define $\mathfrak{B}$ as

$$\mathfrak{B} = L_{\mathfrak{B}} \cap \mathfrak{M} .$$

$\mathfrak{B}$ is the convex set, reduced by F and Pr to the minimum dimension necessary to contain all probability mass.

Corollary 1. *$\mathfrak{B}$ is a polytope.*

Proof. Since $\mathfrak{M}$ is a polytope, and $L_{\mathfrak{B}}$ a linear space, the result follows because the intersection of a linear space with a polytope is a polytope (Weyl, 1949).

Corollary 2. *$\mathfrak{B}$ is the convex hull of a finite number of unique extreme profiles $\lambda^1, \ldots, \lambda^K$, where $\lambda^k = ((\lambda_{jl}^k;\, l = 1, \ldots, L_j),\ j = 1, \ldots, J)$ for $k = 1, 2, \ldots, K$.*

Proof. The proof follows from Lemma 1 using the duality of convex sets (Weyl, 1949). Corollary 2 states that for any profile $\mathbf{p}$ in $\mathfrak{B}$ there exists a vector $\mathbf{g} = (g_k, k = 1, 2, \ldots, K)$, such that $\mathbf{p}$ can be represented as a convex combination of λ^k. Explicitly,

$$p = \sum_{k=1}^{K} g_k \lambda^{(k)} , \tag{2.7}$$

where $\mathbf{g}_k \geq 0$ for all k and $\Sigma_g g_k = 1$. Thus, the λ^k, $k = 1, \ldots, K$, are linearly combined with nonnegative weights g_k^i to represent $\mathbf{p}^i$ sample vectors. The λ^k for all k are determined by $\mathfrak{B}$ and *not* by the distribution of cases in the polytope.

A consequence of corollary 2 is the identifiability of parameters. Representing $\mathbf{p}^i$ as a convex combination of vectors is not, in general, unique. However, if the λ^k in (2.7) are extreme points of $\mathfrak{B}$, the parameterization *is* unique (see Weyl, 1949).

The representation of $\mathbf{p}^i$ as a convex combination of vectors is unique only if the λ^k used in (2.7) are extreme profiles of $\mathfrak{B}$ (i.e., vertices of the convex set B). A random sample of N profiles (assembled into $\mathbf{P}_N$) defines, with probability 1 asymptotically, the linear space. Whether the probability mass in $\mathfrak{M}$ is contained in K points, as assumed in a LCM, or distributed continuously, $\mathbf{P}_N$ has rank n where n is the dimension of $L_{\mathfrak{B}}$ with probability arbitrarily close to 1 when N exceeds n sufficiently. The convex hull of profiles in $\mathbf{P}_N$ is a random set in $\mathfrak{B}$ (since profiles are randomly sampled).

Thus, in summary, if the linear space $L_{\mathfrak{B}}$ spanned by the probability measure that generates the heterogeneity in the p_{ijl} values over individuals intersects with the polytope $\mathfrak{M}(L_{\mathfrak{B}} \cap \mathfrak{M})$, the resulting set $\mathfrak{B}$ is a polytope with a smaller number of extreme points [see Woodbury et al. (1994) for proofs]. Any realization of p_{ijl} is represented by a convex combination of the extreme points λ_{kjl} of $\mathfrak{B}$. The extreme points of $\mathfrak{B}$ correspond to what we referred to previously as extreme profiles. Thus, Conditions I and II assure that there are extreme profiles which represent any vector of multinomial probabilities for an individual as a convex combination of the λ_{kjl}. In addition, the extreme points of the polytope $\mathfrak{B}$ are unique. Thus, we may think of the extreme profiles as a natural basis for the multinomial probabilities p_{ijl} estimated for the J variables with discrete outcomes y_{ijl}. That is, that the location of the extreme profiles are fixed by $\mathfrak{M}$ and $L_{\mathfrak{B}}$, which is generated by the definition of the problem, i.e., selecting the J variables for analysis.

In conclusion, if the definition of the likelihood includes the requirement that the λ_{kjl} values define "extreme" profiles, the likelihood formulation is identifiable since these extreme points are unique. In this book, reference to the extreme profiles λ_{kjl} refers to the extreme points of a convex polytope $\mathfrak{B}$ defined by selection of J discrete variables (to form $\mathfrak{M}$ for $L_{\mathfrak{B}} \cap \mathfrak{M}$) and are *not* simply synthetic constructs for the likelihood model. In Chapter 3 we shall show the necessary theorems by which the p_{ijl} are constructed from the

y_{ijl} in the computations and provide demonstrations of their properties (e.g., the conditions for sufficiency).

2.1.4. The Poisson Likelihood Model

Often it is convenient to approximate the likelihood function by assuming that the probability that an individual has a response in any cell is given by the Poisson distribution. So far we have assumed that any random variable Y_{ijl} can take on only one of two responses, zero or unity. The Poisson assumption, on the other hand, implies that the random variable can assume any nonnegative integer value. There are two justifications for the Poisson likelihood. First, as we will see, one can modify the GoM model to describe aggregate data. In the model for aggregate, or grouped, data there are multiple cases with the same profile of frequencies. A natural model for this likelihood is to assume counts are generated by a Poisson process (see Chapter 4). A second reason for using a Poisson likelihood is as a computational device to estimate parameters of the likelihood. Clearly, for small Poisson rates, essentially only the outcomes zero and unity are likely to occur. Using this justification, if the likelihood is based on the Poisson distribution of counts, rather than the multinomial distribution, the estimation of the parameters is simplified (i.e., there is no need to impose boundary constraints such as in the use of Lagrangian multipliers). We have found that there appears to be little, if any, difficulty in numerically maximizing the Poisson likelihood. This is also true when we consider a specific form for the λ_{kjl} values in the development of an empirical Bayesian estimation scheme (see Chapter 6).

The Poisson likelihood, after the constraints to 0, 1 values are imposed, may be written,

$$L = \prod_i \prod_j \prod_l \left(\frac{\sum_k g_{ik} \cdot \lambda_{kjl}}{\sum_k g_{ik} \cdot \sum_{l=1}^{L_j} \lambda_{kjl}} \right)^{y_{ijl}} \tag{2.8}$$

Asymptotically, (2.1) and (2.8) will give similar parameter values, though they will vary in the stochasticity represented. Specifically, an additional heterogeneity component is represented in the Poisson likelihood, which can reduce the overall uncertainty, i.e., improve the fit of the model to the data. This is discussed in detail in later sections on the empirical Bayes model, where we will discuss the substantive meaning of the extra parameters in the Poisson model and how those parameters represent an additional set of weights for variables which can be used to select informative variables in item analysis.

In summary, the conditional likelihood for the fuzzy partition is given by (2.1). For the Poisson conditional likelihood, the form is given in (2.8). Methods for estimating these conditional likelihoods are discussed in Chapter 3. For the remainder of the chapter, we will discuss changes in the sample design and assumptions and illustrate how the likelihood is modified by these charges.

2.2. MODIFICATIONS OF THE LIKELIHOOD

2.2.1. Incomplete Observations

An incomplete observation arises when the realization y_{ijl} is not observed for some values of j for certain values of i. Using the paradigm above, the case of incomplete observations arises when there are some individuals who do not give responses for all J questions. Such incomplete observations are often referred to as *missing data*. In this section we will examine how different types of missing data points affects the form of the likelihood in (2.1).

Missing data can be of several types. The simplest is where the lack of observations is generated by a random mechanism operating independently of the g_{ik} scores for the individual. Thus, the one or more individuals for whom data are missing are selected at random from the sample. By Assumption 3, the missing observation is conditionally independent of observations given the Grades of Membership of the individuals that are present—even for observations made on the same individual. Consequently, the missing observation can be treated simply as an unobserved, independent observation. In this case, we simply observe that $y_{ijl} = 0$ for L_j responses and the corresponding term drops from the likelihood function (2.1) to produce the correct marginal likelihood form (see Little and Rubin, 1986). This likelihood can be estimated using methods described below.

A more complex situation is when selected variables for cases in a sample are missing *non*randomly (i.e., systematic item nonresponse). An example of this was found by Berkman et al. Certain items had a high rate of missing data on specific latent fuzzy classes. This turned out to result from the inability to respond to specific items by a person being generated by deficiencies in cognitive processes, e.g., dementia. To deal with this problem one may increase the dimensionality of the measurement space by adding a "data missing" category (i.e., an $L_j + 1$ category) for each variable. This modifies the likelihood in equation (2.1) to include a category of response for "missing" for each question. Then dependencies between $L_j + 1$ categories can be identified and possibly (if manifesting a systematic pattern) isolated in certain of the K profiles of λ_{kjl} (or, possibly, providing information to estimate additional classes). In the Berkman et al. (1989) example the missing data pattern defined a class that had a high prevalence of cognitive dementia.

An alternate form of this type of missing data is that of parallel items in questionnaire construction. Specifically, often two questionnaires, applied to different populations but intended to represent the same universe of behavior (e.g., health of the same sampled population), contain items that are worded differently (i.e., have *non*parallel construction). These may, at least, partially elicit the same information about the behavior of interest. If so, the nonparallel items should have similar relations to the other, presumably parallel, items in the questionnaire. The likelihood that this is true increases as J increases. In the analysis of the pooled observations, both items are contained in an analysis with the items not assessed being identified as missing in the appropriate population. Thus, all nonparallel questions are included for both samples with each question including a "missing response" category. The overall likelihood is formed as the product of the two resulting sample likelihoods.

A third type of missing data arises from the systematic censoring (left or right) of whole cases or observations. That is, for certain values of i there are *no* responses. This may be due to a sample defect, to a problem in field procedure, or to attrition correlated with health processes of interest (e.g., Hoem, 1985, 1989; Manton et al., 1991). In cross-sectional data one generally does not have adequate information to resolve this problem unless special data is collected through an independent, ancillary process. For example, NCHS (1966) conducted a study of survey nonresponse by linking nonresponse records to hospital records. From the independent hospital record study they found that the 5% of the sample not responding used 15% of hospital services. However, we seldom have such independent data collection mechanisms in cross-sectional data.

In longitudinal studies these problems are generally more easily resolved because we, in a sense, have replications of the individual experiments (i.e., the cross-time realizations of the process assuming process parameters are constant, or at least reasonably stable) from which to draw information. This is true in the case where we draw our sample from an administrative list (e.g., Medicare eligibility records) and persons are continuously tracked. This is more problematic in area probability samples because we lack an independently generated list; i.e., the sample list is generated in initial enumeration stage in the field work (Corder and Manton, 1991). We deal with both left and right censoring problems in the presentation of GoM models for longitudinal data in Chapter 5.

2.2.2. Complex Sample Designs

An observational design strategy other than random selection by which data is sampled may require modification of the likelihood equation in (2.1). The use of different nonrandom sampling designs will require the inclusion of "weights" with the data so that predicted values of the cell probabilities will be relevant to the population of interest. Thus, different sample designs may change the interpretation of the underlying cell probabilities λ_{kjl}, but need *not* change the representation given in (1.1). The key to identifying the effects of sample design on the GoM model parameters, is in how the likelihood is modified. If the likelihood, conditional on the g_{ik}, can still be factored on i, j, and l (e.g., Hoem, 1985; 1989), then the model is estimated as before. In this case, however, the *distribution* of the g_{ik} changes according to the sampling design. If, however, the sample design causes any of the Assumptions 1–5 to be violated, then the likelihood may not be factorable, and the estimation process must be modified to respond to the violation of the relevant assumption.

To illustrate, suppose that the sample design is such that the ith individual has a probability of being selected in a particular study is proportional to $1/w_i$. A simple random sample will implicitly set w_i to be a constant for all i. More complex designs will require different values of w_i reflecting different probabilities of sampling a case of a given type. The generalized log likelihood function for this design is given by (Woodbury and Manton, 1985).

$$\ln L(\{w_i\}) = \sum_i w_i \left\{ \sum_j \sum_l y_{ijl} \ln\left[\sum_n g_{ik} \cdot \lambda_{kjl} \right] \right\} \tag{2.9}$$

An examination of (2.9) shows that the optimal (i.e., most statistically efficient) weighting for each observation is that already obtained from the maximum likelihood function *assuming*, of course, that the likelihood function as specified is appropriate to the observational or sampling plan. Thus, using any other than the maximum likelihood weights will decrease the efficiency of parameter estimates. Therefore, unequal weights will decrease efficiency, i.e., have greater parameter variance.

Whether or not the survey observational plan fits, or is consistent with, the likelihood function may be a complicated analytic issue of substantive relevance. In some survey designs, cases are geographically clustered in Primary Sampling Units (PSU) to reduce travel and other field costs. The spatial clustering of cases in PSUs, however, may induce empirical intraclass correlation of the p^r_{jl} parameters. This intraclass correlation means that, within PSUs, the draws of individuals in the sample are not independent—a condition violating Assumption 1. A typical substantive basis for the intraclass correlation in a PSU is that persons living in the same geographic area (e.g., a county or group of counties) tend to be more homogeneous in socioeconomic status, in housing choice, and in life stage (e.g., age and number of children or career status) than a comparable number of persons randomly drawn from the entire spatial field. A number of numerical procedures have been proposed to deal with this effect. In general, the variance of estimators must be inflated (or equivalently the number of independent degrees of freedom reduced) to represent the loss of information resulting from the intraclass correlation. Procedures used to adjust the variances of estimators are Taylor series linearizations (Shah et al., 1977), various repeated sampling procedures (e.g., Balanced Repeated Replicates), the direct modeling of the intraclass correlation using an Aitken weighted estimator, or the calculation of effective sample size (Potthoff et al., 1992; 1993). Some of these procedures may become problematic (i.e., unstable) when certain sample domains are small (O'Brien, 1981).

Specifically, the presence of significant intraclass correlations implies that (2.9) is no longer appropriate [nor (2.1)] because the distribution of cases is contaminated, i.e., conditional independence of cases no longer holds due to to the intra-class correlations generated by clustering. If this is true, the factorization in (2.1) no longer is valid and we need to modify the likelihood function to represent a more general contaminated distribution of some form (e.g., the negative binomial; Kendall and Stuart, 1967). A general form representing contamination that does not rely on an assumed distribution function is,

$$L = \prod_i \prod_j T! \prod_l \left(\sum_k g_{ik} \cdot \lambda_{kjl} \right)^{y_{ijl}} \Big/ y_{ijl}! \tag{2.10}$$

where T is a grouping parameter representing membership in, say, a PSU, e.g., $\Sigma_l \, y_{ijl} = T$, which is the number of persons in the PSU (or any set of cases in a defined group or set where there is a within-class correlation not resolved by conditioning on the g_{ik} derived from the J measured variables). To test for contamination we can set $T =$ PSU size in one solution and $T = 1$ in a second solution and see if the likelihood function is significantly altered by the constraint over the sets of T cases in each group.

Specified in this way, constraints on the parameters associated with the Tth set function costs information and thus increases the error variance in the same manner as should be done to reflect the loss of information from the intraclass correlation. The function (2.10) is nested within the function $T = 1$, which is equivalent to (2.1) with the set of J variables. If the J variables contain sociodemographic, economic, and housing factors, then the basis of the intraclass correlation for the PSU may already be explained by the deterministic model (i.e., correlation of all $y_{ijl} = 0$ given g_{ik}), and a test of (2.1) versus (2.6) may show that the intraclass correlation is zero conditionally on the information already present in the model.

An alternate approach is to use an unconditional form of the model where an assumption is made about the distribution from which the g_{ik} are drawn. Specifically, in applying GoM to surveys where there is clustering (i.e., samples are not random in the sense that any sample is as probable as any other sample). The factorization postulated at the individual level in (2.1) from Assumption 3; i.e.,

$$\Pr[Y_i = y_i] = \prod_{j=1}^{J} \Pr[Y_{ij} = y_{ij}] \tag{2.11}$$

still holds regardless of the particular set of individuals in the sample.

However, the expression for

$$\Pr[Y_i = y_i, i = 1, 2, \ldots, N]$$

cannot be factored as

$$\prod_{i=1}^{N} \Pr[Y_i = y_i]$$

and must be written

$$\prod_{m=1}^{M} \prod_{i \in m} \Pr[Y_i = y_i \mid m] , \tag{2.13}$$

where the cases are divided into M clusters. Substituting (2.11) in (2.13) we get

$$\prod_{m=1}^{M} \prod_{i \in m} \prod_{j=1}^{J} \Pr[Y_{ij} = y_{ij} \mid m] .$$

The form chosen to model $\Pr[Y_{ij} = y_{ij} \mid m]$ depends on whether the profiles (i.e. the λ_{kjl}) are chosen to vary over m. Since the λ_{kjl} structure is the coordinate system for the analysis, it is apparent that the λ_{kjl} should be chosen to be the same for *all* clusters or PSUs.

If it happens that more classes (i.e., a larger K) are required to describe the data, then that is the price in information paid for a λ_{kjl} structure (i.e., the space $\mathfrak{B}$) adequately general to describe *all* sample clusters. If that choice is made (i.e., the way we wish to use our information instead of, say, using (2.10)), we may write,

$$\Pr[Y_i = y_i \mid m] = \prod_{j=1}^{J} \Pr[Y_{ij} = y_{ij} \mid m] = \prod_{j=1}^{J} \prod_{l} \left(\sum_{k} g_{ik}^{(m)} \lambda_{kjl} \right)^{y_{ijl}} \tag{2.14}$$

where the $g_{ik}^{(m)}$ are evaluated within the cluster. Equation (2.10) can be implemented in one of two ways. In conditional GoM (i.e., where the g_{ik} are directly estimated) the $g_{ik}^{(m)}$ are chosen by maximizing the likelihood in (2.14) given the λ_{kjl} structure directly. All that is required is to keep the m clusters separated in order to identify the ways in which the structure of the g_{ik} is altered.

If the unconditional form of GoM is used (2.2), where a distribution is assumed for the g_{ik} which does not violate the Jth-order moment constraint or constraint resulting from the number of distinct observations, the sample clusters will have reduced within cluster variability. This will restrict the number of moments identifiable within clusters. The most extreme form of clustering is where all cases in the cluster have the *same* g_{ik}, as in (2.10). This does not necessarily imply the same outcome vector, y_{ijl}. In this case one can aggregate the counts by outcome to obtain $y^j_{ml} = \Sigma_{i \in m} y_{ijl}$ and write

$$L = \prod_m \prod_j \prod_l \left(\sum_k g_{ik} \lambda_{kjl} \right)^{y^j_{ml}}, \tag{2.15}$$

where factorial terms are ignored.

To test whether there is a residual effect (namely, the g_{ik} are distributed within spatial clusters), we can model the within cluster distribution of the g_{ik} as, say, a Dirichlet distribution, namely,

$$f_m(g_{i1}, g_{i2}, \ldots, g_{ik}) = \Gamma(\alpha_{i\cdot}^{(m)}) \prod_k \frac{g_{ik}^{\alpha_{ik}^{(m)}-1}}{\Gamma(\alpha_{ik}^{(m)})} \tag{2.16}$$

so that

$$L = \prod_{m=1}^{M} \prod_{i \in m} \Gamma(\alpha_{i\cdot}^{(m)}) \prod_k \left(\frac{g_{ik}^{(m)\alpha_{ik}^{(m)}-1}}{\Gamma(\alpha_{ik}^{(m)})} \right) \prod_l \left(\sum_k g_{ik}^{(m)} \lambda_{kjl} \right)^{y_{ijl}}. \tag{2.17}$$

To evaluate (2.17) numerically, we need to integrate out the $g_{ik}^{(m)}$ since they are unobservable, Thus, this approach to dealing with survey spatial clustering has certain computational advantages, but would entail analytical complexity [i.e., the need to deal with (2.17)] and the need to make a distributional assumption for the g_{ik}.

Intraclass correlations in PSUs are not the only source of distributional contamination where the case independence assumption of the multinomial likelihood (2.1) could fail. If we were to examine monozygotic (MZ) twins, or populations with a higher than expected degrees of genetic homogeneity (e.g., family groups, isolated villages with high rate of intermarriage), then clusters may be formed due to shared genetic (or other) characteristics. In these cases the nature of the analytic problem is the same; i.e., they all represent some form of distributional contamination that may be resolved using one of the strategies presented above.

In cases where dependency is complex, with multiple levels of groupings, one could also generalize (2.10) to a second-order model where the dependencies among the g_{ik} are used to generate second-order fuzzy sets—in the sense of second-order factors in factor analysis (Jöreskög, 1969)—or shared commonalities among the K classes. Specifically, the matrix $\mathbf{G} = \{g_{ik}\}$ may possess a factor structure with a rank less than K. This would permit the representation of $\mathbf{G}$ as a low-rank matrix of rank $M < K$:

$$L = \prod_i \prod_j \prod_l \left[\sum_{k=1}^{K} \left(\sum_{m=1}^{M} \gamma_{im} \varphi_{mk} \right) \lambda_{kjl} \right]^{y_{ijl}} \tag{2.18}$$

At first look this might seem reducible by the substitution $\varphi_{mjl} = \Sigma_{k=1}^{K} \varphi_{mk} \lambda_{kjl}$, but this replaces at least some of the extreme profiles of $\mathfrak{B}$ by interior points if there is any $\mathfrak{M}$ with two or more nonzero values of φ_{mk}. If there are not, then $\boldsymbol{\varphi} = \mathbf{I}$ and there is no reduction. If there exists such a nontrivial pair of matrices $\boldsymbol{\varphi}$ and $\boldsymbol{\Gamma}$ that replace $\mathbf{G}$ with $\boldsymbol{\Gamma\varphi}$, a demonstrated rank of less than K can be obtained. The set of m second-order factors is calculated from the interdependency (i.e., the joint moments) of the g_{ik}. This postulates a set of second-order factors that cannot be expressed simply by increasing K. This would occur if I cases somehow were clustered to have differing relations within subsets of cases. This could be due to measurement (e.g., persons who are interviewed by the same person or by using a different mode of interview such as telephone versus in-person or individual versus proxy) or by complex subgroup interactions (e.g., outcomes of a surgical procedure is associated with a physician's skill level, usually an unobservable trait, in addition to measured prognostic factors). It would seem possible to represent this heterogeneity by expanding the number of dimensions rather than expecting K to increase, but this might generate less meaningful dimensions; i.e., such effects may be better expressed phenomenologically as second-order factors in explaining the interdependence of the first-order factors, rather than as additional, independent, first-order dimensions. This might also occur in the use of (2.18) to analyze longitudinal panel data which is characterized by R groups ($R \neq K$) of individuals with common trajectories, i.e., subgroups with characteristic tracking (Woodbury et al., 1994).

We could, in addition, use the unconditional form of the model if we were willing to assume that, say, genetic factors could be described by a Dirichlet. This would lead to a model formulated like that in (2.17) but with multiple levels of mixing distributions.

In dealing with the issue of sample weights, further questions arise in (a) choosing between superpopulation versus finite populations sampling theory (e.g., Cassel et al., 1977; Fienberg, 1989; Hoem, 1985, 1989) and (b) adjustment for nonresponse. These issues are taken up in greater detail in Chapter 4.

A second issue of practical importance is the ability to translate from the population structure implied by the sample weights (e.g., population composition differs due to stratification and proportional oversampling) to that in the actual population. This can be done in GoM by using sample weights applied to the estimated g_{ik}. For example, let Y_{ijl} be a variable whose prevalence in the population we wish to know. Let us also assume that a GoM analysis has been conducted and that $\hat{g}_{ik}$ and $\hat{p}_{ijl}$ are estimated. The prevalence rate p^*_{jl} of Y_{ijl} can be calculated from

$$p^*_{jl} = \sum_i p_{ijl} \cdot w_i \Big/ \sum_i w_i \,, \tag{2.19}$$

where $\Pr(Y_{ijl} = 1.0) = \Sigma_k\, g_{ik} \cdot \lambda_{kjl} = p_{ijl}$ and w_i is the sample weight (i.e., inverse of probability of selection). This can be written

$$p_{jl}^* = \sum_i w_i \left(\sum_k g_{ik} \lambda_{kjl} \Big/ \sum_i w_i \right) \tag{2.20}$$

by substitution for p_{ijl}, and then reordered, i.e.,

$$p_{jl}^* = \sum_k \left(\sum_i w_i g_{ik} \Big/ \sum_i w_i \right) \lambda_{kjl} , \tag{2.21}$$

which reduces to the expression

$$p_{jl}^* = \sum_k \bar{g}_k \lambda_{kjl} . \tag{2.22}$$

In this scheme the w_i, and their selection, can be viewed as a problem *independent* of the estimation of g_{ik} and λ_{kjl}, i.e., poststratification weighting (Potthoff et al., 1992; 1993). This separation of weights does not occur for other multivariate procedures such as factor analysis or principal components analysis because those procedures do not estimate the individual scores simultaneously with the structural parameters. That is, the covariance matrix, which is the basic input to principal component analysis, is altered by sample weighting. Since population composition is represented by the distribution of the g_{ik}, which are estimated independently for each person i, the structural parameters in GoM λ_{kjl} are not affected by the sample weighting.

2.3. COMPARISONS TO OTHER MULTIVARIATE MODELS

To contrast the GoM structure to more standard multivariate procedures for analyzing categorical data, in this section, we compare the GoM model with several of the more commonly used strategies for maximum likelihood factor analysis of discrete variables and with the latent class model.

2.3.1. Discrete Factor Analysis

Muthen (1978, 1987, 1989), Christoffersson (1975), Muthen and Christofferson (1981), and Jöreskög and Sörbom (1986) considered estimation of parameters for the factor analysis model from discrete response data using a modification of continuous variable maximum likelihood factor analysis (e.g., Lawley and Maxwell, 1971). Their approaches begin with the standard parameters of a latent factor model (e.g., Muthen, 1978):

$$x_{ij}^* = \Lambda_k f_{ik} + e_i , \tag{2.23}$$

where x_i^* is assumed to be discretely measured. If f_{ik} is continuous, then x_{ij}^*

should be similarly scaled. For discrete variables x_{ij}^* this is an inconsistency. This is resolved by postulating that x_{ij}^* is an indicator of a latent variable representing the dichotomous case as

$$x_{ij} = \begin{cases} 0 & x_{ij}^* < \tau \\ 1 & x_{ii}^* > \tau . \end{cases} \tag{2.24}$$

Thus persons at different distances from the threshold τ have the same observed response x_{ij}. This requires x_{ij} and x_{ij}^* be nonlinearly related; i.e., if x_{ij}^* is normally distributed with means linear in f_{ik}, the relation of the means of x_{ij} and f_{ik} is nonlinear., This can be expressed

$$\Pr(x_{ij} = 0 \mid f_{ik}) = \Pr(x_{ij}^* < \tau) = F\left(\tau - \sum_k \lambda_{jk} f_{ik}\right), \tag{2.25}$$

where F is assumed to be some appropriate response function, e.g., the standard normal (probit) or logistic. Often such a response function is provided without considering its substantive implications but is selected primarily for mathematical and computational convenience.

Estimation of the standard maximum likelihood equations for factor analysis requires that the distribution of variables is normal, or more precisely, that the sample covariances are sufficient statistics for the data (Lawley and Maxwell, 1971). For multiple discrete response variables this is a questionable assumption.

Alternatively, it may be assumed that the discrete density function p of the observed discrete variables derives from a multivariate normal density of K continuous latent variable [e.g., see (2.25)]. The relation of the J discrete variables x to K dimensional normally distributed latent factor scores Y is

$$p(x) = \int_{a_1}^{b_1} \int_{a_a}^{b_2} \cdots \int_{a_K}^{b_K} f(y)\, dy , \tag{2.26}$$

where $f(y)$ is the K dimensional multivariate normal density with covariance Σ and the limits of integration are the threshold parameters h_j defined as follows: if $x_j = 1$, then $a_j = h_j$ and $b_j = \infty$; if $x_j = 0$, then $a_j = -\infty$ and $b_j = h_j$ [see (2.24)]. Thus the likelihood involves three sets of parameters: the threshold parameters h, the factor loading parameters λ, and factor correlations ϕ. The likelihood equation is

$$\begin{aligned} L(h, \lambda, \phi) &= \sum_{i=1}^{I} p(x_i) \\ &= \sum_{i=1}^{I} \left[\int_{a_1}^{b_1} \int_{a_2}^{b_2} \cdots \int_{a_K}^{b_K} f(y)\, dy \right]. \end{aligned} \tag{2.27}$$

Computations based on (2.27) are difficult because of the multiple integrals—indeed they are possibly even more difficult than in the GoM model. The limited information procedure (Christoffersson, 1975) most often used requires only the marginal frequency M for each item j, or

$$M_j = \Pr(x_j = 1) = \int_{h_j}^{\infty} \frac{1}{\sqrt{2\pi}} \exp\left(-\frac{1}{2} y^2\right) dy \tag{2.28}$$

and the joint occurrence of item pairs, or

$$M_{jl} = \Pr(x_j = 1; x_k = 1) = \int_{h_j}^{\infty} \int_{h_l}^{\infty} \frac{1}{2\pi|\Sigma|^{1/2}} \exp\left(\frac{1}{2} y^T \Sigma^{-1} y\right) dy \tag{2.29}$$

Equation (2.29) is based on the multivariate normal distribution and extracts information only from the marginals and the second-order interactions of the discrete responses. This is equivalent to applying maximum likelihood (ML) factor analysis to tetrachoric or polychoric correlations. If the distribution deviates from normality, pathologies can result; i.e., the tetrachoric correlation matrix may not be positive definite.

The discrete variable factor analysis model presented above allows nonnormality only in observed variables. Latent variables must be normally distributed. This assumption was addressed by Muthen (1978), who related the second-order factors to first-order factors by a linear regression. If there are a large number of binary items, their sum can be used as a proxy for second-order factors. If the second-order factors are nonnormally distributed, then, because of the linear regression, the first-order factors are also nonnormal; i.e.,

$$x_{ij}^* = \Pi_j \cdot s_i + \delta_{ij} , \tag{2.30}$$

where $\Pi = \{\Pi_j\}$ is the regression coefficient, $s_i = \Sigma_{j=1}^J x_{ij}$ (assuming x_{ij}^* are dichotomous), and δ_{ij} is a normally distributed residual. Connecting, the observed second- and first-order factors is a probit function that yields a quadratic function that may be evaluated; i.e.,

$$V(x^*) = \Pi V(s) \Pi^T + \mathbf{\Omega} \tag{2.31}$$

where $\mathbf{\Omega}$ is the covariance matrix of δ, $V(s)$ is the variance of the summary scores, and the diagonals of $\mathbf{\Omega}$ are standardized to 1.0. Inserting $\mathbf{\Omega}$ and $V(s)$ sample estimates in (2.31) gives the estimated covariance matrix of the first-order factors $V(x^*)$. This produces estimates of nonnormal tetrachoric correlations of x^*. Factor scores are calculated conditionally from model parameters by Bayes' theorem using a multivariate normal as a prior distribution. Clearly, the approach involves sequential use of sample data and a variety of functional form and distributional assumptions. The fact that factor scores are calculated by application of Bayes' rule *after* estimation indicates the dependency of the solution on the assumed form of the multivariate score distribution. The GoM model is clearly different because no specific form is assumed for the moment distribution $H(\mathbf{x})$ of the latent variables g_{ik} and because the two sets of parameters (i.e., g_{ik} and λ_{kjl}) are simultaneously estimated.

Anderson and Amemiya (1988) suggest that we not be too concerned about distributional assumptions in confirmatory maximum likelihood factor analysis and in the use of ML in estimating linear structural relationships. That is, the assumption of normality is not required for estimation of latent factors. However, it *is* required that all relevant information reside in the covariance matrix (i.e., is resident in second-order moments). If higher-order moments are important, information is lost (i.e., the covariance matrix does not represent sufficient statistics for the data unless the data are normally distributed). It is always possible to use the multivariate normal as the maximum entropy, minimum information representation of a more general multivariate distribution. That is, it is the most conservative (i.e., has largest error bounds), nontrivial distribution that can be used to parametrically summarize the data. Note, however, that much information generated from bounded random variables, as in the current case, is often contained in higher-order moments.

An alternate approach to analyzing contingency tables using the bivariate normal approximation is due to Kendall and Stuart (1967), among others. In the usual case the L_j categories of a two-way contingency table are analyzed using canonical correlation procedures. The extraction of each canonical correlation corresponds to the extraction of information represented by a bivariate correlation surface. The set of eigenvalues associated with each canonical correlation represents a Hermite–Tchbycheff polynomial series that expresses the joint relation of the two marginal variables as a function of those eigenvalues. This decomposition into independent bivariate normal response surfaces leads to the development of the χ^2 statistic and its partitioning for contingency table analyses. The limitation of this approach is in not generating unique correlation measures when three or more variables are involved (Lancaster, 1969). Put in other words, there is an intrinsic lack of identifiability in determining correlation measures in the three-way or higher table (see Kettenring, 1982). This is manifest in multiway contingency table analysis when the results of tests are not invariant to the order of extraction of different interactions.

2.3.2. Latent Class Models

Lazarsfeld and Henry (1968) developed procedures for discrete response data *without* assuming that latent variables are normally distributed. Specifically, they assumed that discrete data may be explained as a weighted combination of K latent tables. On initial examination, the LCM and the GoM models may seem similar in theory and implementation. However, as we will show, LCM and GoM are different conceptually.

First, let us show how LCM generalizes log-linear or other maximum likelihood procedures for multiway contingency table analyses (Bishop et al., 1975; Haberman, 1974). In our notation the classical contingency table model entails estimation of the cell probabilities of J-way tables, where there are $\Pi_j L_j$ such cell probabilities to be estimated from the set of discrete responses y_{ijl}. Such a table can be formulated for each of K independent, observed populations or groups, assuming that the K set assignment functions are known and measured without error. In the classical formulation, the g_{ik} are assumed observed for each of K populations and take the value 0 or 1 for each person.

LCM generalizes the contingency table analysis for K groups in that the g_{ik} are *not* observed and must be estimated from the data. They are, however, still assumed to be 0 or 1 in all cases. Thus, LCM generalizes multiway contingency table models by identifying K latent sets of persons with cell probabilities which can be computed as products of the λ_{kjl}. There are other less statistically rigorous, ad hoc procedures for identifying K latent discrete sets. One that is commonly used, is the *autogroup procedure*, which optimizes a discrete partitioning based on a least squares criteria (Carmelli et al., 1991). However, recursive partitioning models, of which autogroup is an example, are sensitive to the order of inclusion of variables. Furthermore, there is no rigorous basis for evaluating their fit to the data. These ad hoc procedures have often been used in identifying patient groups for reimbursement, e.g., the diagnosis related groups (DRGs) used to reimburse acute hospital stays for persons eligible for the Federal Medicare program. The advantage of LCM over these recursive partitioning procedures is that estimation in LCM uses full information maximum likelihood techniques to estimate a truly multivariate model. Hence, inference is well defined from likelihood principles, as is the basis for accepting a given solution, i.e., determining the true value of K.

In LCM, the latent state variable $C = C_i$ takes only one value in the range $C_i = 1, \ldots, K$. Here the variable $C = C_i$ identifies the latent class from which the ith individual is drawn. This corresponds to a restricted form of GoM in which $g_{ik} = 1$ for $C = k$ and $g_{ik} = 0$ for $C \neq k$; i.e., heterogeneity in the observations is assumed to be generated by a mixture of observations from K latent crisp sets. This difference is evident primarily in terms representing within cell heterogeneity associated with the g_{ik} scores in GoM. We review LCM in detail to make clear its mathematical differences with GoM.

The distribution of responses in LCM is a finite mixture of K discrete distributions, each satisfying local independence (Suppes and Zanotti, 1981). To compare models we will use common notation. Let $P_{ab\cdots e}^{AB\cdots E}$ be the joint probability that variables $A, B, \ldots, E$ have outcomes $a, b, \ldots, e$, i.e.,

$$P_{a,b,\ldots e}^{A,B,\ldots E} = \Pr(A = a, B = b, \ldots, E = e)\,. \tag{2.32}$$

P_k^C is the probability that latent variable C has outcome k; i.e.,

$$P_k^C = \Pr(C = k)\,, \tag{2.33}$$

and $P_{ak}^{\bar{A}C}$ is the conditional probability of outcome a on variable A, given $C = k$; i.e.,

$$P_{ak}^{\bar{A}C} = \Pr(A = a \mid C = k)\,. \tag{2.34}$$

Using this notation LCM can be expressed as,

$$P_{a,b,\ldots e}^{A,B,\ldots E} = \sum_k P_{ak}^{\bar{A}C} P_{bk}^{\bar{B}C} \cdots P_{ek}^{\bar{E}C} \cdot P_k^C\,, \tag{2.35}$$

where each factor on the of (2.35) is a parameter to be estimated. Note that (2.35) assumes that given $C = k$, the variables $A, B, \ldots, E$ are independent.

This is standard notational usage in LCM (e.g., Clogg, 1981). Consistent notation, using the posterior probability vector p_{ik}, is obtained from (2.35) using Bayes formula as

$$p_{ik} = \Pr(C_i = k \mid A_i = a_i; B_i = b_i, \ldots, E_i = e_i) \tag{2.36}$$

where $a_i, b_i, \ldots, e_i$ are data for individual i. The structural parameters λ_{kjl} in GoM satisfy

$$\lambda_{kjl} = P_{lk}^{\overline{A(j)C}}\,, \tag{2.37}$$

where $A(j)$ refers to the jth question. This in turn can be rewritten as

$$\lambda_{kjl} = \Pr(X_{ij} = l \mid k_i = k)\,. \tag{2.38}$$

Put in words these two formulas express λ_{kjl} as the probability that the ith individual will give response l to the jth question given that the individual is a member of the crisp latent set indexed by k. The conditional probability of response $l_1, \ldots, l_J$ for the ith individual, given that the individual is a member of the (latent) crisp set indexed by k is

$$P_{l_1 \cdots l_J k}^{\overline{A(1)\cdots A(J)C_i}} = \prod_j \lambda_{kjl_j}\,. \tag{2.39}$$

Using the binary indicator variable y_{ijl} and the set assignment parameter g_{ik}, equation (2.39) becomes

$$P_{l_1 \cdots l_J k}^{\overline{A(1)\cdots A(J)C}} = \prod_j \left(\sum_k g_{ik} \lambda_{kjl_j}\right)^{y_{ijl}}. \tag{2.40}$$

This is exactly the conditional probability under the GoM model. However, implicit in equation (2.40) is the fact that the set assignment parameter g_{ik} is either 0 or 1 since a crisp set membership was assumed, though it is unknown. Put in words, equation (2.40) states that the conditional probability for LCM and for GoM is the same provided that each individual is a crisp member of *one* latent class. It is this crucial assumption that distinguishes the two models., Under GoM, the membership in sets can be fuzzy. In *both* models the sets are latent.

Without the ability to model individual heterogeneity, the LCM provides very different marginal probabilities.

In GoM notation, the LCM marginal probabilities are

$$P_{l_1 l_2 \cdots l_J}^{A(1)A(2)\cdots A(J)} = \sum_k \lambda_{k1l_1} \lambda_{k2l_2} \cdots \lambda_{kJl_J} P_k \,. \tag{2.41}$$

Note that when put in this form it is clear that $P_{l_1 l_2 \cdots l_J}^{A(1)A(2)\cdots A(J)}$ is a weighted average of the λ_{kjl} probabilities with the weights given by P_k. P_k in turn is the proportion of the population which are members of the crisp latent class indexed by k. Thus,

$$P_k = \lim_{I \to \infty} \sum p_{ik} / I \,, \tag{2.42}$$

where p_{ik} is as defined in (2.36).

Continuing with the notation above, using the indicator variables y_{jkl}, equation (2.41) becomes

$$P_{l_1 l_2 \cdots l_J}^{A(1)A(2)\cdots A(J)} = \sum_k P_k \prod_j \lambda_{kjl_j}^{y_{jl}} \,. \tag{2.43}$$

Compare this to the GoM model for this probability given in (2.2) and reproduced here for a single individual,

$$P_{l_1 l_2 \cdots l_J}^{A(1)A(2)\cdots A(J)} = \int \prod_j \prod_l \left(\sum_k g_k \lambda_{kjl} \right)^{y_{jl}} dH(g_k) \,. \tag{2.44}$$

Thus, LCM is a special (nested) case of GoM where in addition to $\Sigma_k g_{ik} = 1$ the g_{ik} assume only the values 0 or 1. With conditional independence we can form likelihood equations with assumptions about processes generating observations.

To illustrate the differences between the g_{ik} parameters and the p_{ik} consider Table 2.1.

The top panel illustrates the computation of $\Pi_{ijl} = \Pr(y_{ijl} = 1)$ for two variables with two latent classes (Manton et al., 1992). The numbers represent the probability of each response (variable 1, 2; yes/no; rows 1–4) for each class (λ_{kjl}; columns 1–2) for person i for LCM and GoM (Π_{ijl}; columns 3–4) using LCM posterior probability $p_i = (.6, .4)$ and GoM scores $g_i = (.6, .4)$. The GoM prediction of Π_{ijl} is obtained from (2.44) assuming a

Table 2.1. Comparisons of LCM and GoM

1. Computation of Individual Level Marginal Probabilities (%)

	Class 1	Class 2	$\Pr(y_{ijl}=1;\text{ LCM})$	$\Pr(y_{ijl}=1;\text{ GoM})$
Variable 1				
Yes	90.0	20.0	62.0	62.0
No	10.0	80.0	38.0	38.0
Variable 2				
Yes	70.0	40.0	58.0	58.0
No	30.0	60.0	42.0	42.0
P_{ik}	60.0	40.0		
g_{ik}	60.0	40.0		

2. Computation of Class Specific Joint Probabilities (%)

	Variable 2, Class 1		Variable 2, Class 2	
	Yes	No	Yes	No
Variable 1				
Yes	63.0	27.0	8.0	12.0
No	7.0	3.0	32.0	48.0

3. Computation of Individual Level Joint Probabilities (%)

	Variable 2, LCM		Variable 2, GoM	
	Yes	No	Yes	No
Variable 1				
Yes	41.0	21.0	36.0	26.0
No	17.0	21.0	22.0	16.0

discrete point mass distribution at g_{ik} and the LCM is $\Pi_{ijl} = \Pr(y_{ijl}=1) = \sum_k p_{ik} \cdot \lambda_{kjl}$.

Since $p_i = g_i$ the Π_{ijl} are identical in columns 3 and 4. Thus, both models produce the *same* marginal probabilities when $g_i = p_i$. The second panel illustrate class specific *joint* probabilities of response for the two models. Both models assume local independence for their classes; i.e., the joint probabilities for Classes 1 and 2 are generated by multiplying corresponding marginals.

The third panel computes individual joint probabilities. The models produce different joint probabilities for persons whose class $i \neq k$, for a k in the set $k = 1, 2, \ldots, K$. For LCM the probabilities $\Pi_{ijl_1j_2l_2}$ are a weighted average of the joint probability for Classes 1 and 2. For GoM, they are a product of corresponding marginal probabilities. Panel 3 shows that LCM and GoM produce different estimates of joint probabilities even though the values of p_{ik} are set equal to the values of g_{ik} and the marginal probabilities are identical.

The difference in the models is that the g_{ik} are state variables; i.e., that observed variables are independent conditional on g_{ik}. In LCM, the variable C is a state variable denoting latent class, but the p_{ik} are not state variables. Thus, the observed variables are not independent (generally) when conditioning on p_{ik}. This produces different behavior as J is increased. In LCM as J is increased, $p_{ik} \rightarrow 1.0$ for the correct class. In GoM as J increases, the vector $\hat{g}_{ik} \rightarrow g_{ik}$, i.e., the true values of the g_{ik}.

The standard likelihood expression for LCM—assuming the g_{ik} are missing data (Everitt, 1984)—is

$$L = \prod_{i=1}^{I} \sum_{k=1}^{K} P_k \prod_{j=1}^{J} \prod_{l=1}^{L_j} p_{kjl}^{y_{ijl}} . \tag{2.45}$$

For the basic model this is

$$E[Y_{ijl}] = \sum_{k=1}^{K} P_k \cdot p_{kjl} \tag{2.46}$$

$$= \sum_{k=1}^{K} [\Pr(g_{ik} = 1) \cdot p_{kjl}] , \tag{2.47}$$

where the individual's class is *not* known. Knowledge of class membership is equivalent to knowing for which class $g_{ik} = 1.0$. The likelihood in (2.45) does *not* explicitly contain g_{ik}. It represents the likelihood over all classes the individual might occupy.

Because membership is *not* known a priori, to use LCM in a way analogous to GoM to generate discrete (as opposed to fuzzy) classifications means reparameterizing LCM with the g_{ik} rather than the P_k. To do this, (2.45) is first modified as

$$L = \prod_{k} \prod_{i \in d_k} \prod_{j} \prod_{l} p_{kjl}^{y_{ijl}} \tag{2.48}$$

where P_k is replaced by $\Pi_{i \in d_k}$, i.e., the product over all persons in class k, defined here as the index set d_k. Since, in LCM, $g_{ik} = 1.0$ for $i \in d_k$, this can be written,

$$L = \prod_{i} \sum_{k} g_{ik} \prod_{j} \prod_{l} p_{kjl}^{y_{ijl}} . \tag{2.49}$$

Since $g_{ik} = 0$ or 1 only, this may be rearranged to obtain

$$L = \prod_{i} \prod_{j} \prod_{l} \left(\sum_{k} g_{ik} \cdot p_{kjl} \right)^{y_{ijl}} , \tag{2.50}$$

which is the same form as GoM with g_{ik} constrained to be either 1 or 0. This form of the likelihood might have been anticipated from the marginal probability in (2.40).

The translation of GoM into LCM proceeds from

$$L = \prod_{i=1}^{I} \prod_{j=1}^{J} \prod_{l=1}^{L_j} \left[\sum_{k=1}^{K} g_{ik} \cdot \lambda_{kjl} \right]^{y_{ijl}} . \tag{2.51}$$

If $g_{ik} = 1$ or 0 and one wishes only to know the proportion of g_{ik} that are unity for class k for a specific k then the likelihood can be written

$$L = \prod_{i} \prod_{j} \prod_{l} \left(\sum_{k=1}^{K} \left[\sum_{m=1}^{I} g_{mk} \right] \Big/ I \right) \cdot (\lambda_{kjl})^{y_{ijl}} , \tag{2.52}$$

where $[\Sigma_{m=1}^{I} g_{mk}]/I = P_k$ when $g_{ik} = 1$ or 0. Further substitution yields

$$L = \prod_{i} \prod_{j} \prod_{l} \left(\sum_{k} P_k \cdot (\lambda_{kjl}^{y_{ijl}}) , \right. \tag{2.53}$$

which, because P_k is independent of i, can be written as the original LCM likelihood.

From (2.52) we see that the LCM is nested in the GoM model structure and that a likelihood ratio test can be formed for fixed K to determine if LCM or GoM better describes a set of data; i.e., whether the within-class heterogeneity explained by the g_{ik} is significant.

2.3.3. A Comparison of the GoM and LCM Structures

How can it be determined when to select LCM versus GoM? When K, the number of extreme profiles or LCM classes, is fixed, a likelihood ratio test of the empirical performance of the two models can be formed using the ratio of the likelihood (2.53) estimated with the g_{ik} constrained to be 0 or 1 to the likelihood (2.53) estimated without this constraint on the g_{ik}. If GoM fits significantly better than LCM (because it is the more general model with an additional heterogeneity component and more parameters associated with the freedom of the g_{ik}), this suggests there is significant information in the g_{ik}. In LCM, that information is necessarily forced into the p_{kjl} (i.e., g_{ik} must be 0 or 1) and will affect their estimated values. Thus, estimation of the g_{ik} filters extraneous "noise" at the individual level from the fundamental patterns (or signals) within the data (i.e., the λ_{kjl}, location or structural, parameters).

If, however, we allow LCM to have more classes (i.e., $M > K$), then it is potentially possible to "fit" the realized data set as well as with GoM by allowing M to be as large as needed. In this case other substantive and theoretical criteria have to be employed to evaluate the models (Singer, 1989).

If the joint occurrence of K subsets of J^* of the J variables is generated by K latent J^* dimensional *processes* with parameters varying over individuals (which generates variation in the subset of the J measures realized for any individual), then GoM, using g_{ik}, is substantively correct. If the set of J^* outcomes associated with a class is generated by a process where parameters do *not* vary over individuals and person i is affected only by the latent process associated with the mth class, then LCM is correct; i.e., it will perform better in independent replications.

Though we may find an M-class LCM that fits data as well as a K-class GoM model, the fact that each class's outcome is generated by different underlying processes means that the stochastic outcomes (e.g., replication in new data) need not (i.e., will generally not) be the same. For one thing the use of discrete classes means that if persons truly vary in outcome, some persons will be close to the boundary of a class and have a high likelihood of misclassification (Rudemo, 1973). Alternatively, two persons close to the same boundary in two adjacent classes may be more alike than persons at extreme points in the same class. Because GoM may locate a person's state anywhere within the state space simplex, this threshold instability does not exist.

Also, one way to evaluate the ability of the model to forecast is to estimate parameters for ancillary variables in a class (i.e., assess their predictive validity; Reuben et al., 1992). Explicitly, classifications are often performed in order to predict prices for medical services. For example, for Medicare, 473 DRG categories were created using a recursive partitioning algorithm (Mills et al., 1976; Pettingill and Vertrees, 1982). As the number of classes increases, so that there are more p_{kjl} (i.e., $K = 473$) to describe variables, the average number of cases in any class declines, and both the p_{kjl} and ancillary parameter estimates (i.e., the costs in each class K) will decline in stability. In GoM, since g_{ik} are estimated for all persons, such instability in the g_{ik} estimates is absent. Furthermore, cost estimates are derived for each of the K fuzzy classes, so that they are likewise precisely estimated. The discrete class procedures used to generate groups for reimbursement are often validated by using independent samples and reestimating measures of model fit. This, however, is misleading as a method of validation for the purpose of reimbursement since the failure to reject a model in the independent sample confirms not the validity of the M groups for setting costs (e.g., 473 DRGs) but only that the groups cannot be rejected given the information in the independent sample.

2.4. SUMMARY

In this chapter we presented the likelihood formulation of the GoM model. We then compared the model to a number of other multivariate techniques that have been applied to discrete response data. The technique closest in purpose to GoM, and which most closely satisfied the theoretical assumptions required in analyzing multivariate discrete response data with latent constructs, was LCM. A careful examination of the mathematical structure of the two models was undertaken to make clear the implications of parameterizing for fuzzy versus crisp sets.

REFERENCES

Anderson, T. W., and Amemiya, Y. (1988). The asymptotic normal distribution of estimators in factor analysis under general conditions. *Annals of Statistics* **16**: 759–771.

Berkman, L., Singer, B., and Manton, K. G. (1989). Black/white differences in health status and mortality among the elderly. *Demography* **26**(4): 661–678.

Bishop, Y. M., Fienberg, S. E., and Holland, P. W. (1975). *Discrete Multivariate Analysis: Theory and Practice*. MIT Press, Cambridge, Massachusetts, and London.

Carmelli, D., Halpern, J., Swn, G. E., Dame, A., McElroy, M., Gleb, A. B., and Rosenman, R. H. (1991). 27-year mortality in the western collaborative group study: construction of risk groups by recursive partitioning. *Journal of Clinical Epidemiology* **44**(12): 1341–1351.

Cassel, C.-M., Sarndal, C.-E., and Wretman, J. H. (1977) *Foundations of Inference in Survey Sampling*. Wiley, New York.

Christoffersson, A. (1975). Factor analysis of dichotomized variables. *Psychometrika* **50**: 5–32.

Clogg, C. C. (1981). New developments in latent structure analysis. In *Factor Measurement in Sociological Research: A Multi-Dimensional Perspective* (D. J. Jackson and E. F. Borgetta, Eds.). Sage, Beverly Hills, California.

Corder, L. S., and Manton, K. G. (1991). National surveys and the health and functioning of the elderly: The effects of design and content. *Journal of the American Statistical Association* **86**: 513–525.

Cox, D. R., and Hinkley, D. V. (1974). *Theoretical Statistics*. Chapman and Hall, London.

Everitt, B. S. (1984). *An Introduction to Latent Variable Models*. Chapman and Hall, London and New York.

Fienberg, S. E. (1989). Modeling considerations: Discussion from a modeling perspective. In *Panel Surveys* (D. Kasprzyk, G. Duncan, G. Kalton, and M. P. Singh, Eds.). Wiley, New York, pp. 566–574.

Haberman, S. J. (1974) *Analysis of Frequency Data*. University of Chicago Press, Chicago and London.

Hahn, D. D., and Rosenthal, D. D. (1948).

Hoem, J. (1985). Weighting, misclassification and other issues in the analysis of survey samples of life histories. In *Longitudinal Analysis of Labor Market Data* (J. Heckman and B. Singer, Eds.). Cambridge University Press, Cambridge, Massachusetts, pp. 249–293.

Hoem, J. M. (1989). The issue of weights in panel surveys of individual behavior. In *Panel Surveys* (D. Kasprzyk, G. Duncan, G. Kalton and M.P. Singh, Eds.). Wiley, New York, pp. 539–559.

Jöreskög, K. G. (1969). A general approach to confirmatory maximum likelihood factor analysis. *Psychometrika* **34**: 183–202.

Jöreskög, K. G., and Sörbom, D. (1986). *LISREL VI: Analysis of Linear Structural Relationships by Maximum Likelihood and Least Square Methods*. Scientific Software, Mooresville, Indiana.

Kendall, M. G., and Stuart, A. (1967). *The Advanced Theory of Statistics, Vol. 2: Inference and Relationship*. Hafner, New York.

Kettenring: *Biometrika*, 1982. To be supplied later.

Lancaster, H. O. (1969). *The Chi Squared Distribution*. Wiley, New York.

Lawley, D. N., and Maxwell, A. E. (1971). *Factor Analysis as a Statistical Method*. Elsevier, New York.

Little, R. J. A., and Rubin, D. B. (1984). *Statistical Analysis with Missing Data*. Wiley, New York.

Manton, K. G., Stallard, E., and Woodbury, M. A. (1991). A multivariate event history model based upon fuzzy states: Estimation from longitudinal surveys with informative nonresponse. *Journal of Official Statistics* (published by Statistics Sweden, Stockholm) **7**: 261–293.

Manton, K. G., Woodbury, M. A., Corder, L. S., and Stallard, E. (1992) The use of Grade of Membership techniques to estimate regression relationships. In *Sociological Methodology, 1992* (P. Marsden, Ed.). Basil Blackwell, Oxford, England, pp. 321–381.

Mills, R., Fetter, R., Reidel, J., and Aderill, R. (1976). AUTOGRP: An interactive computer system for the analysis of health care data. *Medical Care* **14**: 603–615.

Muthen, B. (1978). Contributions to factor analysis of dichotomous variables. *Psychometrika* **43**: 551–560.

Muthen, B. (1987). LISCOMP: Analysis of Linear Structural Equations with a Comprehensive Measurement Model. Scientific Software, Mooresville, Indiana.

Muthen, B. (1989). Dichotomous factor analysis of symptom data. *Sociological Methods and Research* **18**: 19–65.

Muthen, B., and Christoffersson, A. (1981). Simultaneous factor analysis of dichotomous variables in several groups. *Psychometrika* **46**: 407–419.

National Center for Health Statistics (1966). *Vital and Health Statistics Data Evaluation and Methods Research*, Computer Simulation of Hospital Discharges. PHS PUb. No. 1000-Series 2, No. 13. USGPO, Washington, DC.

O'Brien, K. F. (1981). Life table analysis for complex survey data. Department of Biostatistics, University of North Carolina at Chapel Hill, Institute of Statistics Mimeo Series 1337 (April, 1981), Chapel Hill, NC.

Pettingill, J., and Vertrees, J. C. (1982). Reliability and validity in hospital case-mix measurement. *HCF Review* **4**(2): 101–128.

Potthoff, R. F., Woodbury, M. A., and Manton, K. G. (1992). Equivalent sample size and equivalent degrees of freedom refinements for inference using survey weights under superpopulation models. *Journal of the American Statistical Association, Theory and Methods* **87**: 383–396.

Potthoff, R. F., Manton, K. G., and Woodbury, M. A. (1993). Correcting for nonavailability bias in surveys by weighting based on number of callbacks. *Journal of the American Statistical Association* **89**: 1197–1207.

Reuben, D. B., Siu, A. L., and Kimpau, S. (1992). The predictive validity of self-report and performance-based measures of function and health. *Journal of Gerontology: Medical Sciences* **47**(4): M106–M110.

Rudemo, M. (1973). State estimation for partially observed Markov chains. *Journal of Mathematical Analysis and Applications* **44**: 581–611.

Shah, B. V., Holt, M. M., and Folsom, R. E. (1977). Inference about regression models from sample survey data. *Bulletin of the International Statistical Institute* **67**: 43–57.

Singer, B. H. (1989). Grade of membership representations: Concepts and problems. In *Festschreift for Samuel Karlin* (T. W. Anderson, K. B. Athreya, and D. Iglehardt, Eds.). Academic Press, Orlando, Florida, pp. 317–334.

Suppes, P., and Zanotti, M. (1981). When are probabilistic explanations possible? *Synthese* **48**: 191–199.

Tolley, H. D., and Manton, K. G. (1994). Testing for the number of pure types in a fuzzy partition. In review at *Journal of the Royal Statistical Society, Series B*.

Weyl, H. (1949). The elementary theory of convex polyhedra. *Annals of Mathematics Study* **24**: 3–18.

Woodbury, M. A., and Manton, K. G. (1985) Grade of Membership analysis of complex sample design data: An approach. Presented at a National Center for Health Statistics Seminar, Sept. 26, 1985, Washington, DC.

Woodbury, M. A., Manton, K. G., and Tolley, H. D. (1994). A general model for statistical analysis using fuzzy sets: Sufficient conditions for identifiability and statistical properties. *Information Sciences*, in press.

CHAPTER 3

Estimation of the Parameters of the GoM Model

INTRODUCTION

In Chapter 2 we discussed the form of the likelihood function for the GoM model. In this chapter we discuss estimation of the parameters of the GoM model using maximum likelihood principles. Although maximum likelihood methods generally require iterative computations, this type of estimator is often preferred both because of its statistical properties and heuristic appeal. For the GoM model, however, the properties of the estimators that maximize equation (2.1) must be considered carefully for two reasons. First, the number of g_{ik} parameters increases as the number of observations increases. Consequently, properties of consistency and asymptotic distribution convergence would not be expected to hold without some type of additional "smoothing" constraints on parameter estimators. These constraints in GoM come from the properties of the convex space which defines the bounds of parameter space and hence are based on fundamentally different principles than the more usual estimators. Second, since parameters are constructed to lie in a bounded region, in practice many of the parameter estimates may lie on the boundaries of this region. As a result, large sample properties of the estimates and test statistics need modification to reflect the effects of these boundary constraints on the distribution of the statistics and consequently the formation of the decision criteria for evaluating model features (see Self and Liang, 1987). In this chapter we briefly present results on both sets of properties. A more extensive development is accessible in published references (e.g., Manton and Stallard, 1988; Tolley and Manton, 1992, 1994; Woodbury et al., 1994).

A second concern in maximizing the likelihood function is its computational complexity. Not only are the parameters constrained to lie in a compact region, but the number of parameters that must be simultaneously considered is very large. As a result, modified Newton–Raphson techniques are required. This chapter presents an iterative method of maximizing the likelihood equation by using the missing information principle (MIP; Orchard and Woodbury, 1971) to modify classical optimization techniques (Cox and Hinckley, 1974). We also discuss some of the practical issues involved in its maximization.

3.1. CONSISTENCY OF MAXIMUM LIKELIHOOD ESTIMATES

There are two statistical properties of maximum likelihood estimates of GoM parameters that are of immediate interest. The first is whether or not parameters are consistently estimated. That is, as the number of observations increase (more generally, the information in the data set), will the variance of the maximum likelihood parameter estimators decrease to zero? The second is the large-sample distributional characteristics of the parameters. For example, often the asymptotic distribution of maximum likelihood estimators, suitably standardized, is normal or Gaussian, producing test statistics and decision functions of a convenient form.

The estimates produced from (2.1) will have some unique statistical properties because GoM parameters (g_{ik}) are estimated for each individual. Heuristically, we would not expect estimates of the individual g_{ik} values to be consistent (given fixed K and J) as the number of individuals I increases. However, the estimates of (a fixed finite number of) parameters of $H(\mathbf{x})$ (the distribution of the ξ_{ik} of which the g_{ik} are realizations) will be consistently estimated. We discuss this, and a second way (based on generalization of the measurement space) that consistency of parameter estimates may be achieved. Once consistency is established, the large-sample distribution properties of the parameters is discussed.

3.1.1. Consistency with J Increasing

We saw in (2.1) that g_{ik} is constrained over j while λ_{kjl} is constrained over i. Thus, while we expect that the λ_{kjl} would be better estimated as I increases (for fixed K and J), the number of g_{ik} increases directly with the sample size I. However, the stability of the g_{ik}, if K and I are fixed, will improve if J increases, assuming that the $(J + 1)$th measure is not increasingly correlated with the first J measures as J increases, i.e., the information on individual i increases as J increases while K is fixed. This can be shown by considering the behavior of a random walk with an infinite number of discrete steps (e.g., Gillespie, 1983; Kushner, 1974). As the number of discrete steps (each step corresponding to the jump in information when a new variable is added to the analysis) increases to infinity, the discrete-step random walk can be shown to better approximate a continuous-time continuous-state stochastic process where, if step size is independently generated, the process is Gaussian. Woodbury (1963) showed this formally for a Wiener process where the number of test items was increased (test items or questions representing the discrete, independent steps in "information") until the "true" test scores indicated by the items were well approximated. Thus, in this case the individual g_{ik} are consistently estimated as K is fixed for increasing J, or, more generally, if J simply increases much faster than K.

This approach to consistency is used both in mental test theory (e.g., Lord and Norvick, 1968) and in models of stochastic processes and time series. In a number of applications of GoM we have found that J is often large enough (e.g., 198 symptoms of schizophrenic disorders in Manton et al., 1993; or the nursing home example presented in Chapter 5 where we have 111 variables and 406 answers) that the g_{ik} appear well estimated for the individual. In such cases one can examine the individual's vector of the g_{ik} and make inferences about outcomes at the individual level using the scores. It can be shown that the g_{ik} (for the correct value of K) contain all significant information in the y_{ijl} and can be useful as summary indices in actuarial underwriting formulas where it is necessary to predict the risk of an individual for a "catastrophic," but rare, loss (Tolley and Manton, 1991). This also has interesting philosophical relations to the physician's task of uniquely characterizing the individual patients manifestation of a disease—and not just the disease. For example, in the extreme, one could always kill tumor cells and eradicate a cancer by increasing the dose in chemotherapy. However, as that dose is escalated, more side effects are generated, and the physician's task of keeping the patient in homeostasis (i.e., "alive") becomes increasingly complex as the therapy affects the function of more systems (liver, kidneys, bone marrow, etc.).

3.1.2. Consistency for I Increasing (J Constant)

For most applications (and data collection designs), the number of measures J is usually fixed (and relatively small), but I can be readily increased. This is the standard condition under which consistency is evaluated (i.e., I increases to ∞ with K and J fixed). In this case we can consider the arguments of Kiefer and Wolfowitz (1956), who provide important "bridging" concepts for inference in the fuzzy set case. Specifically, they deal with inference in the case of infinitely many "nuisance" parameters by integrating across the nuisance parameters to get a restricted set of moments to represent the distribution of the nuisance parameters for which information does increase as I increases (Heckman and Singer, 1984a,b; Manton and Stallard, 1988; Woodbury et al., 1993).

To modify the arguments of Kiefer and Wolfowitz (1956) to apply to GoM, we assume that the distribution $H(\mathbf{x})$ is completely specified (up to an equivalence class; see Section 3.2.1) by the moments of order J or less. This means two distributions that differ only in moments of order greater than J are considered equivalent. Mathematically, this amounts to creating an equivalence class of distributions for $H(\mathbf{x})$. Provided that the data are such that there are no statistical identifiability problems (discussed below in Section 3.2) the consistency of λ_{kjl} and *moments* of $H(\mathbf{x})$ can be established using the results of Kiefer and Wolfowitz (1956) for this equivalence class of distributions. Note that, in this case, (2.2) is actually the likelihood being maximized, and not (2.1). Inferences and large sample properties of a fitted GoM model must, in these analyses, be limited to the estimators of λ_{kjl} and of the low-order moments of $H(\mathbf{x})$. In this definition the order of moments, J or less, is identical to the number of "questions" asked of, or measurements made for, each individual (Tolley and Manton, 1992, 1994).

3.2. STATISTICAL IDENTIFIABILITY

3.2.1. Theoretical Considerations: Necessary Conditions for Parameter Identifiability

We discussed identifiability in Section 2.2 with regard to the unique *mathematical* specification of the likelihood. In this section we discuss the "statistical" identifiability of parameter estimates (i.e.) the ability to generate unique parameterization from the available data. Even though the likelihood and the probability density function may be identifiable in that they are uniquely specified by the probability distribution of the observable random variables, in practice we may not have enough data to estimate all of the structural parameters λ_{kjl}. In the fuzzy partition case the data "shortfall" may be because the number of extreme profiles K in the model is too large, the number of questions J asked each individual in the sample is too small, or the number of individuals I queried is too small. We begin by recalling that the true likelihood is given as L^* in equation (2.2). Since the y_{ijl} are discrete, we can rearrange terms. After interchanging the order of the finite sums and integrals, we get the marginal density of Y_{ijl} given the "parameters" of $H(\mathbf{x})$. From the likelihood formulation given in equation (2.2) we have the density function (Tolley and Manton, 1992, 1994),

$$h(y_{ijl} \mid \gamma) = \sum_{k_1=1}^{K} \sum_{k_2=1}^{K} \cdots \sum_{k_j=1}^{K} \left(\prod_{j=1}^{J} \prod_{l=1}^{L_j} \lambda_{k_j jl}^{y_{ijl}} \right) \times \left(\int \cdots \int \prod_{j=1}^{J} \prod_{l=1}^{L_j} \xi_{ikj}^{y_{ijl}} \, dG_i(\xi_{i1}, \ldots, \xi_{ik}) \right) \tag{3.1}$$

Since, for any i and j, y_{ijl} has only one value equal to 1, (3.1) involves only raw moments of order J or less. Thus the density in (3.1) can be reduced to

$$h(y \mid \gamma) = \sum_{k_1=1}^{K} \sum_{k_2=1}^{K} \cdots \sum_{k_j=1}^{K} a(k_1, \ldots, k_j) \cdot b(k_1, \ldots, k_j) \tag{3.2}$$

where $a(k_1, \ldots, k_j)$ is a function of the λ_{kjl} and $b(k_1, \ldots, k_j)$ is a function of moments of the random variables $(\xi_1, \ldots, \xi_k)$ up to order J.

Recall that no specification of the form of $H(\mathbf{x})$ is made except for its moment space to order J. The equations require that only moments to order J be specified. As a result of this constraint, when we "estimate" $H(\mathbf{x})$, we are only estimating its first J-order moments. Thus any two distributions $H(\mathbf{x})$ and $H^*(\mathbf{x})$ with the same first J moments will be considered equivalent as far as statistical estimation and identifiability of parameter estimates is concerned. Since J is fixed, we will not be able to estimate more moments as I increases. Our ability to estimate more moments of $H(\mathbf{x})$ increases only if J increases.

Equation (2.22) describes the number of nonlinear functions available in the likelihood function to estimate unique values of parameters. The maximum number of equations for J variables each with L_j responses is $\Pi_{j=1}^{J} L_j - 1$. For a solution with K classes, the number of λ_{kjl} parameters to estimate is $K\,\Sigma_{j=1}^{J}(L_j - 1)$. The number of moments in $H(\mathbf{x})$ of degree J or less is $[(J+K-1)!/J!(K-1)!] - 1$. Thus in order for there to be an adequate number of equations to estimate the parameters, a *necessary* condition for identifiability is

$$\prod_{j=1}^{J} L_j > K \sum_{j=1}^{J} (L_j - 1) + \frac{(J+K-1)!}{J!(K-1)!} - 1 \tag{3.3}$$

This compares the number of equations formed by varying the y_{ijl} outcomes with the number of λ_{kjl} and moments for a model with K classes.

3.2.2. Sufficient Conditions for Identifiability in Estimation

In Section 2.1.3 we discussed the mathematical identifiability of the GoM model. By this, we mean the uniqueness of the specified probability model. In this section we use some of the notation and results developed in that section to examine statistical estimation of model parameters. A question that must be addressed is, "Given the application of a GoM model, to a set of data, can the parameters be uniquely estimated?" Of particular importance here are the moments of H. In this section we examine this problem by first showing that the dual space of B, denoted B^*, consists of all linear functionals on L_B that are nonnegative. B^* is a polytopal cone, and the extreme rays of B^* are $R \geq n$ in number. These rays define R linear functions $\mu_r(p)$, $r = 1, \ldots, R$, each of which is zero on a particular facet of B. These $\mu_r(p)$ are used to define the grade of membership for a profile p_i and also provide a representation of the moments of H. This is discussed more formally below.

Consider real-valued weights w_i^k, one for each individual for each of K extreme points of B, assembled in a matrix $\mathbf{W}$ that is $N \times K$ where $\Sigma_i w_i^k = 1$ for $k = 1$ to K. A matrix $\mathbf{W} = (w_i^k)$ exists since $\mathbf{\Lambda} = (\boldsymbol{\lambda}^k, k = 1, 2, \ldots, K) \subset$

B, which produces $\mathbf{P}_N\mathbf{W} = \Lambda$ where $\mathbf{\Lambda}$ is the $M \times K$ matrix with extreme $\boldsymbol{\lambda}^k$ for its columns.

To relate vertices of B to the distribution $\mathbf{P}_N$ write

$$\mathbf{P}_N = \mathbf{\Lambda}\boldsymbol{\phi}^{\dagger}\mathbf{\Gamma}_N , \tag{3.4}$$

where $\mathbf{P}_N$ is the set of N individual case profiles, $\mathbf{\Lambda}$ is the matrix of extreme profiles of B, and $\mathbf{\Gamma}_N$ is a matrix of "extreme" functionals for rows of $\mathbf{P}_N$. The matrix $\boldsymbol{\phi}^{\dagger}$ is the pseudo-inverse of the matrix of probabilities $\boldsymbol{\phi}$, relating B and B^*; i.e., there exist linear functionals (μ_r, $r = 1, 2, \ldots, R$) that are "extreme" elements of the convex polytope B^* dual to B. The set B^* is a cone and consists of all linear functionals that are nonnegative on B. We choose (below) an appropriate normalization of each ray in B as needed in (3.4).

Using the definition of linear functionals, B^* is convex for an arbitrary B and is a polytope with the roles of extreme points and extreme functionals interchanged if B is a polytope. The set of all profiles that are nonnegative for all functionals in B^* is the intersection of half spaces defined by $H_\mu(\mathbf{p}) = \{\mathbf{p} \mid \mu(\mathbf{p}) \geq 0\}$ and hence is convex. The intersection of the half spaces is convex, contains B, and is the closed convex hull in general. Since B is convex the intersection is B.

The extreme functionals of B^* are 0 on a facet of B and nonnegative in B. Since nonnegative multiples of such functionals are nonnegative on B, and 0 on a facet, we choose multipliers so that $\Sigma_{k=1}^{K} \mu_r(\lambda^k) = 1$. We take $\gamma_r^i = \mu_r(\mathbf{p}^i)$ and $\phi_k^r = \mu_r(\boldsymbol{\lambda}^k)$ for all $r = 1, 2, \ldots, R$ and $k = 1, 2, \ldots, K$. Since μ_r is linear,

$$\mu_r(\lambda^k) = \mu_r\left(\sum_{i=1}^{N} w_i^k \mathbf{p}^i\right). \tag{3.5}$$

Equation (3.5) can be rewritten as

$$= \sum_{i=1}^{N} w_i^k \mu_r(\mathbf{p}^i) \tag{3.6}$$

and

$$\sum_{i=1}^{N} w_i^k \mu_r(\mathbf{p}^i) = \sum_{i=1}^{N} w_i^k \gamma_r^i = \phi_r^k \tag{3.7}$$

i.e. $\mathbf{\Gamma W} = \boldsymbol{\phi}$. Since $\boldsymbol{\phi}$ is unique, we choose $\mathbf{W} = \Gamma^{\dagger}$ the pseudo-inverse of Γ to produce a unique $\mathbf{W}$.

To illustrate, let $K = R = 5 \neq n = 3$, and let $\boldsymbol{\phi}^{\dagger}$ be the pseudo-inverse of $\boldsymbol{\phi}$. This defines a B with five extreme points and five bounding facets (edges) that intersects a cube M with three dimensions corresponding to $J = 3$ dichotomous variables. Though $R = 5$, all $\mathbf{p}^i$ fall in the intersecting linear space L_B of $(n=)\,3$ dimensions, so $\boldsymbol{\phi}^{\dagger}$, the pseudo-inverse of $\boldsymbol{\phi}$, relates the five types on the boundary of $\mathfrak{M}$ to cases in the three dimensional L_B containing λ^k. When $K = R = n$ and $\boldsymbol{\phi}^{\dagger} = \boldsymbol{\phi} = \mathbf{I}$, L_B must pass through three of the edges of $\mathfrak{M}$ (e.g., 000 to 100, 000 to 010, and 000 to 001), so $R = K = 3$. The metric is *fixed* by the intersection of M and the linear space L_B, so profiles in B are *uniquely* defined by distances to extreme half-space boundaries whether on the boundary or not.

More generally (i.e., $K \neq R$; $\boldsymbol{\phi}^{\dagger} \neq \mathbf{I}$), B can be defined by factoring $\mathbf{P}$ as $\mathbf{\Lambda G}$, assuming $\mathbf{\Lambda}$ is the matrix of extreme profiles of the event space B, *and* $\mathbf{G}$ are boundary points of M. $\mathbf{P}$ is defined by $\mathbf{\Lambda}, \mathbf{\Phi}$, and $\mathbf{\Gamma}$. Factoring $\mathbf{G}$ as $\mathbf{\Phi}^{\dagger}\mathbf{\Gamma}$, with $\boldsymbol{\phi}^{\dagger} \neq \mathbf{I}$ requires additional constraints. That is, $\mathbf{\Gamma}$ represents the profile of coordinates on the variables $1 \rightarrow 1$ with the extreme rays of B^*. When $R \neq K$, $\boldsymbol{\phi}$ relates the K extreme types to R extreme functionals on the variable space. When $\boldsymbol{\phi} = \boldsymbol{\phi}^{-1} = \mathbf{I}$, then $\boldsymbol{\phi}^{\dagger}\mathbf{\Gamma} = \mathbf{I\Gamma} = \mathbf{\Gamma} = \mathbf{G}$, so that $\mathbf{G}$ defines the relation of each case i to each of R extreme functionals and matches exactly the K vertices of B. The set of $\mathbf{\Lambda}$ columns is unique modulo reordering the columns and the rows of $\mathbf{\Gamma}$. The matrix $\mathbf{W}$, however, need not be unique if $\mathbf{P}$ is not full rank modulo its linear constraints $(\Sigma_l\, p^i_{jl} = 1)$. Nor is $\mathbf{W}$ necessarily nonnegative. However, $\mathbf{\Lambda}$, being the set of extreme points of B, is unique.

For $R \neq K$ and $\boldsymbol{\phi}^{\dagger} \neq \mathbf{I}$, restrictions are needed to define $\boldsymbol{\phi}^{\dagger}$. To define $\mathbf{W}$ uniquely we need vectors for which $\Sigma_i\, p^i_{jl} w^k_i = \lambda^k_{jl}$ for all k, j, and l to form a convex set that may be unbounded. If we focus on small $\mathbf{W}$s, we can examine $\hat{\lambda}^k_{jl} = \Sigma_i\, y^i_{jl} w^k_i$ whose true value is λ^k_{jl} since $E(Y^i_{jl}) = p^i_{jl}$. The variance of $\hat{\lambda}^k_{jl}$ is to be minimized by choice of $\mathbf{W}$. The constraint on $\mathbf{W}$ is to produce an expected value which produces a constrained mean square and uniquely identifies $\mathbf{W} = \mathbf{\Gamma}^{\dagger}$ for $K \neq R$, $\boldsymbol{\phi}^{\dagger} \neq \mathbf{I}$.

The p^i_{jl}, the true probabilities, have to be related to outcomes y^i_{jl} (which are 0 and 1). For $\boldsymbol{\phi}^{\dagger} = \mathbf{I}$ and $K = R$, to identify the space containing the "true" p_{jl}, $\mathfrak{M}$ contains images of the Radon–Nicodym measures in (Ω, F, P_r) for the mapping.

$$\text{Pr}\colon \omega \rightarrow ((p_{jl}(\omega), l = 1, 2, \ldots, L_j)\,,\; j = 1, 2, \ldots, J)$$

The image on $\mathfrak{M}$ need not be absolutely continuous and can contain singular components as required by dimensional restrictions. The integral producing the expected value is

$$E[\cdot] = \int_{\mathfrak{M}} f(\mathbf{p})[\cdot]\,\text{Pr}[dV] \tag{3.8}$$

where dV is a volume element in M, $\text{Pr}[dV]$ is the probability that a random profile is in dV, and $f(\mathbf{p})$ is the density for $\mathbf{p}$. The "centroid" of profile distributions in $\mathfrak{M}$ is

$$\mathbf{p}^0 = \lim_{N\to\infty} \frac{1}{N} \sum_{i=1}^{N} \mathbf{p}^i = E[\mathbf{p}]\,, \tag{3.9}$$

with probability one. Expected cross products define $\mathbf{\Pi}$, an $M \times M$ grammian matrix, as

$$\mathbf{\Pi} = \int_{\mathfrak{M}} f(\mathbf{p})\,\mathbf{p}\mathbf{p}^{\mathrm{T}}\,\Pr[dV]\,. \tag{3.10}$$

With these definitions, we can state

Theorem 1. *The rank of* $\mathbf{\Pi}$ *is the dimension of the closed convex set* B.

Proof. Since n is the dimension of B, there exists basis vectors $\mathbf{V}^1, \mathbf{V}^2, \ldots, \mathbf{V}^n$, such that for every $\mathbf{p} \in B$ there is an unique $\beta_m(\mathbf{p})$ such that $\Sigma_{m=1}^{n} \beta_m(\mathbf{p})V^m = \mathbf{p}$. On coordinates defined by $V = (V^1, V^2, \ldots, V^n)$, $\mathbf{p} \leftrightarrow (\beta_k(\mathbf{p}),\ k = 1, 2, \ldots, K)$. Also $\mathbf{V}_k$ is an element in $B \subseteq \mathfrak{M}$. Thus

$$\lim_{N\to\infty} \frac{1}{N} \sum_{i=1}^{N} p_{jl}^i = p_{jl}^0 = \lim_{N\to\infty} \frac{1}{N} \sum_{i=1}^{N} y_{jl}^i\,, \tag{3.11}$$

which exists with probability 1, is consistently estimated, and defines a unique point (the centroid) in B. The grammian matrix with elements

$$\lim_{N\to\infty} \frac{1}{N} \sum_{i=1}^{N} p_{jl}^i p_{rs}^i = \Pi_{jr}^{ls} \tag{3.12}$$

exists with probability 1. The $M \times M$ matrix,

$$\mathbf{\Pi} = (\mathbf{\Pi}_{jr}^{ls})$$

has rank n if $\mathbf{p}_i \in B$ with probability 1 for all i $(i = 1, 2, \ldots)$. Hence the result.

If $j \neq r$, under conditional independence,

$$\lim_{N\to\infty} \frac{1}{N} \sum_{i=1}^{N} y_{jl}^i y_{rs}^i = \Pi_{jr}^{ls} \tag{3.13}$$

with probability 1. If $j = r$, then

$$\lim_{N\to\infty} \frac{1}{N} \sum_{i=1}^{N} y_{jl}^i y_{rs}^i = p_{jl}^0 \delta_{ls} \neq \Pi_{jj}^{ls}\,. \tag{3.14}$$

In partitioned form $\mathbf{\Pi}$ is

$$\mathbf{\Pi} = \begin{pmatrix} \Pi_{11} & \Pi_{12} & \cdots & \Pi_{1J} \\ \Pi_{21} & \Pi_{22} & \cdots & \Pi_{2J} \\ \vdots & \vdots & & \vdots \\ \Pi_{J1} & \Pi_{J2} & \cdots & \Pi_{JJ} \end{pmatrix} \tag{3.15}$$

where $\Pi_{jr} = (\Pi_{jr}^{ls})$ $(l = 1, 2, \ldots, L_j;\ s = 1, 2, \ldots, L_r)$. Define the matrix $\mathbf{\Pi}^*$ using $\mathbf{\Pi}$ where $\mathbf{\Pi}_{jj}$ is replaced by a matrix with p_{jl}^0 on the diagonal (i.e., ${}^*\Pi_{jj}^{ls} = \delta_{ls} p_{jl}^0$). Then we get, with probability 1,

$$\Pi_{jr}^{*ls} = \lim_{N\to\infty} \frac{1}{N} \sum_{i=1}^{N} y_{jl}^i y_{rs}^i .$$

$\mathbf{\Pi}$ is rank n and defines B uniquely with probability 1. Given $\mathbf{V}$ and β_i^m defined in the proof of Theorem 1 for all i, m we replace (3.11) by

$$p_{jl}^i = \sum_{k=1}^{n} \beta_k^i V_{jl}^k \tag{3.16}$$

$$p_{rs}^i = \sum_{m=1}^{n} \beta_m^i V_{rs}^m , \tag{3.17}$$

so that (3.12) is

$$\frac{1}{N} \sum_{i=1}^{N} p_{jl}^i p_{rs}^i = \sum_{k=1}^{K} V_{jl}^k \sum_{m=1}^{n} V_{rs}^m \frac{1}{N} \sum_{i=1}^{N} \beta_k^i \beta_m^i . \tag{3.18}$$

Since the limit on the left for $\mathbf{P}$ exists as $N \to \infty$, so does the limit on the right. Hence, $\mathbf{\Pi} = \mathbf{VBV}^{\mathrm{T}}$ where B is symmetric and nonnegative definite. Consequently, $\mathbf{\Pi}$ is symmetric and nonnegative definite. Furthermore, $\mathbf{\Pi}$ is of rank at most n since $\mathbf{V}$ is rank n. If the rank of B were less than n, conditions defining B would be violated (i.e., K could be reduced). The preceding showed that $\mathbf{\Pi}^*$ was *not* equal to $\mathbf{\Pi}$ on the *diagonal* because of the orthogonality of observations $y_{jl}^i y_{jl'}^i = 0$ for $l \neq l'$. The data, however, are used to estimate $\mathbf{\Pi}^*$, not $\mathbf{\Pi}$. Specifically, the crucial theorem for estimating the individual profiles is

Theorem 2. *A* sufficient *condition for* $\mathbf{\Pi}$ *to be estimable (and identifiable) from* $\mathbf{\Pi}^*$ *is that* $K < J/2$ *and a nonsingular submatrix of* $\mathbf{\Pi}$ *of size* $K \times K$ *exists and contains no elements in the diagonal blocks (i.e., where* $\mathbf{\Pi}^*$ *differs from* $\mathbf{\Pi}$*).*

This is the essential "bridging" theorem in relating probability spaces, discrete outcome measures, and fuzzy state measures (Woodbury et al., 1994).

Proof. Assume **A**, a nonsingular $n \times n$ submatrix of **Π**, and **Π*** exist. Let ψ_r map rows of **A** into rows of **Π**, and ψ_c map columns of **A** into columns of **Π**, where mappings are $1 \rightarrow 1$ and the partial inverses, to map rows and columns of **Π** into rows and columns of **A**, exist. Then rows of **A** in **Π** do not have elements common with rows of $\mathbf{A}^T$ (otherwise an element of **A** is in a diagonal block where $\mathbf{\Pi}^* \neq \mathbf{\Pi}$). Further, no two rows of **A** are in the same block of rows of Π, and no two columns of **A** are in the same block of columns of **Π** (i.e., rows of $\mathbf{A}^T$ in **Π**) for the same reason.

We may permute variables so the row blocks of **Π** containing rows of **A** come first. Similarly, column blocks of **Π** are arranged so columns of **A** are last in lexicographical order. If $n < J/2$, there is no contradiction. We assume that the rows of **A** are the first rows in the blocks of **Π** containing rows of **A** and the columns of **A** are the last columns of the blocks of columns of **Π**. Consider the following illustration of **Π**. Blank cells are not involved in the proof.

$\mathbf{\Pi}(M \times M)$: $M = 5$ $(n = 2)$

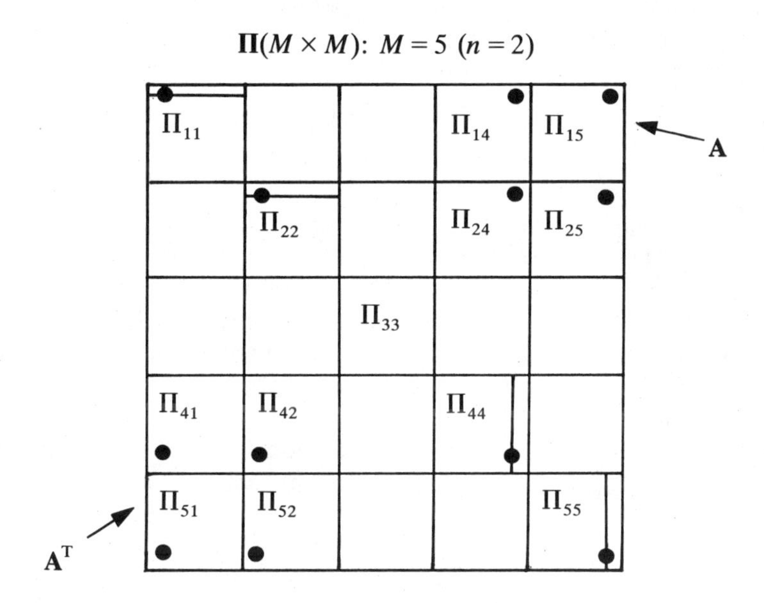

The submatrices **A** and $\mathbf{A}^T$ indicated by boxes with dots are assumed nonsingular, rank n. We wish $\{\Pi_{ij}\}$, $i \neq j$, to define Π_{jj} uniquely, given the resulting matrix is rank n.

Select an element in Π_{jj} containing no rows of **A** and no columns of $\mathbf{A}^T$. Enlarge **A** to $\mathbf{A}^+$ where the element in Π_{jj} defines an additional row and column of $\mathbf{A}^+$. The matrix $\mathbf{A}^+$ contains exactly one element in Π_{jj}. The rest are *not* in Π_{jj} by construction. $\mathbf{A}^+$ is a rank n submatrix of **Π** of size $n + 1 \times n + 1$ and must be *singular* by assumption. All elements of $\mathbf{A}^+$ are known except a_{rc}^{++}.$\mathbf{A}^{+}$ extracted from $\mathbf{\Pi}$ satisfies

$$\det\begin{bmatrix} a_c^+ \mathbf{A} \\ a_{rc}^{++} a_r^+ \end{bmatrix} = 0 \quad \text{but} \quad \det\begin{bmatrix} a_c^+ \mathbf{A} \\ a_{rc}^{++} a_r^+ \end{bmatrix} = \det \mathbf{A} \det(a_{rc}^{++} - a_r^+ \mathbf{A}^{-1} a_c^+)$$

and $\det \mathbf{A} \neq 0$, so

$$a_{rc}^{++} = a_r^+ \mathbf{A}^{-1} a_c^+ .$$

This can be done for any element of Π_{jj} that is neither in a row or column of $\mathbf{A}$ in $\mathbf{\Pi}$. This defines all elements in Π_{jj} except for n rows, one in each of the Π_{jj} corresponding to rows of $\mathbf{A}$, and n columns in Π_{jj}, which are in columns of $\mathbf{A}$ in $\mathbf{\Pi}$. But $\Pi_{ij} = \Pi_{ij}^{\mathrm{T}}$ so there are at most $2n$ elements in Π_{jj} not evaluated. If other $n \times n$ matrices satisfy the conditions for $\mathbf{A}$, the number of unidentified elements is reduced. The elements in Π_{jj} in a row or column of $\mathbf{A}$ are indicated by a dot in the diagonal blocks of the figure. The principal minor $\mathbf{\Pi}$ generated by the diagonal elements is a matrix $2n \times 2n$ where $\mathbf{C}_r$ and $\mathbf{C}_c$ are diagonal matrices

$$\mathbf{A}^* = \begin{bmatrix} \mathbf{C}_r \mathbf{A} \\ \mathbf{A}^{\mathrm{T}} \mathbf{C}_c \end{bmatrix}$$

$\mathbf{A}^*$ is rank K and size $2n \times 2n$ with only diagonals $\mathbf{C}_r$ and $\mathbf{C}_c$ unknown.

Albert (1944) shows that a nonnegative definite $M \times M$ matrix can be transformed to a matrix of rank$(M+1)/2$ or less with unique diagonal values. But $M = 2n$, so diagonal values can be found to make the rank of $\mathbf{A}^*$ be at most $(2n+1)/2$. But $\mathbf{A}^*$ is of rank at least n since it contains the nonsingular submatrix $\mathbf{A}$ of rank n. Hence the result.

Given these results, note that the conditional likelihood can be written where data are functions of parameters of the convex set as

$$L_N^+ = \prod_{i=1}^{N} \prod_{j=1}^{J} \sum_{k_j=1}^{K} g_{k_j}^i \lambda_j^{k_j}(x_j^i) . \tag{3.19}$$

This is transformed to an unconditional likelihood using the law of total probability and integrating the g_k^i with respect to $H(\mathbf{x})$, the distribution function of the g_k^i.

Tolley and Manton (1992, 1994) replaced realizations of the $\mathbf{g}_k^i$ by variables of integration ξ_k^i with the same constraints as the g_k^i, i.e., $\xi_k^i \geq 0$, $\Sigma_k \xi_k^i = 1$, and

$$L_N = \prod_{i=1}^{N} \int_{S_i} \sum_{k_1=1}^{K} \sum_{k_2=1}^{K} \cdots \sum_{k_J=1}^{K} \prod_{j=1}^{J} (\xi_{k_j}^i) \prod_{j=1}^{J} \lambda_j^{k_j}(x_j^i) \, dH(\boldsymbol{\xi}) \tag{3.20}$$

where $\boldsymbol{\xi}^i = (\xi_1^i, \ldots, \xi_K)^{\mathrm{T}}$. If S_i is a regular simplex defined by constraints on ξ_k^i, the moments μ_k of $H(\mathbf{x})$ are

$$\mu_k = \int_{S_i} \prod_{j=1}^{J} \xi_{k_l}^i \, dH(\boldsymbol{\xi}) \tag{3.21}$$

where $k = (k_1, k_2, \ldots, k_J)$. The array of moments $\mu = (\mu_k, (k_1 = 1, 2, \ldots, K), (k_2 = 1, 2, \ldots, K), \ldots, (k_J = 1, 2, \ldots, K))$ is invariant over the complete symmetry group on J. The equation for μ_k is replaced by

$$L = \prod_{i=1}^{N} \sum_{k} \mu_k \boldsymbol{\Lambda}^k(\mathbf{x}^i) \tag{3.22}$$

where $\boldsymbol{\Lambda}^k(\mathbf{x}^i) = \lambda_i^{k_1}(x_1^i) \otimes \lambda_2^{k_2}(x_2^i) \otimes \cdots \otimes \lambda^{k_J}(x_J^i)$ is the Kroneker product of J matrices $\lambda_j^k(x_j^i)$, $(j = 1, 2, \ldots, J)$. $\boldsymbol{\Lambda}$ is dimension K^J by $L = L_1 \times L_2 \times \cdots \times L_J$. The rank of $\boldsymbol{\Lambda}$ is the product of the ranks of separate Λ_j where $\Lambda_j = (\lambda_{jl}^k \ (l = 1, 2, \ldots, L_j), (k = 1, 2, \ldots, K))$. Denote the rank of Λ_j as R_j, where necessarily $R_j \leq K$ and $R_j \leq L_j$. The rank of $\boldsymbol{\Lambda}$ is the product $R_1 \times R_2 \times \cdots \times R_J$.

The fact that the g_k^i and λ_{jl}^k are unique dual representations (for $\boldsymbol{\phi}^+ = \mathbf{I}$) means that, with known $\boldsymbol{\Lambda}$, the moments μ_k can be estimated *without* bias. Specifically, consider the decomposition of $\Lambda_j = (\lambda_{jl}^k, (l = 1, 2, \ldots, L_j), (k = 1, 2, \ldots, K)$.

$$\begin{aligned}
\boldsymbol{\Lambda} &= (\mathbf{U}^{(1)}\mathbf{V}_1^{\mathrm{T}}) \otimes (\mathbf{U}^{(2)}\mathbf{V}_2^{\mathrm{T}}) \otimes \cdots \otimes (\mathbf{U}^{(J)}\mathbf{V}_J^{\mathrm{T}}) \\
&= (\mathbf{U}^{(1)} \otimes \mathbf{U}^{(2)} \otimes \cdots \otimes \mathbf{U}^{(J)})(\mathbf{V}_1^{\mathrm{T}} \otimes \mathbf{V}_2^{\mathrm{T}} \otimes \cdots \otimes \mathbf{V}_J^{\mathrm{T}}) \\
&= \mathbf{U}\mathbf{V}^{\mathrm{T}}
\end{aligned}$$

where $\mathbf{U}$ and $\mathbf{V}^{\mathrm{T}}$ have maximum possible rank $R_1 R_2 \cdots R_J = R$. The vector of moments μ_k is $1 \times K^J$ but has symmetries, i.e., at most only $R^* = (J + K - 1)!/J!(K-1)!$ values of μ_k exist among the K^J values of k (Tolley and Manton, 1992a,b). Define $\boldsymbol{\mu}^*$ as a vector of distinct moments of order J. Define $\boldsymbol{\Delta}$ as an $R^* \times K^J$ matrix with only one nonzero value in each column of $\boldsymbol{\Delta}$ (equal to 1). The rank of $\boldsymbol{\Delta}$ is at most R^*. To prove that nonbiased estimates of the moments exist, consider

Theorem 3. *Assuming conditional independence, the unconditional likelihood* (3.22) *with the decomposition for* $\boldsymbol{\Lambda}^k$ *substituted is*

$$L_i = \mu^* \{\boldsymbol{\Delta}\mathbf{U}\mathbf{V}^{\mathrm{T}}\}(y_i) . \tag{3.23}$$

Proof. Define statistic T as a product (using the multinomial variable x_{ij}

to simplify notation),

$$T(\mathbf{x}^i) = \prod_{j=1}^{J} T(x_j^i) . \tag{3.24}$$

Compute the expectation of T by "repeated sampling," namely,

$$E(T) = \sum_{x_1^i=1}^{L_1} \sum_{x_2^i=1}^{L_2} \cdots \sum_{x_J^i=1}^{L_J} \prod_{j=1}^{J} p_j^i(x_j^i) \prod_{j=1}^{J} T_j(x_j^i) \tag{3.25}$$

$$= \prod_{j=1}^{J} \sum_{x_j^i=1}^{L_i} p_j^i(x_j^i) T(x_j^i)$$

$$= \prod_{j-1}^{J} \sum_{x_j^i=1}^{L_j} \sum_{k_j=1}^{K} g_k^i \Lambda_j^{(k_j)}(x_j^i) T_j(x_j^i)$$

$$= \sum_k \left(\prod_{j=1}^{J} g_{k_j}^i \sum_{x_j=1}^{L_j} \Lambda_j^{(k_j)}(x_j^i) T_j(x_j^i) \right) \tag{3.26}$$

$$= \sum_k \left(\prod_{j=1}^{J} g_{k_j}^i \right) \sum_x \Lambda_k(\mathbf{x}) T(\mathbf{x}) .$$

We can extract any order moment d using the conditional model, i.e.,

$$E_x(T^d) = \sum_k \left(\prod_{j=1}^{J} g_{k_j}^i \right) \sum_x \Lambda_k(\mathbf{x}) T^d(\mathbf{x}) . \tag{3.27}$$

For the unconditional model, integrate (compute the expected value of) $\Pi_{j=1}^{J} g_{k_j}^i$, denoted μ_k,

$$E_g(E_x(T_x^d \mid g)) = \sum_k \mu_k \sum_k \Lambda_k(\mathbf{x}) T^d(\mathbf{x}) . \tag{3.28}$$

The statistic has the form

$$T(\mathbf{x}_i) = \prod_{j=1}^{J} T_j(x_{ij}) . \tag{3.29}$$

Any statistic is expressible as the sum of such statistics, i.e., where

$$T_w = \prod_{j=1}^{J} T_{w_j}(x_J^i) \tag{3.30}$$

and

$$T_{w_j} = \begin{cases} 1 & \text{if } x_i^j = w_j \\ 0 & \text{otherwise}. \end{cases}$$

Call T_{w_j} an "outcome" variable. A basis for T can be constructed from the $L_1 \cdot L_2 \cdots L_j$ outcome variables T_{w_j}. Thus, $E_g(E_x(T))$ is estimated by selecting statistics forming a matrix $\mathbf{T}$ of dimension $L_1 \cdot L_2 \cdots L_j$ by M. Rewriting L_i with $\mathbf{T}$ substituted in produces

$$E_g E_x(T_x \mid g) = \mu^* \mathbf{\Delta U V}^{\mathrm{T}}(y_i)$$

where T_x represents the xth element of $\mathbf{T}$, which is, using this expansion of $\mathbf{\Lambda}$,

$$\mathbf{T} = \mathbf{V}(\mathbf{V}^{\mathrm{T}}\mathbf{V})^{-1}(\mathbf{\Delta U})^{\mathrm{T}}((\mathbf{\Delta U})(\mathbf{\Delta U})^{\mathrm{T}})^{-1}$$

Thus,

$$\begin{aligned} E_y(E_x(T_x \mid g) &= \mu^* \mathbf{\Delta U V}^{\mathrm{T}}([\mathbf{V}(\mathbf{V}^{\mathrm{T}}\mathbf{V})^{-1}(\mathbf{\Delta U})^{\mathrm{T}}(\mathbf{\Delta U}(\mathbf{\Delta U})^{\mathrm{T}})^{-1}] \\ &= \mu^* \mathbf{\Delta U}([\mathbf{V}^{\mathrm{T}}\mathbf{V}(\mathbf{V}^{\mathrm{T}}\mathbf{V})^{-1}](\mathbf{\Delta U})^{\mathrm{T}}(\mathbf{\Delta U}(\mathbf{\Delta U})^{\mathrm{T}})^{-1} \end{aligned} \tag{3.31}$$

which reduces to

$$\begin{aligned} &= \mu^*[\mathbf{\Delta U}(\mathbf{\Delta U})^{\mathrm{T}}(\mathbf{\Delta U}(\mathbf{\Delta U})^{\mathrm{T}})^{-1}] \\ &\equiv \mu^* . \end{aligned}$$

Thus the μ^* (distinct moments of the distribution of the ξ_k^i) are estimated *un*biasedly for the distribution H. $R_* \leq R$ is *sufficient* for existence of a statistic T to estimate μ^*, and the theorem is proved (Woodbury et al., 1994).

3.2.3. Practical Aspects of Statistical Identifiability: A Geometric Example of the Dual Spaces

The necessary condition for identifiability establishes a theoretical upper bound for the number of moments that may be extracted from a data set. The sufficient conditions show that n must be less than $J/2$. It is also important to realize how the likelihood function ensures that the parameters generated are the correct estimates and are unique. This is illustrated in Figure 3.1.

In the figure we consider two conditions for a $K = 3$ solution. One is for the "correct" solution. The other is for a solution where the position of vertices of the simplex have been moved out by a distance Δ. In Figure 3.1(a), for the maximum likelihood solution, we assume for simplicity that at

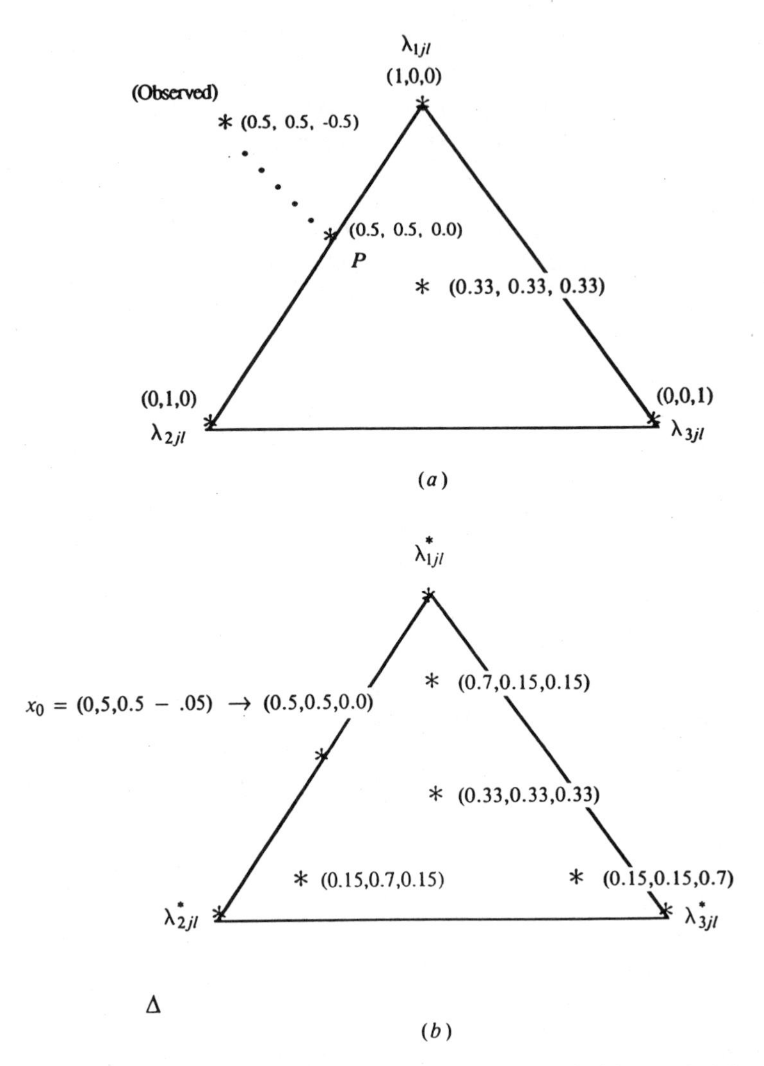

Figure 3.1 (*a*) MLE (x indicates location of case in g_{ik} space ($K = 3$). (*b*) Extended simplex to incorporate exterior points.

least one case falls at each of the vertices. These cases c_1, c_2, and c_3 have the g_{ik} (or set assignment functions) vectors $(1,0,0)$, $(0,1,0)$, and $(0,0,1)$, respectively. Now the tendency of any likelihood solution (or at least the portion dealing with the score functions in the solution for λ_{kjl}) will be to maximize the size of the simplex to encompass all points exterior to the simplex that, because they are constrained at the boundary, are "penalizing" the likelihood; i.e., maximizing the size of the simplex will help encompass all cases in the simplex and maximize the likelihood associated with the λ_{kjl} parameters.

We can see in Figure 3.1(a) that there is one case, x, beyond the boundaries of the fitted simplex. Presumably many cases are on the boundary because there is an exterior point in one or more dimensions that is beyond the boundary of the $K-1$ dimensional simplex. However, given the assumptions of the model, these values are fixed at the boundary, and a penalty in the likelihood for the λ_{kjl} is taken when the score functions corresponding to these terms are optimized in an iteration. In estimation this point is moved to the boundary (dotted line) to provide a permissible estimate.

In Figure 3.1(b) we present the situation where we have allowed the simplex to increase to encompass this exterior point (this would imply altering, in a compensating fashion, the simultaneously evaluated likelihood score equations associated with both the λ_{kjl} and g_{ik}). In this case the g_{ik} vectors for cases 1–3 have to change, e.g., in the example to $(0.7, 0.15, 0.15)$, $(0.15, 0.7, 0.15)$ and $(0.15, 0.15, 0.7)$, respectively, because of the space constraint $\Sigma_k\, g_{ik} = 1.0$. Thus, as the simplex increases to include exterior (boundary) cases, the amount of heterogeneity represented by the g_{ik} not on the boundary must be decreased, which will tend to reduce the contribution (say, measured by a weighted series of squared distances) to the likelihood of those parameters.

Ultimately, without restrictive conditions on the likelihood, the g_{ik} vectors would all approach a singularity [i.e., they would all go to (0.33, 0.33, 0.33)], in which case there would be little heterogeneity of the data explained by the g_{ik} (i.e., the likelihood associated with the $g_{ik} \rightarrow 0.0$). The "force" or feature of the likelihood that prevents the boundary from expanding "too" far (e.g., having the g_{ik} distribution "collapse") is the fact that the λ_{kjl} are constant over i, the g_{ik} are constant over j, and they are simultaneously maximized. Thus, if there is variation associated with i that operates differently over individuals within an extreme profile K, for a variable j, then its variation cannot be represented by the λ_{kjl} (i.e., only the g_{ik} can represent heterogeneity across cases). Thus, after reaching some point that would encompass most points (there is a penalty in terms of a component of likelihood associated with the λ_{kjl} *not* gained if the simplex is too small), there will be a constraint on the λ_{kjl} from moving further because the loss of explained heterogeneity due to the g_{ik} more than balances further gains from a continuing expansion of the boundaries of the simplex B by changing the K extreme profiles (i.e., coordinates in the J-variable measurement space defining the vertices of B). This is a result of the "quadratic loss" function implied in the likelihood leading to the likelihood ratio χ^2. Furthermore, in addition to balancing distances in the parameter span that are quadratically weighted, the boundaries of B are ultimately bounded by M, the "possible" response space.

Thus, there is an "optimal" balance of the total heterogeneity explained in a data set between that due to λ_{kjl} by expansion [Figure 3.1(b)] and the heterogeneity *lost* as the expansion of the λ_{kjl} reduces the effect of the g_{ik}

faster than the expansion of the simplex gains in explaining heterogeneity. This "balance" is enforced by the form of the likelihood [i.e., (2.1) and (2.3) and (2.4)]. That is, one can see in Figure 3.1(*b*) that the heterogeneity of the g_{ik} has decreased, i.e., $(1, 0, 0)$, $(0, 1, 0)$, $(0, 0, 1) \rightarrow (0.7, 0.15, 0.15)$, $(0.15, 0.7, 0.15)$, $(0.15, 0.15, 0.7)$. Thus the algorithm will have a tendency to increase the size of the simplex (to capture exterior points), which is balanced by the discounting that the expansion costs the g_{ik} effects in explaining the data points. Indeed, in most solutions we will have many parameters on the boundary because the heterogeneity attributable to the g_{ik} is highly significant and because the possible response space $\mathfrak{M}$ represents a set of a priori established boundaries. The effects of $\mathfrak{M}$ will also be to make estimates of parameter values near boundaries "hyperefficient" because of the a priori measurement space constraints. This will be empirically shown later, in the example of the ECA data on psychiatric disorders where the "nested" test of LCM and GoM models show the additional heterogeneity at the *individual* level is highly significant. Otherwise the LCM model would be superior to the GoM model and all g_{ik} heterogeneity would be zero with all cases located at the vertices (i.e., all $g_{ik} = 0$ or 1).

The failure to use the heterogeneity in higher-order moments in discrete mixture models, produces identifiability problems (Everitt, 1984). For example, in Heckman and Singer (1984a,b) a discretely mixed hazard function is presented and a proof of the consistency of the parameters given. Unfortunately, standard likelihood estimation does not implement one of the assumptions of the theoretical proof of consistency, i.e., that the tail of the distribution function does not decrease too slowly. This can be implemented in the estimation algorithm by imposing a penalty function for the addition of the $K + 1$ class (Behrman et al., 1991). In GoM, the g_{ik} represent a model with a natural parameterization where the likelihood estimates are balanced by an increased penalty in adding the g_{ik} estimates for the $K + 1$ fuzzy class.

A less consequential question of identifiability is the order of extraction of the K classes. That is, depending upon the initial partition used, the vector of parameters for a profile 1 (e.g., $\boldsymbol{\gamma}_1$) and 6 ($\boldsymbol{\gamma}_6$) in an analysis could be permuted in order of extraction; i.e., $\boldsymbol{\gamma}_6$ could emerge as the first vector of parameters and $\boldsymbol{\gamma}_1$ as the sixth vector of parameters. This is not important in GoM,because the maximum likelihood calculation solves for all parameter vectors simultaneously. In addition the characteristics of the profiles are described by the distribution of the ξ_{ik} (of which the g_{ik} are realizations). Permutation of the profiles causes a corresponding permutation of these distributions. Confusion sometimes arises because, in least-squares factor-analysis models (unconstrained), factors, for identifiability, must be extracted in some order, e.g., in decreasing size of the variance explained by the eigenvalues for the factor (Harman, 1967).

3.3. NUMERICAL ESTIMATION

We now come to a basic issue with regard to analyzing data, that of numerical estimation of the two types of parameters in (2.1). Clearly, maximization of (2.1) or (2.2) entails iterative numerical estimation techniques. Though many readers may be familiar with iterative and nonlinear numerical estimation techniques, care should be taken in maximization because of the constraints contained in Condition I and Condition II. We first discuss the maximization of (2.2) through maximizing (2.1). We then briefly present techniques for numerically maximizing (2.1). Details of computation are discussed in Woodbury and Clive (1974) and Woodbury et al. (1978).

3.3.1. Maximizing Equation (2.2) by Maximizing Equation (2.1)

As described above, the maximization of (2.2) when no specification of $H(\mathbf{x})$ has been made entails determining estimates of λ_{kjl} and estimates of μ (i.e., moments of $H(\mathbf{x})$) that maximize (2.2) subject to Conditions I and II of Chapter 1. Note that Condition I applies to the g_{ik} values themselves and *not* to the moments of $H(\mathbf{x})$. In other words, the constraints on the moments of $H(\mathbf{x})$ are those implicitly imposed by Condition I on the realizations of the random variable ξ_{ik} with the distribution $H(\mathbf{x})$. This implicit set of constraints must be imposed on estimates of the moments of $H(\mathbf{x})$. Thus, even if one were to attempt to estimate moments of $H(\mathbf{x})$ directly, one could not avoid the computations for individuals to ensure $H(\mathbf{x})$ fulfilled the convexity constraints on g_{ik}.

Now suppose that (2.1) is maximized instead of (2.2). The estimated g_{ik} parameters satisfy Condition I by definition. They can also be used to form estimates of the moments of $H(\mathbf{x})$. As noted above, the g_{ik} can form estimates of moments only to order J. These estimated moments satisfy the implicit constraints imposed by Condition I by construction. In general, the set of parameters that maximize the likelihood given by equation (2.1) will not maximize the likelihood given by equation (2.2). However, they do maximize (2.2) asymptotically, if $H(\xi)$ is replaced by the empirical distribution function $H_N(g)$. In this case since the integrand in (2.2) contains mass only at "observed values" of g, and maximizing (2.1) maximizes this integrand at these mass points, the maximization of (2.1) maximizes (2.2). In the limit, as N increases (2.2) with H replaced by H_N converges to (2.2) under general conditioning. Therefore, maximizing (2.1) and using the estimates of the g_{ik} to form moment estimates of $H(\mathbf{x})$ provides "maximum likelihood" estimates asymptotically of (2.2). The results in Section 3.2.3 show these estimates are unbiased asymptotically and that a sufficient condition (3.2.2) for estimation is $K \leq J/2$. When we refer to maximizing (2.2) we mean maximizing (2.1) and substituting this value in (2.2).

3.3.2. Computational Formulas for Maximizing Equation (2.1)

The method used for maximization is to iteratively optimize (2.1) with respect to one or the other of the sets of g_{ik} or λ_{kjl} parameters keeping the other set constant. The method estimates parameters for models with missing data. The basic theory for missing data, the "Missing Information Principle" (MIP), is discussed in Orchard and Woodbury (1971) as is the general nature of computational algorithms for missing data. The MIP suggests that calculations be conducted iteratively, exchanging maximization steps between sets of parameters using modified, restricted Newton–Raphson procedures.

The method of solving for g_{ik} and λ_{kjl} is to first equate the derivatives of (2.1) to 0, subject to Conditions I and II. This can be done using Lagrange multipliers, yielding two sets of equations, each from solving the resulting expression for one set of parameters with the second set held fixed. The iterative solution of these two sets of equations (and their cross-derivatives) is what numerically imposes the balance of the λ_{kjl} expansion and the g_{ik} variation discussed in geometric terms in Section 3.2.3. These sets of equations or "score functions" (with modifications), are given by Woodbury and Clive (1974). The first set of score functions (functions that determine the value of the parameter set that maximizes the likelihood, i.e., $\partial L/\partial g_{ik}$, or the partial derivatives), which is used to estimate g_{ik}, is

$$\hat{g}_{ik} = \frac{1}{y_{i++}} \sum_{i=1}^{I} \sum_{l=1}^{L_j} y_{ijl} \frac{g_{ik}^* \cdot \lambda_{kjl}^*}{p_{ijl}} \tag{3.32}$$

where $y_{i++} = \Sigma_j \Sigma_l y_{ijl}$ and $p_{ijl} = \Sigma_k g_{ik}^* \cdot \lambda_{kjl}^*$ for estimating the set assignment function parameters (g_{ik}) for individuals.

The complementary set of score functions, used to determine the structural or location parameters λ_{kjl}, is

$$\hat{\lambda}_{kjl} = \frac{\sum_{i=1}^{I} y_{ijl} \dfrac{g_{ik}^* \cdot \lambda_{kjl}^*}{p_{ijl}}}{\sum_{i=1}^{I} y_{ij+} \sum_{l=1}^{L_{ij}} \dfrac{g_{ik}^* \lambda_{kjl}^*}{p_{ijl}}}\,. \tag{3.33}$$

In both equations, g_{ik}^* and λ_{kjl}^* represent estimates of the parameters from the prior iteration.

Equations (3.32) and (3.33) are used in an iterative manner as follows. First, establish an initial discrete classification from an a priori selected categorical variable by setting all the g_{ik} to a value of either 0 or 1. Using these values of g_{ik} the distribution of the J variables for the I set assignment functions (here $g_{ik} = 0, 1$) can be tabulated for the K classes and used to form preliminary estimates of the λ_{kjl} using (3.33). Note that Condition II must be satisfied for these initial estimates. Second, the g_{ik} are determined by holding these estimates of λ_{kjl} "fixed," and maximizing (2.1) using (3.32). This set of parameters must satisfy Condition I. The λ_{kjl} are then re-estimated using (3.33) with this new estimate of g_{ik} held fixed. This is continued, alternating between (3.32) and (3.33) until no increase in the likelihood function is obtained. Each set of estimates should adjust parameters so that the likelihood is increased *and* so that Condition I and Condition II are satisfied.

Parenthetically, the equations provide the basis for imposing substantively interesting constraints on the solution space. For example, in certain examples to be presented, we will wish to choose sets of "internal" variables and sets of "external" variables. "Internal" variables are those for which both (3.32) and (3.33) are simultaneously solved. "External" variables are those for which we wish λ_{kjl} estimates conditioned on the g_{ik} parameters fixed to the values obtained for the "internal" variables. Thus, to estimate parameters for exterior variables, an additional "loop" can be added to the algorithm so that, after (3.32) and (3.33) are solved for J "internal" variables, the score functions of both types for the J variables are then "fixed" and the score function (3.33) is maximized for each of J^+ "external" variables separately. This gives maximum likelihood estimates of λ_{kj*l} for external variables conditional on the g_{ik} constructed from the J internal variables only.

In solving for $\hat{g}_{ik}$ and $\hat{\lambda}_{kjl}$ in equations (3.32) and (3.33) any estimates outside the admissible parameter space would have to be forced to the boundary before the next iteration is made and the likelihood at that iteration is calculated with the parameter at the boundary but free to move off it. Thus estimates are constrained to be in the defined admissible regions and not simply projected into the region after the likelihood is maximized. Parameter estimates constrained to be on the boundary in one iteration are free to move into the interior of the admissible region in subsequent iterations, i.e., the set of boundary parameters is not fixed during the iterations. This iterative solution between parameter sets is continued until (2.1) reaches a maximum (or the portion of (2.1) associated with each of the J^+ external variables reaches a conditional maximum).

Recall that the likelihood was unique, provided that the λ_{kjl} values used in the parameterization are the extreme profiles of the convex polytope B that contains the sample space of the individual probability profiles p_{ijl} (generated from the discrete responses y_{ijl} as shown in Section 3.2.2). What this means in practice is that the likelihood estimators obtained by the above procedure define only one point in the span of possible estimators. The final step is to chose the estimate in this span that has the greatest number of λ estimates equal to 0, to represent the maximum number of "extreme" elements; i.e., the maximum value of K is bounded by M. Elements of the span can be generated from a single point by postmultiplying the g_{ik} matrix by $\mathbf{A}$ and premultiplying the matrix of λ estimates by $\mathbf{A}^{-1}$ (see equations

(2.3) and (2.5) in Chapter 2). Alternatively, we may take the singular-value decomposition of **P** and project this onto the boundary of B using the Kullback–Leibler metric. This will give a unique maximum likelihood procedure.

The information matrix can be determined from standard likelihood methods once a maximized set of parameter estimates are obtained. Computationally direct methods of obtaining the elements of the information matrix are presented in Woodbury and Clive (1974).

In (3.32) and (3.33), we see that the parameters are dependent across score functions because $p_{ijl} = ((\Sigma_k g_{ik}\lambda_{kjl})$ appears in both expressions. Thus the g_{ik} *cannot* be factored as can the p_{ikl} in the EM implementation of LCM (McLachlan and Basford, 1987), and thus they are *not* sufficient statistics for the data as are the posterior probabilities of classification p_{ikl}. It is important to note that the EM algorithm (Dempster et al., 1977), often employed in situations with missing data (e.g., Heckman and Singer, 1984a,b), is a special case of the MIP. The MIP, however, involves a direct parameterization of both g_{ik} and λ_{kjl} and does not have the convergence problems (either theoretical or practical) that have been identified for the EM algorithm (e.g., Wu, 1983). For example, not only has EM been found to be slow (i.e., computationally cumbersome because the information matrix is not evaluated) but also that the EM solution is not guaranteed to converge (i.e., Wu, 1983) unless additional conditions are satisfied.

3.4. PRACTICAL ESTIMATION ISSUES

We have described the theory of GoM and compared it with logically related model structures to highlight its unique features. In this section we deal with several fundamental issues in its application to data.

In prior sections, we showed how consistency could be shown in one of two ways—each with different implications. The first is when J increased (and K and I fixed). In that case, consistency of the g_{ik} is relatively straightforward to demonstrate. However, in this framework, the sample distribution fraction is not necessarily consistent (since sample size is fixed). Where J is small (~20), we appeal to a model relying on moment constraints applied to the g_{ik}. That model does not use moments of $H(\mathbf{x})$ over degree J. In this case, the moments of the distribution function are consistently estimated but not necessarily the g_{ik}. Of course, if both J and I increase for fixed K consistency of both types could be produced. Thus, which argument for consistency is relevant is in part dependent on the analyses. Independent of the consistency issue, the GoM numerical algorithm directly estimates the g_{ik} parameters. It was shown (Section 3.2.3) that the g_{ik} also serve as a numerical tool that can reproduce, asymptotically without bias, the moments of the distribution $H(\mathbf{x})$ of the random variables ξ_{ik} when J is of relatively low degree. Thus, the same algorithm (maximizing the g_{ik}) is applicable no matter which form of consistency is assumed. The reasons for doing this are numerical, specifically, if all raw moments up to degree J are defined by the vector $\boldsymbol{\mu}$, $H(\mathbf{x})$ is the distribution of the g_{ik} characterized by these moments. The λ_{kjl} and $\boldsymbol{\mu}$ can be determined using the estimation scheme employing the λ_{kjl} and g_{ik} by choosing the λ_{kjl} that fulfill the boundary constraints. The sheer size of the moments problem (i.e., the size of H) makes direct estimation of μ_k difficult. The $\hat{\boldsymbol{\mu}}$ must be chosen so that $H(\mathbf{x})$ does *not* imply any g_{ik} that are negative, greater than 1, or sum to any number but 1 with nonzero probability. In other words the $\hat{\boldsymbol{\mu}}$ must maximize the expectations of (2.1) with respect to the $H(\mathbf{x})$ generated by $\hat{\boldsymbol{\mu}}$ *and* satisfying the convexity constraints on g_{ik} ($g_{ik} \geq 0$ and $\Sigma^K g_{ik} = 1$).

Numerical optimization of such moments is difficult because, for the value of any moment selected, one has to insure that the conditions on $H(\mathbf{x})$ are also not violated. This makes estimation of the μ_k impractical (i.e., after making an estimate of $\boldsymbol{\mu}$ to increase the likelihood, the estimate must be used to calculate $H(\mathbf{x})$ to see if the step keeps all the g_{ik} in bounds). However, if we maximize (2.1), and not its expectation, with respect to λ_{kjl} and g_{ik} jointly, then the g_{ik} estimated will provide an approximation of $H(\mathbf{x})$ with appropriate constraints which will have the correct moments.

To demonstrate, we formed the statistic T in Section 3.2.3. Conditional on the λ_{kjl}, the $\hat{g}_{ik}$ are specific to i. If $\hat{\boldsymbol{\mu}}$ are the moments estimated for $H(\mathbf{x})$ by estimating the g_{ik}, then the $\hat{\boldsymbol{\mu}}$ satisfy the constraints on the parameter space *and* maximize the integral used in taking the expectation of (2.1) with respect to $H^*(\mathbf{x})$ for each value of i. The distribution defined by these moments is in the same Jth-degree equivalence class as represented by $H^*(\mathbf{x})$. Thus the estimate [i.e., the $\hat{\boldsymbol{\mu}}$ derived from the $\hat{g}_{ik}$ produced from maximizing the expected value of (2.1)] with respect to the subset of the equivalence class represented by $H^*(\mathbf{x})$ will have the same moments as the sample moments. The $\hat{\boldsymbol{\mu}}$ so derived, as shown in Theorem 3, are conditional maximum likelihood estimates of the moments of $H(\mathbf{x})$, and the moments of the distribution $H(\mathbf{x})$ can be estimated unbiasedly by estimating g_{ik}. Thus, the estimation of g_{ik} fulfills the properties required for consistency and unbiasedness of estimators in the second case (Woodbury et al., 1993).

Heuristically, $\hat{\boldsymbol{\mu}}$ has the same large sample consistency and distributional properties (with the possible exception of differences in asymptotic relative efficiency) as produced by maximizing (2.2). The g_{ik} in this case are similar to the individual parameters implicitly estimated in a parametric empirical Bayes model (Morris, 1982), except that a specific distribution function is not selected. This is also a resolution of the problem identified by Gill (1992) of estimating a semimetric likelihood function for use in hazard and other modeling.

If J is small, the individual g_{ik} are not consistently estimated. However, using them as a group (i.e., averaging information over them) to estimate parameters of the empirical distribution generates estimates of the moments that are consistent. Since the g_{ik} are independently optimized for each person, they are required by the constraints to fall into the region of M intersecting the linear space L_B bounded by the λ_{kjl}. The moments of the g_{ik} will be the moments of some actual unknown distribution, though that distribution need not be restricted to a specific parametric form. Any distributional family uniquely identified by the Jth-degree moments is estimable. When J is large and I is large, the constraints on the distribution $H(\mathbf{x})$ will be "tight" if the density is smooth, and the distribution $H(\mathbf{x})$ will be well estimated as will the g_{ik}.

Another issue is the possibility that the maximum likelihood procedure will not find the global maximum. This is a possibility with any nonlinear maximum likelihood procedure (Cox and Hinkley, 1974). Asymptotically, if the parameters are consistent, then the global maximum will be found with probability 1 (by virtue of the definition of consistency). In practice, with finite samples, however, several problems can arise. Depending on the initial parameter values (i.e., a $K-1$ dimensional discrete partition of the data where partitions are defined by setting appropriate $g_{ik}=1$ (and 0) and the λ_{kjl} are the empirical distribution of the J attributes within each partition) after an iterative search a *local* maximum may be obtained in any sample. This problem can be evaluated by determining the sensitivity of the solution to alternate starting points using different starting partitions based on certain of the J variables in the analysis. This approach has the limitation that (a) if J is large, only a few options can be explored, and (b) the use of different variables to "find" a best partition is an ad hoc exploitation of the data, generating unknown bias.

A preferred approach is to a priori select an "informed" set of start values. If the start values that are a priori selected are closer to the maximum than a randomly selected set of start values, then the probability of getting to the true global maximum is increased. If the trial partition is based on a variable theoretically likely to be correlated with the resultant fuzzy classes (e.g., age to health; income to consumption behavior), the resultant likelihood is usually better than for an alternate solution generated from a randomly selected partition. Even if the solution so obtained is "biased," the direction of bias may be known. As in ridge regression, such substantively "biased" parameters may have better properties than unbiased unrestricted parameters.

Another approach is to analyze the convergence behavior of the algorithm. This requires that we define the "domain of attraction," i.e., the set of all starting values that converge to a specific "maximum." If the domain of attraction is larger for higher likelihood values, as information increases, there will be a greater distance between the likelihood associated with local solutions and the influence of the initial partition declines; i.e., "prior information" in an initial set of starting values is dominated by increasing sample information. Asymptotically, the initial partition has no effect on the solution and the global maximum is obtained. This procedure is justified if start values define a specific discrete prior distribution because an informed prior, if available, should dominate an uninformed prior, as in Bayesian inference.

If an initial partition is informative, the selection of it, rather than an arbitrary partition, may guard against incorrect inferences in samples with limited information; i.e., a biased solution may be restricted to a region of the total parameter space that, on substantive grounds is "known" to contain the solution. That is, despite the possible existence of a "higher" likelihood value for an unconstrained model, weighting the direction of the search by prior knowledge suggests a decreased probability that the true maximum is in an a priori unacceptable region of the parameter space (Singer, 1989). This principle is obvious in modeling, say, hazard rates for mortality, which must be nonzero. In the GoM model, which is multidimensional, the principles are the same but are applied more complexly, i.e., though the limits imposed by the matrix **M**.

One special case of interest for constraining the solution is that where a "null" class or extreme profile is defined. By "null" class we mean an extreme profile where *none* of the J attributes are present. This obviously cannot occur for variables where a person must have an attribute (e.g., age, sex). However, in a study of disease, certain groups may manifest none of the symptoms of the disease. This identifies a natural origin for the fuzzy state space if, it is only with the onset of a disease that symptoms begin to be accumulated. Thus the existence of a "null" fuzzy class (which often naturally emerged in many GoM analyses of health-related variables) provides additional structure to a solution and may help exclude unreasonable solutions. Other "origins" in the fuzzy state space could be generated depending upon the problem. The "null" class is a natural origin. Likewise, one could form both a "null" class and a "saturated" class (i.e., an extreme profile where *all* attributes are manifest). This could help align other fuzzy classes along prespecified dimensions.

In addition, information can be gleaned about solutions from the behavior of the solution as it progresses and by the robustness of the solution to data conditions. In the search algorithm we examine the numerical value of derivatives for the λ_{kjl} and g_{ik}, the Newton–Raphson step size (α), and the distribution of individual likelihood elements. Each of these factors, in addition to the rate of change of the total likelihood R, provides information about the performance of the search and the topology of the solution space. In addition, we can examine this cross-moments and correlations of the g_{ik}. The information matrix for small problems may be calculated. An examination of the information matrix will help the analyst decide if a global solution is reached. For large problems (e.g., $K > 4$; $I > 1000$) this is computationally difficult.

Usually the algorithm moves rapidly from its start values to a region near a solution and slows down and often makes a number of corrections in direction. These corrections, with associated increases and decreases in step size, suggest that the algorithm is able to move freely through the convex, multidimensional parameter space and is not "hung" up on local solutions. Indeed, under conditions violating the properties of the model, the algorithm can retreat, select a new direction and reject a local solution if the signs of the derivatives in the "Newton–Raphson" search are wrong. The intensity of search behavior increases in the region of the solution. As a consequence, the most effort is expended in searching the region near the global solution. Perturbation analysis of one, or multiple, parameters suggests that the solution returns to the same parameter values most of the time. When it returns to a different solution, it differs only on a small number of parameters (usually due to boundary constraints) and not in general structure. In addition, the distribution of the likelihood terms tells us if there are subsets of cases that are not well described by a solution.

We will now examine how to formally test the hypothesis that the order of the model, the number of extreme profiles, is K_{I} versus the hypothesis that the number of extreme profiles is K_{II}, where we assume that $K_{\mathrm{I}} < K_{\mathrm{II}}$. Let θ denote the vector of extreme profile parameters λ and the moments μ of the distribution H_n. Usually in testing a hypothesis about the order of the model one fits the model under the null hypothesis which constrains a number of the elements of θ to be zero and compares the fit to the fit obtained using a higher order model where specific elements of θ are not constrained to be zero. The problem in GoM modeling is that the exact extreme profiles are implicitly defined by the data. This means that when we wish to determine if, say, a model with four extreme profiles fits nearly as well as a model with five extreme profiles, we cannot specify which four of the five profiles make up the constrained model. The estimation process will choose the four "best." This makes the formation of a test procedure more difficult.

To resolve this problem we define a set of null hypothesis associated with a single alternative hypothesis. Explicitly, let the alternative hypothesis denote the model with K_{II} extreme profiles. Let $\mathbf{P}$ be an index set whose elements consists of the $K_{\mathrm{II}}!/[K_{\mathrm{I}}!(K_{\mathrm{II}} - K_{\mathrm{I}})!]$ sets of indices that can be obtained from all possible combinations of K_{I} extreme profiles selected from the set of K_{II} extreme profiles. Note that we have implicitly assumed a numbering of the profiles so that we can determine an index. For any particular element i of $\mathbf{P}$ there is a corresponding set of indexes that specifies a null hypothesis of $K_{\mathrm{II}} - K_{\mathrm{I}}$ specific extreme profiles whose parameters are constrained to be zero. Denote this specific null hypothesis as H_i and let $l_i(\theta)$ be the likelihood associated with this hypothesis. Then the likelihood specified by the GoM null hypothesis that there are K_{I} extreme profiles is $l_{\mathrm{I}}(\theta)$, where

$$l_{\mathrm{I}}(\theta) = \max_{i \in \mathbf{P}} l_i(\theta) , \tag{3.34}$$

To determine the distribution of the likelihood ratio of $l_{\mathrm{I}}(\theta)$ to $l_{\mathrm{II}}(\theta)$ we will first determine the distribution of ratio $l_{\mathrm{I}}(\theta)/l_{\mathrm{II}}(\theta)$. The distribution of the ratio above for testing the number of extreme profiles in GoM is then the distribution of the maximum of the individual ratios over values of i.

The likelihoods under both a specific null hypothesis H_i and the alternative hypothesis H_A are maximized subject to the constraints that the estimates of θ are in a closed region. This usually results in several parameter estimates being on the boundary of the parameter space. This complication has been mentioned before and affects the distribution of the likelihood ratio used to test the hypothesis (see, e.g., Chernoff, 1954; Moran, 1971; or Chant, 1974). Below we determine the asymptotic distribution of the ratio $l_i(\theta)/l_{II}(\theta)$ when some of the estimated parameters are on the boundary.

Lemma 1. *Under H_i, $\hat{\theta}$ converges in probability to the true parameters, θ_0. In addition, $N^{1/2}(\hat{\theta}-\theta_0)=0_p(1)$.*

Proof. The proof follows from Tolley and Manton (1992).

From this lemma and the results of Self and Liang (1987), we can show

Theorem: *Let Z be a multivariate Gaussian distribution with mean θ and information matrix $I(\theta_0)$. Let C_{Ii} and C_{II} be nonempty cones approximating the parameter space under H_i and H_A, respectively. Then the asymptotic distribution of $l_i(\theta)/l_{II}(\theta)$ is the same as the distribution of the likelihood ratio test of $\theta \in C_{Ii}$ versus $\theta \in C_{II}$ based on a single realization of Z when $\theta = \theta_0$.*

This theorem provides a basis for calculating the distribution of the log likelihood. Explicitly $-2\log(l_i(\theta)/l_{II}(\theta))$ is asymptotically equivalent to

$$Q_i = \inf_{\theta \in C_{II}} \{(Z-\theta)'I(\theta_0)(Z-\theta)\} - \inf_{\theta \in C_{Ii}} \{(Z-\theta)'I(\theta_0)(Z-\theta)\} \quad (3.35)$$

when H_i obtains. If the infimum was not over a bounded region in which boundary values were likely to occur, then (3.35) resembles a chi-square random variable. However, since values are likely to occur on the boundary the large sample distribution, in general, is not chi-square. As noted by Self and Liang (1988), this distribution with just two random variables on the boundary can be very involved. When more parameter estimates are on the boundary, calculation of the exact large sample distribution may be intractable. In the next paragraphs we suggest an approximate distribution.

To approximate the distribution for (3.35) we define five regions of Z_j, the jth element of Z based on the value of θ_j which maximizes

$$q = \inf(Z-\theta)'I(\theta_0)(Z-\theta)\,, \qquad \theta \in C \quad (3.36)$$

These regions are defined as

$$Z_j \in R_{\mathrm{I}} \text{ if } \theta_j = Z_j \text{ when } C = C_{\mathrm{II}} \text{ and } C = C_{\mathrm{I}i}$$

$$Z_j \in R_{\mathrm{II}} \text{ if } \theta_j = Z_j \text{ when } C = C_{\mathrm{II}} \text{ and } \theta_j = 0 \text{ when } C = C_{\mathrm{I}i}$$

$$Z_j \in R_{\mathrm{III}} \text{ if } \theta_j = 0 \text{ when } C = C_{\mathrm{II}} \text{ and when } C = C_{\mathrm{I}i}$$

$$Z_j \in R_{\mathrm{IV}} \text{ if } \theta_j = Z_j \text{ when } C = C_{\mathrm{II}} \text{ and } \theta_j = Z_j^* \neq Z_j$$

$$\text{when } C = C_{\mathrm{I}i}, \text{ where } Z_j^* \neq 0$$

$$Z_j \in R_{\mathrm{V}} \text{ if } \theta_j = Z_j^* \text{ when } C = C_{\mathrm{II}} \text{ and } \theta_j = Z_j^{**} \text{ when } C = C_{\mathrm{I}i}\,,$$

$$\text{where } Z_j \neq Z_j^* \neq 0 \text{ and } Z_j \neq Z_j^{**} \neq 0\,.$$

Note that if $I(\theta_0)$ is a diagonal matrix then R_{IV} and R_{V} are empty. This is the situation examined by Chant (1974) and Moran (1971). However, as noted by Self and Liang (1987), when $I(\theta_0)$ is not diagonal then, depending on the null and alternative hypotheses these last two regions are not empty.

It can be shown, in this case, that replacing Z^* in region R_{IV} with Z_j results in a conservative test statistic (one that is smaller in probability). Although the same cannot be said for region R_{V}, it is approximately true for situations where $I(\theta_0)$ tends to be diagonally dominant. In practice we have found that $I(\theta_0)$ tends to be strongly diagonal. Hence, an approximately conservative statistic for q_i is the following.

Define the column vectors Z_{I} and Z_{II} as

$$Z_{\mathrm{I}} = \{Z_j : Z_j \in R_{\mathrm{II}}\}$$

$$Z_{\mathrm{II}} = \{Z_j : Z_j \in R_{\mathrm{III}}\}$$

Let

$$I^*(\theta_0) = \begin{pmatrix} I_{11} & I_{12} \\ I_{21} & I_{22} \end{pmatrix}$$

be the submatrix of $I(\theta_0)$ corresponding to

$$Z^* = \begin{pmatrix} Z_i \\ Z_{\mathrm{II}} \end{pmatrix}.$$

Lemma. *Under the assumption that Z_j can be used in place of Z_j^* and Z_j^{**} in region R_{V},*

$$\Pr(q_i \le x) \ge \Pr(z_{\mathrm{I}}' I_{11} Z_{\mathrm{I}} + Z_{\mathrm{I}}' I_{12} Z_{\mathrm{II}} + Z_{\mathrm{II}}' I_{21} Z_{\mathrm{I}} \le x) \quad \text{for all } x \qquad (3.37)$$

The proof of this lemma is given in Tolley and Manton (1994).

To apply this result we note that if I_{12} is close to zero the righthand side of (3.37) is a chi-square random variable. In more general cases, the probability can be determined by simulation methods.

Another problem of practical consequence is counting the degrees of freedom for specific solutions. This is determined by the structure, or implied structure, of the likelihood function. The actual degrees of freedom are determined by the rank of $I(\theta)$. In practice, for large problems, though many of the information matrix elements are calculated, they may not be stored. In this case, we may set some heuristic conservative bounds on the degrees of freedom. Specifically, the degrees of freedom consumed in a solution are equal to the number of λ_{kjl} and g_{ik} parameters estimated. Specifically, for GoM the degrees of freedom are

$$df_{\text{GoM}} = K \sum_{j=1}^{J} (L_j - 1) + \frac{(J + K - 1)!}{J!(k-1)!} \tag{3.38}$$

where the first term on the right of (3.38) represents the number of λ_{kjl} estimated and the second represents the number of $\hat{\boldsymbol{\mu}}$ (of the ξ_{ik}) estimated from the g_{ik}. Naturally, since the second term increases factorially, there will be a limit to the number of $\hat{\boldsymbol{\mu}}$ that can be estimated. The theoretical upper bound is the number of data points in a data set, which is $I \times (\Sigma_{j=1}^{J} L_j - 1)$. Thus the maximum number of moments that can be estimated is

$$[(K-1) \times I] \leq I \times \left(\sum_{j=1}^{J} (L_j - 1) - \left[K \sum_{j=1}^{J} (L_j - 1) \right] \right. . \tag{3.39}$$

In fact, the number of g_{ik} that can be estimated (3.39) will usually be more of a restriction than (3.38).

For LCM the number of parameters estimated when not making explicit class assignments (i.e., only the p_{ik} are estimated) is

$$dF_{\text{LCMI}} = K \sum_{j=1}^{J} (L_j - 1) + (K - 1) . \tag{3.40}$$

The LCM model has as many p_{kjl} as the GoM model has λ_{kjl}, (for the same K); however, in the usual implementations of the solution using the EM algorithm, only $K - 1$, P_k, or class frequencies, are estimated. When using an LCM model to explicitly make classifications (i.e., estimate g_{ik}), additional degrees of freedom must be consumed. Since only one $g_{ik} = 1$, all other g_{ik} must be 0. Thus there is only one degree of freedom used in each LCM classification (i.e., one degree of freedom in the set assignment function for a crisp set model). Thus the degrees of freedom count is

$$dF_{\text{LCMII}} = K \sum_{j=1}^{J} (L_j - 1) + 1 \tag{3.41}$$

where I represents the one degree of freedom used in classifying each case. In fact, when using LCM to make a classification, there emerges a numerical issue; i.e., though the likelihood can be formulated for the classification form of the LCM, with the g_{ik} and λ_{kjl} having the required constraints, there is a numerical problem in getting direction vectors for the solution of the g_{ik} derivative equations because the crisp set model generates singularities in the set assignment function.

There are a large number of clustering algorithms that could be used to make discrete set assignments (using different approximations; Hartigan, 1975), but we used a different numerical strategy. Specifically, in estimating parameters for the standard LCM model the P_k are the averages of the p_{ik} estimated for each case. Though the p_{ik} are random variables, and not true parameters, one can assign a person to class K where p_{ik}/P_k is a maximum. This will increase the likelihood over any other class assignments. The p_{ik} are still the probability that the assignment in a class is correct, but we have forced a discrete choice to get a unique set assignment function which contain the true state parameters. An alternative is to take a GoM solution and cause it to move its g_{ik} values to 0 or 1. This will decrease the likelihood value but allows use of the partial derivatives in the MIP algorithm to generate a solution. In either case the number of degrees of freedom is given by (3.41).

A likelihood ratio test of the GoM and LCM model for a particular K can be calculated by taking minus twice the difference in the associated likelihoods with $[(J+K-1)!/J!(K-1)!]-I$ used as the difference in degrees of freedom when J is "small" and $[K-1\times I]-I$ when J is large (and the g_{ik} are consistent). The likelihood ratio is asymptotically χ^2 distributed when there are no boundary constraints. When there are boundary constraints and the null hypothesis is assumed to hold, the likelihood ratio is proportional to a χ^2 distribution with degrees of freedom conservatively defined above. When K is increased past the critical test value, the χ^2 can be affected, in some examples with limited data, by the dependence of the K extreme profiles, which may affect the estimated sums of squares. In the examples below this is not a problem since K is held fixed and we are comparing χ^2 between the two nested models.

To understand how the GoM and LCM degrees of freedom change for different size problems we present in Table 3.1 calculations for 2 to 30 binary variables and 2 to 30 variables with four response levels, the number of parameters used for 2 to 10 classes in each type of model (Manton et al., 1991).

The number of degrees of freedom consumed for LCM increases linearly with the number of classes. GoM has a factorial increase of degrees of freedom consumed as K increases. $\Pi_{j=1}^{J} L_j$ represents the number of categories, or unique parameters, that can be estimated. We see that it becomes quite large—especially when $L_j = 4$. Clearly, LCM should perform well with small numbers of variables, e.g., up to 6. GoM's performance will

Table 3.1. The Number of Parameters Required for Identifiability of GoM and LCM with *K* Classes

J	2	3	4	5	10	20	30	
					All Binary Variables			
$\Pi_{j=1}^{J} L_j$	4	8	16	32	64	1,024	1,048,576	1,073,741,824
K					*LCM*			
2	5*	7	9	11	13	21	41	61
3	—	11*	14	17	20	32	62	92
4	—	—	19*	23	27	43	83	123
5	—	—	—	29	34	54	104	154
6	—	—	—	35*	41	65	5	185
7	—	—	—	—	48	76	146	216
8	—	—	—	—	55	87	167	247
9	—	—	—	—	62	98	188	278
10	—	—	—	—	69*	109	209	309
					GoM			
2	6*	9*	12	15	18	30	60	90
3	—	—	26*	35*	45	95	290	585
4	—	—	—	—	107*	325	1,850	5,575
5	—	—	—	—		1,050*	10,725	46,525
6	—	—	—	—		—	53,249	324,811
7	—	—	—	—		—	230,369	1,948,001
8	—	—	—	—		—	888,189	10,295,711
9	—	—	—	—		—	3,108,284*	48,903,761
10	—	—	—	—		—	—	211,915,431
J	2	3	4	5	6	10	20	30
					All Variables with 4 Categories			
$\Pi_{j=1}^{J} L_j$	16	64	256	1,024	4,096	1,048,576	1.1×10^{12}	1.15×10^{18}
K					*LCM*			
2	13*	19	25	31	37	61	121	181
3	20*	29	38	47	56	92	182	272
4	—	39	51	63	75	123	243	363
5	—	49	64	79	94	154	304	454
6	—	59	77	95	113	185	365	545
7	—	69*	90	111	132	216	426	636
8	—	—	103	127	151	247	487	727
9	—	—	116	143	170	278	548	818
10	—	—	129	159	189	309	609	909
K					*GoM*			
2	14	21	28	35	42	70	140	210
3	23*	36	50	65	81	155	410	765

Table 3.1. (*Continued*)

$\Pi_{j=1}^{J} L_j$	16	64	256	1,024	4,096	1,048,576	1.1×10^{12}	1.15×10^{18}
4	—	55	82	115	155	405	2,010	5,815
5	—	79*	129	200	299	1,150	10,925	46,825
6	—	—	197	341	569	3,182	53,489	325,171
7	—	—	293*	566	1,049	8,217	230,649	1,948,421
8	—	—	—	911	1,859	19,687	888,509	10,296,191
9	—	—	—	1,421*	3,164	44,027	3,108,644	48,904,301
10	—	—	—	—	5,184*	92,677	10,015,604	211,916,031

* *Number of parameters* $\geq \Pi_{j=1}^{J} L_j$; hence model not identified.

improve as J increases. Operating characteristics of GoM appear to become good when $J - K \geq 10$. The argument for consistency of the g_{ik} would appear to be reasonably applied when $J - K \geq 50$.

3.5. AN EXAMPLE

We compared GoM with LCM in an analysis of 33 psychiatric symptoms representative of six psychiatric disorders (i.e., simple depression, chronic, severe depression with somatic symptoms, social phobia, simple phobia, panic) in the Johns Hopkins Epidemiological Catchment Area (ECA) study ($N = 3{,}349$; Eaton et al., 1989) and the Duke ECA study ($N = 3{,}834$; Blazer et al., 1989). These data were chosen because the symptoms had known clinical relations to the six disorders. The number of classes was set at six ($K = 6$) because six fuzzy classes were sufficient to describe the data in the GoM analysis.

We present two sets of results (Woodbury and Manton, 1989). First we present the λ_{kjl} and p_{kjl} for the 33 psychiatric symptoms. These are in Table 3.2 for the Johns Hopkins data described above ($N = 3{,}349$).

The p_{kjl} are diffuse, because the $g_{ik} = 1$ or 0, so all within cell heterogeneity is forced into the structural parameters. The λ_{kjl} for the GoM model, in contrast, form distinct patterns that are strongly related to the clinical criteria for the diagnoses.

We also selected a subset of cases to see how altering the marginals by systematic sampling affected the solution. In the first analysis the GoM χ^2 (20, 571.6) is higher than for LCM ($\chi^2 = 6{,}243.8$). The same starting partitions were used for both models. Because the models are nested, the 14,507.8 χ^2 point increase for GoM (with $4 \times 3{,}349 = 13{,}396$ additional degrees of freedom) is highly significant ($t = 6.7$).

When we deleted all cases where *no* symptoms were reported, we find that 1,661 persons had at least one symptom. The χ^2 for GoM dropped from 20,751.6 to 12,262.4 or 8,489 χ^2 points—within 0.5% of the decline in degrees of freedom (i.e., $3{,}349 - 1{,}661 = 1{,}688$; $1{,}688 \times (6 - 1) = 8{,}440$).

Table 3.2. LCM (p_{kjl}) and GoM (λ_{kjl}) Estimates for 33 Variables 6 Classes: Johns Hopkins ECA Data ($N = 3{,}349$)

	Frequency	Classes											
		I		II		III		IV		V		VI	
	%	p_{1jl}	λ_{1jl}	p_{2jl}	λ_{2jl}	p_{3jl}	λ_{3jl}	p_{4jl}	λ_{4jl}	p_{5jl}	λ_{5jl}	p_{6jl}	λ_{6jl}
Sad for 2 weeks	4.1	37.9	100.0	34.8	0.0	2.4	0.0	2.6	0.0	16.5	0.0	1.0	0.0
Sad for 2 years	1.2	13.1	39.6	10.4	0.0	0.7	0.0	0.0	0.0	12.9	0.0	0.3	0.0
Fainting	0.3	3.0	0.0	0.9	30.8	3.5	0.0	0.0	0.0	0.0	0.0	0.2	0.0
Shortness of breath	1.2	6.1	0.0	16.0	100.0	17.1	0.0	0.9	0.0	0.0	0.0	0.4	0.0
Palpitations	2.1	12.6	0.0	16.6	100.0	19.6	0.0	1.9	0.0	22.1	0.0	0.8	0.0
Felt dizzy	2.3	12.4	0.0	19.9	100.0	15.2	0.0	2.8	0.0	0.0	0.0	1.1	0.0
Feel weak	1.4	6.1	0.0	18.5	100.0	16.9	0.0	2.5	0.0	1.9	0.0	0.5	0.0
Nervous person	23.6	67.9	100.0	86.3	100.0	68.7	100.0	34.1	0.0	39.0	0.0	17.0	0.0
Fright attack	1.8	7.3	0.0	30.9	100.0	100.0	0.0	0.0	0.0	0.0	0.0	0.0	0.0
Phobias													
Eating in public	1.1	1.9	0.0	25.1	100.0	0.9	0.0	5.3	0.0	0.0	0.0	0.2	0.0
Speaking in small group	1.5	2.4	0.0	18.9	100.0	15.9	0.0	8.1	0.0	0.8	0.0	0.5	0.0
Speaking to strangers	1.9	2.9	0.0	39.7	100.0	4.7	0.0	10.5	0.0	0.0	0.0	0.5	0.0
Fear of being alone	1.4	4.5	0.0	32.0	80.5	5.1	0.0	3.9	0.0	5.4	37.9	0.3	0.0
Tunnels and bridges	3.6	1.8	0.0	50.9	0.0	17.2	0.0	30.0	100.0	0.0	0.0	0.6	0.0
Crowds	2.7	4.4	0.0	48.4	100.0	22.8	0.0	16.0	0.0	0.0	0.0	0.5	0.0
Public transportation	4.0	2.8	0.0	55.6	100.0	17.9	0.0	30.6	100.0	0.0	0.0	0.9	0.0
Outside house alone	1.3	3.7	0.0	23.5	100.0	0.0	0.0	6.8	0.0	0.0	0.0	0.4	0.0
Heights	7.7	5.2	0.0	54.3	0.0	17.1	0.0	47.9	100.0	0.0	0.0	3.4	0.0
Closed place	2.9	2.6	0.0	43.7	100.0	2.9	0.0	17.7	100.0	6.5	0.0	0.9	0.0
Storms	4.7	3.9	0.0	46.0	0.0	23.0	0.0	26.5	100.0	6.1	0.0	1.9	0.0
Water	4.4	4.0	0.0	35.4	0.0	32.7	0.0	30.5	100.0	6.5	0.0	1.4	0.0
Bugs	9.0	13.0	0.0	55.5	0.0	23.9	0.0	52.4	100.0	13.2	0.0	3.7	0.0
Animals	2.0	2.9	0.0	25.6	0.0	0.7	0.0	11.5	91.6	0.0	0.0	0.8	0.0
Crying spells	4.9	22.5	0.0	32.1	0.0	9.2	0.0	5.7	0.0	100.0	100.0	2.2	0.0
Felt hopeless	4.0	29.5	0.0	26.5	0.0	5.6	0.0	3.1	0.0	100.0	100.0	1.1	0.0
Change in weight or appetite	6.1	32.9	100.0	48.4	0.0	26.2	0.0	6.5	0.0	11.0	0.0	2.8	0.0
Sleeping more or less	10.7	52.6	100.0	64.9	0.0	24.9	0.0	16.9	0.0	36.0	0.0	5.2	0.0
Talking or moving slower	5.7	37.8	100.0	72.9	0.0	5.1	0.0	6.4	0.0	5.4	0.0	1.7	0.0
Interest in sex much less	2.2	13.6	63.0	23.3	0.0	9.2	0.0	3.6	0.0	1.4	0.0	0.9	0.0
More tired	7.4	48.7	100.0	61.6	0.0	5.2	0.0	6.8	0.0	2.8	0.0	2.8	0.0
Feel worthless, sinful, or guilty	2.8	25.0	100.0	32.8	0.0	0.6	0.0	1.6	0.0	16.8	0.0	0.7	0.0
Difficulty in concentrating or thinking	5.2	38.9	100.0	53.9	0.0	6.6	0.0	4.2	0.0	11.8	0.0	1.5	0.0
Thoughts of death or suicide	9.3	41.9	0.0	63.3	0.0	24.7	0.0	15.1	0.0	36.4	100.0	4.5	0.0
"Prevalence"		7.1	3.6	1.9	0.9	0.8	18.3	7.3	3.7	0.5	2.4	82.4	71.1

Source: Johns Hopkins Epidemiological Catchment Area Study

Thus, eliminating cases with no information produced a change in the GoM likelihood almost exactly at expectation.

In LCM the χ^2 declined from 6,243.8 to 3,110.8. The drop of 3,133 is *greater* than expectation, i.e., the change in the marginal significantly affected the prediction of individual class probabilities. Using the usual degrees of freedom count [i.e., equation (3.18)] in the LCM produced a t of 29.4. Even without penalizing the LCM for any degrees of freedom for setting the $g_{ik} = 1$, the difference between the two models was $t = 5.4$ and highly significant.

In Table 3.3 we present the results for 554 cases selected from the total samples in both ECA sites where there were enough symptoms reported by the individual to make at least one of the six diagnoses based on clinical algorithms. Again the λ_{kjl} from the GoM are more informative than the p_{kjl} produced by the LCM. The LCM actually loses some diagnoses it had represented before, while the GoM profiles become more clearly defined.

Thus the GoM model had superior operating characteristics to LCM in this data set. We conducted the same comparison of LCM and GoM in four other data sets with very different characteristics [e.g., 198 patients assessed on 778 psychiatric symptoms (Blazer et al., 1989); 550 subjects evaluated for 68 allergic antigen extracts and 25 clinical diagnoses (Buckley and Woodbury, 1989); 130 subjects in a study of depression (Davidson et al., 1989); and 4,525 nursing home residents, 65+ assessed on 111 clinical characteristics (Manton et al., 1993; see Chapter 5)]. Thus the theoretical and practical advantages of GoM (its greater generality and ability to describe individual heterogeneity) has been confirmed in a direct comparison with LCM in at least five data sets with different characteristics and content.

3.6. SUMMARY

We examined the statistical properties of maximum likelihood estimates of the GoM model. The property of consistency was established for two cases: (i) as J increases and (ii) as I increases. In the first case the g_{ik} values are fixed unknown parameters to be estimated. In the later case the g_{ik} values are realizations of a set of random variables ξ_{jk}. Only the moments of the distribution function $H(\mathbf{x})$ of the random variables were consistently estimated and not the g_{ik} values themselves. We also considered the necessary and sufficient number of questions and observations to provide data to estimate the model parameters. Next we reviewed methods of obtaining maximum likelihood estimates from observations. In practice the iterative procedures require attention in their implementation. As with all nonlinear routines, it is possible to reach local maxima or nonsense results from "canned" methods of estimation.

We illustrated the GoM model with data on mental disorders. The analysis was compared with an LCM analysis of the same data. The different

Table 3.3. Discrete Mixture Model (p_{kjl}) and GoM (λ_{kjl}) Values for 33 Variables by 6 Classes: Johns Hopkins and Duke ECA Data on Diagnosed Cases ($N = 554$)

	Frequency	Classes											
		I		II		III		IV		V		VI	
	%	p_{1jl}	λ_{1jl}	p_{2jl}	λ_{2jl}	p_{3jl}	λ_{3jl}	p_{4jl}	λ_{4jl}	p_{5jl}	λ_{5jl}	p_{6jl}	λ_{6jl}
Sad for 2 weeks	23.8	64.4	100.0	24.1	0.0	4.8	0.0	5.6	0.0	4.4	0.0	4.6	15.3
Sad for 2 years	8.4	24.4	58.2	8.5	0.0	1.7	0.0	0.0	0.0	0.0	0.0	2.4	0.0
Fainting	3.6	4.5	0.0	9.2	38.5	0.8	0.0	0.0	0.0	0.0	0.0	0.0	0.0
Shortness of breath	15.0	18.2	0.0	34.4	100.0	1.3	0.0	13.4	0.0	2.3	0.0	0.7	0.0
Palpitations	17.5	21.2	0.0	36.1	100.0	11.5	0.0	5.7	0.0	2.0	0.0	4.5	0.0
Felt dizzy	18.4	27.2	0.0	33.9	100.0	6.3	0.0	0.0	0.0	8.8	0.0	4.5	0.0
Feel weak	14.8	20.1	0.0	27.9	100.0	8.9	0.0	0.0	0.0	5.7	0.0	1.8	0.0
Nervous person	64.4	85.4	100.0	72.7	100.0	68.5	50.0	23.5	100.0	52.8	26.8	39.7	56.4
Fright attack	19.6	26.4	45.9	16.1	0.0	100.0	41.5	0.0	32.0	0.0	0.0	0.0	0.0
Phobias													
Eating in public	7.1	14.6	0.0	1.0	0.0	5.8	0.0	48.1	100.0	5.4	0.0	1.1	0.0
Speaking in small group	11.8	13.5	0.0	3.1	0.0	12.3	0.0	71.0	100.0	14.1	0.0	7.1	0.0
Speaking to strangers	11.6	18.3	0.0	1.1	0.0	11.9	0.0	97.5	100.0	11.0	0.0	1.5	0.0
Fear of being alone	11.6	25.1	0.0	5.2	0.0	6.7	0.0	2.0	100.0	16.0	0.0	2.2	0.0
Tunnels and bridges	23.9	17.6	0.0	1.5	0.0	12.2	0.0	1.7	100.0	65.1	100.0	31.7	0.0
Crowds	23.5	30.4	0.0	7.2	0.0	29.8	0.0	18.4	100.0	49.5	0.0	3.7	0.0
Public transportation	26.7	21.5	0.0	3.3	0.0	19.9	0.0	26.3	100.0	67.7	100.0	24.2	0.0
Outside house alone	9.2	14.1	0.0	2.6	0.0	1.7	0.0	10.0	100.0	15.3	0.0	8.6	0.0
Heights	33.8	26.2	0.0	7.5	0.0	18.5	0.0	13.3	98.1	71.8	100.0	55.7	0.0
Closed place	19.6	22.4	0.0	4.5	0.0	13.4	0.0	18.3	100.0	38.2	64.2	21.8	0.0
Storms	26.8	29.2	0.0	5.0	0.0	18.3	0.0	2.1	0.0	32.3	100.0	71.0	0.0
Water	25.7	18.7	0.0	3.9	0.0	33.9	0.0	25.4	100.0	43.7	93.3	49.1	0.0
Bugs	38.3	37.8	0.0	8.3	0.0	30.3	0.0	28.0	100.0	46.4	100.0	94.8	0.0
Animals	10.0	11.8	0.0	0.9	0.0	2.6	0.0	5.0	50.6	10.9	36.4	31.6	0.0
Crying spells	25.8	37.4	0.0	44.3	0.0	10.0	0.0	9.7	0.0	10.8	0.0	7.2	100.0
Felt hopeless	28.6	56.1	100.0	48.1	0.0	5.4	0.0	1.6	0.0	5.3	0.0	3.5	100.0
Change in weight or appetite	23.5	50.0	100.0	19.4	0.0	23.6	0.0	10.1	54.2	15.4	0.0	4.5	0.0
Sleeping more or less	38.9	79.4	100.0	36.3	73.5	31.9	0.0	14.3	100.0	26.2	0.0	8.4	23.1
Talking or moving slower	29.0	75.1	100.0	20.8	0.0	9.1	0.0	9.6	0.0	17.8	0.0	4.8	0.0
Interest in sex much less	8.3	16.7	31.4	2.6	0.0	9.4	0.0	1.3	41.2	11.4	6.4	2.9	0.0
More tired	32.7	70.8	100.0	38.5	100.0	10.5	0.0	18.8	0.0	11.9	0.0	8.4	0.0
Feel worthless, sinful, or guilty	14.6	45.4	100.0	11.0	0.0	2.5	0.0	1.9	0.0	2.9	0.0	1.9	0.0
Difficulty in concentrating or thinking	25.6	71.7	100.0	18.8	0.0	7.5	0.0	2.4	0.0	11.1	0.0	4.8	0.0
Thoughts of death or suicide	34.6	76.9	100.0	31.0	100.0	19.5	0.0	2.0	100.0	22.4	0.0	10.2	56.4
"Prevalence"		23.2	15.8	28.3	9.2	8.9	31.5	3.8	5.6	21.8	21.7	14.1	16.1

Source: Johns Hopkins Epidemiological Catchment Area Study.

operating characteristics of GoM and LCM were compared. GoM both described the data better and produced substantively more meaningful results in terms of describing symptom patterns. Similar conclusions were reached in analyses of four other data sets not presented in this chapter.

REFERENCES

Behrman, J. R., Sickles, R., Taubman, P., and Yazbeck, A. (1991). Black–white mortality inequalities. *Journal of Econometrics* **50**: 183–203.

Blazer, D., Woodbury, M. A., Hughes, D., George, L. K., Manton, K. G., Bachar, J. R., and Fowler, N. (1989). A statistical analysis of the classification of depression in a mixed community and clinical sample. *Journal of Affective Disorders* **16**: 11–20.

Buckley, C. E., and Woodbury, M. A. (1989). Poster Presentation at Duke Medical Center Symposium.

Chant, D. (1974). On asymptotic tests of composite hypotheses in nonstandard conditions. *Biometrika* **61**: 191–198.

Chernoff, H. (1954). On the distribution of the likelihood ratio. *Annals of Mathematics and Statistics*, **25**: 573–578.

Cox, D. R., and Hinkley, D. V. (1974). *Theoretical Statistics*. Chapman and Hall, London.

Davidson, J. R., Woodbury, M. A., Zisook, S. and Giller, E. L. (1989). Classification of depression by Grade of Membership: A confirmation study. *Psychological Medicine* **19**: 987–998.

Dempster, A. P., Laird, N. M., and Rubin, D. B. (1977). Maximum likelihood from incomplete data via the EM algorithm (with discussion). *Journal of the Royal Statistical Society Series B* **39**(1): 1–38.

Eaton, W. W., McCutcheon, A., Dryman, A., and Sorenson, A. (1989). Latent class analysis of anxiety and depression. *Sociological Methods & Research* **18**: 104–125.

Everitt, B. S. (1984). *An Introduction to Latent Variable Models*. Chapman and Hall, London and New York.

Gill, R. D. (1992). Multistate life-tables and regression. *Mathematical Population Studies* **3**(4): 259–276.

Gillespie, D. T. (1983). The mathematics of simple random walks. *Naval Research News* **35**: 46–52.

Harman, H. H. (1967). *Modern Factor Analysis*. University of Chicago Press, Chicago.

Hartigan, J. A. (1975). *Clustering Algorithms*. Wiley, New York.

Heckman, J., and Singer, B. (1984a). The identifiability of the proportional hazards model. *Rev. Economic Studies* **51**: 231–241.

Heckman, J., and Singer, B. (1984b). A method for minimizing the impact of distributional assumptions in econometric models for duration data. *Econometrics* **52**: 271–320.

Kiefer, J., and Wolfowitz, J. (1956). Consistency of the maximum likelihood estimator in the presence of infinitely many parameters. *Annals of Mathematical Statistics* **27**: 886–906.

Kushner, H. J. (1974). On the weak convergence of interpolated Markov chains to a diffusion. *Annals of Probability* **2**: 40–50.

Lord, F. M., Norvick, M. R. (1968). *Statistical Theories of Mental Test Scores*. Addison-Wesley, Reading, Massachusetts.

Manton, K. G., and Stallard, E. (1988). *Chronic Disease Modeling: Measurement and Evaluation of the Risks of Chronic Disease Processes*. Charles Griffin Ltd., London.

Manton, K. G., Stallard, E., and Woodbury, M. A. (1991). A multivariate event history model based upon fuzzy states: Estimation from longitudinal surveys with informative nonresponse. *Journal of Official Statistics* **7**: 261–293 (published by Statistics Sweden, Stockholm).

Manton, K. G., Cornelius, E. S., and Woodbury, M. A. (1992). Methodological study of nursing home residents. Duke University Center for Demographic Studies, Working Paper 425, Durham, North Carolina.

Manton, K.A G., Korten, A., Woodbury, M. A., Anker, M., Jablansky, A. (1993). Symptom profiles of psychiatric disorders based on graded disease classes: An illustration using data from the WHO International Study of Schizophrenia. *Psychological Medicine*, 1993.

McLachlan, G. J. and Basford, K. E. (1987). Mixture Models: Inference and Clustering Applications. (Statistics and Monographs Series: No. 4).

Moran, P. A. P. (1971) Maximum likelihood estimators in non-standardized conditions. *Proceedings of the Cambridge Philosophical Society*, **70**: 441–450.

Morris, C. N. (1982). Natural exponential families with quadratic variance functions. *Annals of Statistics* **10**: 65–80.

Orchard, R., and Woodbury, M. A. (1971). A missing information principle: Theory and application. In *Proceedings of Sixth Berkeley Symposium on Mathematical Statistics and Probability* (L. M. LeCam, J. Neyman, and E. L. Scott, eds.). University of California Press, Berkeley, pp. 697–715.

Self, S. G., and Liang, K. Y. (1987). Asymptotic properties of maximum likelihood estimators and likelihood ratio tests under nonstandard conditions. *Journal of the American Statistical Association* **82**: 605–610.

Singer, B. H. (1989) Grade of membership representations: Concepts and problems, In *Festschreift for Samuel Karlin* (T. W. Anderson, K. B. Athreya and D. Iglehardt, Eds.). Academic Press, Orlando, Florida, pp. 317–334.

Tolley, H. D., and Manton, D. G. (1991). A Grade of Membership method for partitioning heterogeneity in a collective. *SCOR Notes*, International Prize in Acturial Science, April, 1991, pp. 121–151.

Tolley, H. D., and Manton, K. G. (1992). Testing for the number of pure types of a fuzzy partition. In review at *Journal of the Royal Statistical Society*, *Series B*.

Tolley, H. D., and Manton, K. G. (1994). Large sample properties of estimates of discrete Grade of Membership model. *Annals of Statistical Mathematics* **44**: 85–95.

Woodbury, M. A. (1963). The stochastic model of mental testing theory and application. *Psychometrika* **28**: 391–394.

Woodbury, M. A., and Clive, J. (1974). Clinical pure types as a fuzzy partition. *Journal of Cybernetics* **4**: 111–121.

Woodbury, M. A., and Manton, K. G. (1989). Grade of Membership analysis of depression-related psychiatric disorders. *Sociological Methods and Research* **18**(1): 126–163.

Woodbury, M. A., Clive, J., and Garson, A. (1978). Mathematical typology: A grade of membership technique for obtaining disease definition. *Computers and Biomedical Research* **11**: 277–298.

Woodbury, M. A., Manton, K. G., and Tolley, H. D. (1993). A general model for statistical analysis using fuzzy sets: Sufficient conditions for identifiability and statistical properties. *Information Science*, forthcoming.

Wu, C. F. J. (1983). On the convergence properties of the EM algorithm. *Annals of Statistics* **11**: 95–103.

CHAPTER 4

A GoM Model for Aggregate Data

INTRODUCTION: THE EFFECTS OF AGGREGATION ON FUZZY SETS

In prior chapters we presented a basic model to analyze multivariate discrete response data for individuals. In this chapter we modify that model to deal with aggregate data, i.e., units of observation are populations with the potential for multiple responses. Such aggregates might be institutions (e.g., hospitals or nursing homes) where a certain number of patients are treated in a unit time, or populations where disease or mortal events (or more generally, emergence of disease natural histories over time) are expressed. The fuzzy set models have advantages in describing aggregates of individuals where additional heterogeneity components are present but there is little a priori information upon which to select a parametric "mixing" distribution.

4.1. LIKELIHOOD FUNCTION FOR AGGREGATE DATA

4.1.1. The Poisson Model

To apply to aggregates, the GoM model requires that we reformulate the likelihood function in (2.1). First, we need to modify notation and define additional terms. Let i refer to a population or institution—not a person. Let j index the frequency count of a type of event. In the model in Chapter 2, the responses for each j were binary responses denoted y_{ijl}. The binary nature was preserved by requiring one of the random variables Y_{ijl}, $l = 1, \ldots, L_j$, to be 1 and the rest to be 0. l indexed the outcome cell. We could consider the present case in the same manner where l indexes the frequency count. For example, $l = 0$ would indicate a zero count, $l = 1$ a count of one, and so forth. In the current application this notation is cumbersome. Instead we use Y_{ij} to denote the random variable of counts of the jth event type for the ith unit. That is, Y_{ij} will take values of the actual count. The realization of Y_{ij} is denoted y_{ij}. For example, $y_{ij} = 5$ would indicate a realization of 5 events. The count of events of the jth type in the ith institution is assumed to be generated by Poisson processes with rate parameters λ_{ij}.

For aggregate data, the λ_{ij} are parameterized similar to the parameterization of probabilities of the Y_{ijl} in (2.1), namely,

$$\lambda_{ij} = \sum_{k=1}^{L} u_{ik} \cdot v_{jk} . \tag{4.1}$$

Although the representation in (4.1) is of the same bilinear form as (2.1), the parameters are of different types with different ranges. The first, u_{ik}, describes the relative frequency of the events for the ith institution associated with the kth fuzzy class or profile of frequencies. The u_{ik} are normed to represent the total frequency of events in all I institutions or aggregates, i.e.,

$$\sum_{i=1}^{I} \sum_{k=1}^{K} u_{ik} = I . \tag{4.2}$$

The u_{ik} are related to Grade of Membership scores, g_{ik}, as

$$g_{ik} = u_{ik} \Big/ \sum_{k=1}^{K} u_{ik} . \tag{4.3}$$

Thus, the u_{ik} are scale adjustment factors (i.e., proportionality factors representing the size of the unit) for the underlying fuzzy set structure. That is, these factors adjust the frequency of event J for type K for the differences in volumes of cases processed by the I institutions or differences in sizes of the populations of interest.

The second set of coefficients, v_{jk}, represents the rate of the jth type (extreme profile) of event for the kth fuzzy set. These parameters serve a role similar to that of λ_{kjl} in Chapter 2. Each v_{jk} represents the Poisson rate parameter for the kth fuzzy set for the jth type of outcome. The v_{jk} are constrained to be positive but are *not* bounded by 1.0 as the λ_{kjl} were in the preceding chapters. Their sum is the average number of cases in the Kth profile.

Under Assumptions 1–4 of Chapter 1, the Y_{ij} follow a product Poisson distribution. This gives the likelihood for observed data $\{y_{ij}\}$ as

$$L = \prod_{i} \prod_{j} e^{-\lambda_{ij}} \cdot \lambda_{ij}^{y_{ij}} / y_{ij}! . \tag{4.4}$$

Reparameterizing with (4.1), we get

$$L = \prod_i \prod_j e^{-[\Sigma_k u_{ik} v_{jk}]} \cdot \left(\sum_{k=1}^{K} u_{ik} v_{jk} \right)^{y_{ij}} \Big/ y_{ij}! \qquad (4.5)$$

The likelihood may be thought of as representing the ith J-way contingency table composed of counts y_{ij} and combining weights for the ith table given as u_{ik} (with scale adjustments).

Equation (4.5) is used to estimate the u_{ik} and v_{jk} parameters of the aggregate model. Note that by construction the model is the likelihood conditional on the u_{ik} parameters. Unlike Chapter 2, where we used the unconditional model, there is no need to remove the u_{ik} parameters here, because as the sample size increases for each unit, estimates of u_{ik} will be consistent (see Chapter 3 for the case of increasing J). In case the units or populations indexed by i are finite sized and the distribution of the u_{ik} is of fundamental interest, we may follow the procedure given for the GoM model in Chapter 3 for increasing I.

4.1.2. The Negative Binomial Model

The aggregate model provides for heterogeneity in responses through the representation of the λ_{ij} using (4.1). On occasion, however, a specific parametric form of the distribution of heterogeneity may be used. One approach is to assume that the λ_{ij} parameters are gamma-distributed random variables with parameters c_{ij} and β. The shape parameters, c_{ij}, are functions of both unit i and event type j. The scale parameter, β, is constant over i and j.

Under the assumption that the λ_{ij} are gamma-distributed random variables, the expectation of Y_{ij} is

$$E(Y_{ij}) = \lambda_{ij} = \beta c_{ij} \,. \qquad (4.6)$$

The β parameter represents the excess variation in Y_{ij} over that generated by a Poisson process, i.e., heterogeneity of individuals' risk of the event or outcome within areas or for other aggregates which is assumed to be independent of i and j. To account for the additional variation (4.5) generalizes to

$$L = \prod_i \prod_j \int_0^\infty \frac{e^{-\lambda_{ii}/\beta}}{\Gamma(c_{ij})} \left(\frac{\lambda_{ij}}{\beta} \right)^{c_{ij}-1} \frac{e^{-\lambda_{ij}} \lambda_{ij}^{y_{ij}}}{\beta(y_{ij}!)} \, d\lambda_{ij} \,. \qquad (4.7)$$

This is the likelihood arising from negative binomial distribution function parameterized for the GoM model. β can be made specific to areas or sets of areas. Equation (4.7) can be used to explicitly introduce empirical Bayesian principles into GoM by using β to represent information from an empirical Bayesian prior distribution of individual differences (see, e.g., Morris, 1982). This also distinguishes the GoM semimetric representation of heterogeneity of individuals from the assumed type distribution of parameters over individuals assumed in empirical Bayes models. A more general model allows β to vary across population groups. If β is allowed to vary across subpopulations, then the subpopulations cannot be simply weighted and added together. Clearly, (4.7) is more complicated to estimate because the integral precludes a convenient direct analytic solution. The expression in (4.7), which we estimate for an example below, could also be used for the survey cluster problem discussed in Chapter 2 or for any other data structure with contamination over individual responses.

In both the Poisson and the negative binomial cases the fuzziness of membership in one set or another of each institution is maintained by the parameterization of the rate parameter λ_{ij}. Estimation of the parameters from either of these models using maximum likelihood follows the same general techniques described in Chapter 3. Thus we also refer to analyses of aggregate data using fuzzy set models as GoM analyses.

4.2. EXAMPLE: STRUCTURAL AND FUNCTIONAL CLASSIFICATION OF MILITARY TREATMENT FACILITIES

As an example, consider the set of 167 Military Treatment Facilities (MTF) maintained by the U.S. Department of Defense (DoD) (Woodbury et al., 1993). By their nature and the role they served the MTFs have no market mechanisms to control costs or obvious ways to value the rarely used, but necessary, standby wartime health services. This makes resource allocation difficult, especially since the services provided by MTFs vary drastically between peacetime and wartime conditions. To solve this problem, the MTFs were characterized using the methods described in Section 4.1. Allocation of resources can be made based on the characteristics of each of the MTFs (see Woodbury et al., 1993).

In this example we show how such resource allocation can be done if the MTFs can be accurately classified by structure and function. The institutional function (or purpose) is indicated by the frequency of patients treated in each of 473 Diagnosis Related Group (DRG) categories (Vertrees and Manton, 1986). We conducted two different GoM analysis.

1. The 167 MTFs were analyzed according to their medical "production" function roles using the aggregate GoM model in (4.1). Institutional function is described by the number of patients in each of the 167 facilities in each of the 473 medically defined DRGs (Woodbury et al., 1993). Thus there is an $I(=167) \times J(=473)$ table of patient counts (y_{ij}). These are analyzed using (4.5).

2. The basic GoM model Chapter 2 was used to classify each facility on the basis of 19 structural characteristics describing general facility features such as whether the hospital has teaching functions, its standby bed capacity, etc. These attributes are analyzed using (2.1), where $I = 167$ and $J = 19$. In the analysis a number of characteristics are continuous. They have been recorded as categorical random variables so that the information lost from the original distribution is minimized.

4.2.1. Structural Analysis

We first applied the standard GoM model. The λ_{kjl} for 19 facility characteristics are in Table 4.1.

Table 4.1 is divided into two parts. Part A refers to the 19 variables used to define the MTF classes. The variables, and their levels, are in the left-most column of the table. The next five columns describe the λ_{kjl} for each of the ($K = 5$) five profiles which the likelihood ratio χ^2 indicated were necessary to explain the structural characteristics of the 167 MTFs (i.e., the y_{ijl}). The 9 external variables in Part B were *not* used to define the K fuzzy classes. That is, they were not used in equation (2.1) to calculate the g_{ik}. Their λ_{kjl} were estimated in an independent maximization using the score function (3.34) with the g_{ik} held fixed in (3.33), i.e., the g_{ik} were not altered. These variables help to validate the five structural classes of MTFs.

The nature of each of the five MTF structural classes is described by comparing the λ_{kjl} to the sample proportion with this response. If the λ_{kjl} are larger than the marginal probability, they "define" characteristics distinctive to that particular MTF class. For example, the first variable is the percentage of the facilities' budget allocated to standby activities. The first response is "missing," a response included to avoid biased results due to missing data (see Section 2.2). The second response, "0," indicates that 1.86% of these facilities did *not* have standby units. No standby services, however, is characteristic of Class 1 in that the probability of an MTF that is 100% like this type (i.e., $g_{i1} = 1$) having zero standby units is 7.42%. The probability for the four other classes is 0 (i.e., $g_{i2} = g_{i3} = g_{i4} = g_{i5} = 0$). Thus a facility with no standby costs will exhibit at least partial membership in Class 1.

The five classes can be characterized (refer to Table 4.1) as follows:

1. Small, rural in-patient MTFs, located in the U.S., serving primarily the medical needs of a local military installation. These facilities have the lowest number of operating beds (70% have fewer than 30 beds), serve areas under 20,000 in population with moderately high levels of active duty personnel and with moderate numbers of admissions. These facilities have the lowest case-mix severity, the fewest disposi-

Table 4.1. Coefficients of Treatment Facilities on 19 MTF Characteristics, Basic GoM Analysis

Range	Sample Proportion	Type Profiles 1	2	3	4	5
A. Internal Variables						
1. % Monies Used to Support Standby Capacity						
Missing	3.59	0.10	0.00	20.40	0.00	0.00
0	1.86	7.42	0.00	0.00	0.00	0.00
<0.002	21.74	0.00	27.12	32.23	0.00	59.14
<0.006	34.16	27.11	31.34	8.81	69.82	40.86
<0.015	26.71	36.69	41.53	12.94	30.18	0.00
≥0.015	15.53	28.78	0.00	46.01	0.00	0.00
2. Number of Operating Beds, Set Up and Staffed						
0–29	25.15	69.53	0.00	0.00	0.00	0.00
30–49	23.35	30.47	45.37	12.77	0.00	0.00
50–99	18.56	0.00	54.63	82.44	0.00	0.00
100–199	19.16	0.00	0.00	0.00	87.73	0.00
200–399	7.78	0.00	0.00	4.79	12.27	44.68
400+	5.99	0.00	0.00	0.00	0.00	55.32
3. Total Catchment Area Population						
Missing	1.20	0.00	1.64	6.85	0.00	0.00
<20,000	31.52	100.00	0.00	43.11	0.00	0.00
20,000–39,999	30.30	0.00	89.90	35.65	0.00	0.00
40,000–59,999	15.76	0.00	10.10	21.24	33.22	27.67
60,000–79,999	12.12	0.00	0.00	0.00	49.27	18.19
80,000+	10.30	0.00	0.00	0.00	17.51	54.14
4. Percent Active Duty Personnel in Area						
Missing	1.20	0.00	1.64	6.85	0.00	0.00
<20	24.85	0.00	60.57	0.00	0.00	46.91
20–29	31.52	48.20	39.43	0.00	15.15	53.09
30–39	23.03	51.80	0.00	0.00	84.85	0.00
40+	20.61	0.00	0.00	100.00	0.00	0.00
5. Percent Retired in Area						
Missing	1.20	0.00	1.64	6.85	0.00	0.00
<4	23.64	0.00	0.00	100.00	0.00	0.00
4–11	19.39	61.06	0.00	0.00	55.27	0.00
12–19	34.55	38.94	44.52	0.00	44.73	56.76
20+	22.42	0.00	55.48	0.00	0.00	43.24
6. Average Age (in years) in Area						
<24	26.95	0.00	0.00	100.00	0.00	0.00
24–29	25.15	100.00	0.00	0.00	73.08	0.00
30–35	33.53	0.00	64.07	0.00	26.92	77.49
36+	14.37	0.00	35.93	0.00	0.00	22.51

Table 4.1. (*Continued*)

Range	Sample Proportion	Type Profiles 1	2	3	4	5
7. Number of Teaching Programs						
Missing	7.78	0.00	0.00	0.00	41.54	0.00
0	88.31	100.00	100.00	100.00	100.00	0.00
1–5	2.60	0.00	0.00	0.00	0.00	22.22
6–17	4.55	0.00	0.00	0.00	0.00	38.89
18+	4.55	0.00	0.00	0.00	0.00	38.89
8. Number Residents + Interns (I Series)						
0	81.44	100.00	100.00	100.00	60.09	0.00
1–30	6.59	0.00	0.00	0.00	39.91	0.00
31–149	6.59	0.00	0.00	0.00	0.00	55.00
150+	5.39	0.00	0.00	0.00	0.00	45.00
9. Total Visits in 1985						
Missing	2.40	2.21	0.00	10.55	0.00	0.00
<100,000	25.15	77.87	0.00	32.09	0.00	0.00
100,000–199,999	33.74	22.13	81.15	51.07	0.00	0.00
200,000–399,999	21.47	0.00	18.85	16.84	53.30	25.49
400,000+	19.63	0.00	0.00	0.00	46.70	74.51
10. Inpatient Service Complexity, 1 = Low, 9 = High						
Missing	2.99	1.01	0.00	15.69	0.00	0.00
1–2	10.49	32.82	0.00	0.00	0.00	0.00
3–4	38.89	67.18	58.19	49.62	0.00	0.00
5–7	40.12	0.00	41.81	50.38	100.00	0.00
8–9	10.49	0.00	0.00	0.00	0.00	100.00
11. Outpatient Service Complexity, 1 = Low, 11 = High						
Missing	2.99	1.01	0.00	15.69	0.00	0.00
1–3	34.57	100.00	31.39	0.00	0.00	0.00
4–5	23.46	0.00	68.61	100.00	0.00	0.00
6–8	30.86	0.00	0.00	0.00	100.00	0.00
9–11	11.11	0.00	0.00	0.00	0.00	100.00
12. Total Service Index						
Missing	2.29	1.01	0.00	15.69	0.00	0.00
3–9	19.14	62.86	0.00	0.00	0.00	0.00
10–13	25.31	37.14	40.04	56.13	0.00	0.00
14–16	17.90	0.00	59.96	43.87	0.00	0.00
17–19	16.67	0.00	0.00	0.00	60.00	0.00
20–24	11.11	0.00	0.00	0.00	40.00	0.00
25–29	9.88	0.00	0.00	0.00	0.00	100.00

Table 4.1. (*Continued*)

Range	Sample Proportion	Type Profiles				
		1	2	3	4	5
13. Number of Admissions						
Missing	1.20	0.00	0.00	6.85	0.00	0.00
<50	22.42	0.00	0.00	100.00	0.00	0.00
50–499	23.64	82.57	6.75	0.00	0.00	28.50
500–999	25.45	17.43	47.83	0.00	26.64	32.50
1,000–2,999	23.64	0.00	45.42	0.00	64.03	12.91
3,000–10,000	4.85	0.00	0.00	0.00	9.33	26.09
14. Region of World						
U.S.	76.05	100.00	100.00	0.00	100.00	95.26
Europe	14.37	0.00	0.00	61.54	0.00	0.00
Other	9.58	0.00	0.00	38.46	0.00	4.74
15. Category						
Overseas	22.75	0.00	0.00	100.00	0.00	0.00
Undeserved Area	11.38	48.66	0.00	0.00	5.31	0.00
Med. Ed.	10.78	0.00	0.00	0.00	0.00	100.00
Operational Force	32.93	51.34	45.13	0.00	54.10	0.00
Combat Train	13.17	0.00	21.74	0.00	40.59	0.00
Other	8.98	0.00	33.14	0.00	0.00	0.00
16. Facility Condition						
Good	16.17	24.34	11.67	4.71	32.65	0.00
Fair	17.37	13.40	25.34	7.09	18.70	20.17
Poor	46.11	40.91	62.99	27.84	48.65	42.35
Replace	20.36	21.35	0.00	60.36	0.00	37.47
17. Total Visits/Normal Beds (in 100's)						
Missing	2.99	1.01	0.00	15.69	0.00	0.00
<80	11.11	3.95	0.00	30.06	17.77	13.09
80–109	11.11	0.00	0.00	7.18	14.70	45.04
110–149	17.28	33.14	0.00	0.00	12.07	36.59
150–199	20.99	12.50	34.29	14.15	35.61	5.28
200–249	17.28	22.04	16.35	24.99	19.86	0.00
250–299	12.96	12.52	29.51	20.19	0.00	0.00
300+	9.26	15.86	19.84	3.42	0.00	0.00
18. Support Base Operations						
Missing	2.99	1.01	0.00	15.69	0.00	0.00
Yes	24.69	0.00	0.00	37.06	25.77	81.62
No	75.31	100.00	100.00	32.94	74.23	18.38
19. Support Military Activities						
Missing	2.99	1.01	0.00	15.69	0.00	0.00
Yes	96.91	96.40	100.00	86.22	100.00	100.00
No	3.09	3.60	0.00	13.78	0.00	0.00

Table 4.1. (*Continued*)

Range	Sample Proportion	Type Profiles				
		1	2	3	4	5
	B. External Variables					
1. Number Residents + Interns						
0	81.44	100.00	100.00	100.00	52.47	0.00
1–78	8.98	0.00	0.00	0.00	47.53	0.00
79+	9.58	0.00	0.00	0.00	0.00	100.00
2. Branch of Service						
Army	29.94	0.00	23.76	43.80	69.39	31.53
Air Force	49.10	92.26	64.52	17.42	16.15	23.14
Navy	20.96	7.74	11.72	38.78	15.46	45.33
3. CHAMPUS Percentage						
Missing	3.59	1.65	0,00	18.20	0.00	0.00
<5	30.43	13.39	0.00	100.00	0.00	52.77
5–19	33.54	72.69	18.95	0.00	62.48	31.00
20–49	24.22	0.00	51.43	0.00	37.52	16.23
50–100	11.80	13.92	29.62	0.00	0.00	0.00
4. Case Mix Index						
Missing	0.60	2.41	0.00	0.00	0.00	0.00
≤0.5	15.66	39.33	7.28	23.40	0.00	0.00
0.6	57.23	60.67	66.05	58.80	71.07	0.00
0.7	20.48	0.00	24.40	17.79	28.93	45.76
0.8–1.0	6.63	0.00	2.27	0.00	0.00	54.24
5. Teaching Index, 0 = None, 10+ = Medical Center						
0	81.44	100.00	100.00	100.00	56.02	0.00
1–9	7.78	0	0	0	43.98	0.00
10+	10.78	0.00	0.00	0.00	0.00	100.00
6. Discharges in 1985						
Missing	2.40	2.21	0.00	10.55	0.00	0.00
<2,000	20.86	60.88	7.03	0.00	0.00	0.00
2,000–4,999	41.10	39.12	89.93	64.14	0.00	0.00
5,000–9,999	20.25	0.00	3.15	35.86	74.98	0.00
10,000+	17.79	0.00	0.00	0.00	25.02	100.00
7. Cost Per Work Unit, Case Mix Adjusted						
Missing	2.99	1.47	0.00	15.02	0.00	0.00
<1000	6.79	25.07	0.00	0.00	0.00	0.00
1,000–1,499	35.19	41.76	44.06	39.32	33.12	0.00
1,500–1,999	37.65	19.11	32.35	60.63	66.30	20.31
2,000–2,999	16.05	11.78	21.31	0.06	0.58	55.13
3,000+	4.32	2.28	2.28	0.00	0.00	24.56

Table 4.1. (*Continued*)

Range	Sample Proportion	Type Profiles 1	2	3	4	5
8. Cost Per Day						
Missing	2.40	2.21	0.00	10.55	0.00	0.00
<300	12.27	6.64	8.26	33.62	16.29	1.97
300–349	24.54	16.49	20.15	15.14	44.09	32.03
350–379	16.56	21.13	3.91	0.00	34.65	27.75
380–429	18.40	12.01	30.67	10.67	0.76	38.25
430–519	17.79	24.09	26.38	23.72	4.21	0.00
520+	10.43	19.64	10.63	16.85	0.00	0.00
9. Cost Per Discharge						
Missing	2.40	2.21	0.00	10.55	0.00	0.00
<110	9.82	34.24	0.00	8.42	0.00	0.00
110–139	25.15	26.54	32.05	29.23	26.08	0.00
140–164	24.54	20.91	28.30	32.01	32.91	0.00
165–184	13.50	3.93	10.03	3.22	36.90	15.70
185–239	16.56	14.38	25.09	15.51	4.11	23.73
240+	10.43	0.00	4.52	11.61	0.00	60.57

tions (discharges), the highest per bed costs, but the lowest cost per case.

2. Slightly larger MTFs located in small urban areas. These MTFs service an older, retired population and are concentrated in the U.S. They have the highest proportion of CHAMPUS reimbursement and a more severe case-mix than Class 1 (the case mix level is similar for the second, third, and fourth MTF classes).
3. MTFs larger than the second class and serving a smaller catchment area, but with a high proportion of active duty personnel. There are few retired personnel in the area served, and the average age of patients is low. There is a small number of admissions. These MTFs are found overseas.
4. MTFs of moderate size having a large catchment area population with a high proportion of active duty personnel and large numbers of patients.
5. MTFs with significant teaching functions. These MTFs have large numbers of beds, a large catchment area population, serve many retired personnel, have many teaching programs and the most serious case mix. Costs are moderate on a per diem basis but high on a per patient episode or stay basis (generally because of greater LOS).

Thus a clear pattern emerges from the description of the five classes of

MTFs. Classes 1 and 3 are MTFs serving active military populations on bases with Class 3 located overseas and Class 1 in the continental U.S. Class 2 provides a large amount of basic medical services to retired populations. Classes 4 and 5 are larger MTFs with Class 5 representing major teaching facilities.

In Table 4.2 we present the g_{ik} distributions to see how the MTFs are described by each class profile.

We see that 53 (i.e., 11 + 15 + 4 + 10 + 13) MTFs have a g_{ik} of 1, i.e., are strongly characterized by the features associated with that class. In Class 5 the average g_{ik} (for MTFs that had $g_{ik} > 0$) was 0.6, indicating that this capability is strongly concentrated in specialized institutions. There are 42 facilities in Class 1 with a $g_{ik} \geq 0.50$, also a strong degree of concentration in a class. Most MTFs were strongly characterized as one class with a partial association of one or two of the remaining classes to indicate admixtures of ancillary services.

4.2.2. Functional Analysis

We can contrast the fuzzy set classes generated in the structural analysis using standard GoM (equation (2.1)) with the fuzzy set classes derived from the frequency-based GoM analysis of MTF function using the Poisson rate likelihood function (equation (4.5)). The functional analysis will better portray the volume of services provided by the 167 MTFs to what is a fairly large patient population ($N \sim 1{,}000{,}000$ for the service year 1985). There is an average of 5500 patients seen at each MTF. In the Poisson rate analysis, the frequency table is $I(=167) \times J(=473)$ for the 473 DRG categories. The

Table 4.2. Number of MTFs with GoM Score in Given Ranges for the Five GoM Profiles, Basic GoM Analysis ($N = 167$)

	Type Profiles				
g_{ik} Intervals	1	2	3	4	5
	93	90	111	105	130
0.01–0.09	2	6	1	1	2
0.10–0.19	3	5	11	11	7
0.20–0.29	6	9	3	8	7
0.30–0.39	12	8	1	4	0
0.40–0.49	9	5	3	7	1
0.50–0.59	8	3	12	8	2
0.60–0.69	7	7	8	3	1
0.70–0.79	5	8	9	5	1
0.80–0.89	4	7	3	4	2
0.90–0.99	7	4	1	1	1
	11	15	4	10	13

data analyzed are the frequency of patients within a DRG served in the ith MTF. To help compare the results of the functional analysis (i.e., the MTF classes generated by patient treatment patterns) with the structural analysis, we again used a five-class solution ($K = 5$). The frequency-based analysis (because it is based on the experience of more than a million patients) could have supported the identification of more MTF classes. This is a general property of the aggregate model, i.e., in an analysis of social structure in 512 census tracts in Baltimore, Maryland, the number of classes that could be estimated was consistent with the large population count in the tracts rather than the number of tracts (Goldsmith et al., 1984).

In Table 4.3 we present the v_{jk}, which are logically similar to the λ_{kjl} except that they represent the *average* frequency of that DRG in the typical MTF in the kth class rather than the probability (i.e., the λ_{kjl}) for the lth response to the jth variable for a class. For parsimony, we presented only 46 of 473 DRGs (all 473 DRGs were in the solution) in Table 4.3. These 46 DRGs were selected because they illustrated typical differences between classes.

In the left-most column are DRG names. The next column (labeled "Sample Frequency") is the *average* number of cases in that DRG in the 167 MTFs. The remaining columns contain the average frequency of cases seen in the DRGs for each of five MTF classes. The v_{jk} describe the treatment role associated with each of those five classes of facilities. Each MTF class can be characterized by the highest-frequency DRGs. The five classes may be described as follows:

1. MTFs providing large amounts of services to pregnant women (military dependents) and children. The most frequent DRGs involve pregnancy and delivery. For example, normal vaginal delivery (DRG = 373) with an average annual rate (in 1985) of occurrence of 936 cases and treatment of normal neonates (DRG = 391) with an average rate of 1025 events are the most frequent cases in these facilities. Babies with medical problems, although born at these MTFs, are transferred for treatment to other MTFs or hospitals with special neonatal services and units. In addition, these MTFs provide a broad range of primary medical services, e.g., treating fractures and sprains, and provide acute care services (e.g., head injuries, etc.). There is also a large volume of simple surgical procedures (e.g., appendectomy) performed.

   The analysis did *not* use facility characteristics as classification variables. Thus, one might not expect that one of the functionally defined classes would be characterized by location in foreign countries, as was found in the structural analysis where location was used. However, Navy and Army MTFs in foreign countries (these are generally larger than Air Force MTFs) tend to have high grades of membership (more than 0.5) in this class along with midsized Army and Navy MTFs in the continental U.S.

Table 4.3. v_{jk} Estimated for 46 DRGs, Extended GoM Analysis of MTFs

DRG Description	Sample Frequency	Type Profiles				
		1	2	3	4	5
Craniotomy age >17 except for trauma	2.84	0.00	0.08	0.00	0.00	18.80
Craniotomyy age <18	1.21	0.00	0.00	0.00	0.00	8.04
Extracranial vascular procedures	3.90	0.00	0.07	0.50	0.00	25.14
Carpal tunnel release	13.40	2.73	8.29	22.93	19.31	25.78
Nervous system neoplasms age >69 and/or C.C.	1.56	0.39	0.49	1.54	0.65	6.31
Seizure & headache age 18–69 W/O C.C.	35.28	25.85	34.30	35.08	63.43	34.56
Traumatic stupor & coma <1 hr age 18–69 W/O C.C.	18.71	36.52	3.90	2.91	40.90	3.78
Major head & neck procedures	1.90	0.00	0.09	2.01	0.00	9.84
Tonsillectomy and/or adenoidectomy only age >17	13.34	5.71	0.00	43.97	0.27	17.27
Tonsillectomy and/or adenoidectomy only age 0–17	14.41	6.15	2.20	49.04	0.00	13.94
Otitis media & URI age 18–69 W/O C.C.	84.05	7.69	14.57	0.00	638.52	9.64
Otitis media & URI age 0–17	31.69	32.48	18.15	18.73	113.49	0.00
Major chest procedures	6.37	1.14	0.33	2.08	0.00	36.63
Respiratory neoplasms	21.48	0.00	5.05	24.71	14.78	90.97
Chronic obstructive pulmonary disease	33.17	1.26	32.70	56.03	49.79	58.50
Simple pneumonia & pleurisy age 18–69 W/O C.C.	26.97	15.21	11.10	14.80	135.04	3.27
Simple pneumonia & pleurisy age 0–17	17.46	20.98	17.02	15.08	32.18	1.91
Coronary bypass W/O cardiac catheter	4.71	0.00	0.00	0.00	0.00	31.32
Circulatory disorder with AMI W/O comp. disch. alive	20.96	5.14	12.26	44.17	24.92	32.52
Circulatory disorders except AMI, with cardiac catheter W/O complex diagnosis	31.71	0.00	0.00	0.00	0.00	210.73
Cardiac arrest, unexplained	2.95	0.03	4.16	6.77	2.93	2.54
Angina pectoris	20.64	0.00	19.60	64.42	21.55	7.14
Hernia procedures except inguinal & femoral age >69 and/or C/C.	1.75	0.00	2.02	3.65	0.96	3.22

Table 4.3. (*Continued*)

DRG Description	Sample Frequency	Type Profiles 1	2	3	4	5
Hernia procedures except inguinal & femoral age 18–69 W/O C.C.	15.06	13.36	16.50	19.88	19.93	6.41
Inguinal & Femoral hernia procedures age 18–69 W/O C/C.	60.46	51.93	58.52	96.44	63.92	30.51
Esophagitis, castroent. & misc. digest, age 18–69 W/O C./C.	143.99	104.27	227.83	13.70	420.78	68.13
Dental extractions & restorations	74.00	0.00	309.21	7.29	79.28	0.00
Malignancy of hepatobiliary system or pancreas	3.41	0.96	2.12	1.98	2.10	13.41
Fractures, sprains, strains, & dislocations of upper arm or lower leg except foot age 18–69 W/O C.C.	13.11	15.62	2.54	21.35	24.92	1.48
Malignant breast disorders age >69 and/or C.C.	3.73	0.00	3.61	3.18	1.03	6.92
Cellulitis age 18–69 W/O C.C.	31.80	30.85	20.05	30.94	89.00	4.94
Kidney transplant	0.32	0.00	0.00	0.00	0.00	2.15
Renal failure	10.71	1.08	3.28	0.00	2.56	62.33
Admit for renal dialysis	0.63	0.00	0.00	0.00	0.00	4.18
Kidney & urinary tract infections age 18–69 W/O C.C.	22.29	20.49	30.93	14.97	41.37	8.94
Sterilization, male	11.21	11.90	0.00	0.00	60.89	0.00
Laparoscopy & incisional tubal interruption	55.56	83.37	96.41	1.14	32.26	32.02
Vaginal delivery W/O complicating diagnoses	488.07	936.28	707.51	203.23	0.00	0.00
External immaturity or respiration distress syndrome, neonate	5.56	7.68	5.27	0.00	0.00	13.33
Normal newborns	570.98	1024.82	912.29	246.11	0.00	26.06
Radiotherapy	3.39	0.00	0.12	0.31	0.26	21.75
Chemotherapy	15.68	1.37	1.30	3.06	2.09	93.72
Viral illness age >17	55.42	40.07	45.78	37.13	212.30	0.00
Alcohol/drug dependence, detox and/or other symptomatic treatment	84.39	57.44	30.25	250.30	51.47	19.99

2. The second MTF class shares some characteristics with Class 1 in that it also deals with a high volume of pregnancies and deliveries. This class also provides dental services that the first MTF class did not, i.e., far more cases in DRG 187 are discharges from the second MTF class. The first MTF class, in contrast, performed more orthopedic procedures. Thus Class 1 appears to be a fuller-service facility (concentrating on pregnancies and deliveries), while the second is oriented to simple inpatient and outpatient services.

   Among MTFs with $g_{i2} > 0.5$, it is striking that they are virtually all (46 of 53) Air Force hospitals located in remote areas of the U.S. Of the 48 MTFs meeting this criterion, only one is a Navy hospital and one is an Army MTF. These MTFs probably perform a role similar to that of small rural hospitals in the civilian sector.

3. The third MTF class handles a mix of cases with large proportions of retirees and dependents. The most important DRG is lens extractions (i.e., for cataracts). They also handle cerebrovascular problems (strokes), transient ischemic attacks (TIAs), chronic obstructive pulmonary disease (COPD), pneumonia, and some circulatory problems. Rhinoplastices, tonsillectomies, and adenoidectomies are also common.

   No MTF had a $g_{ik} > 0.75$ in Class 3, implying that facilities may be using their standby capacity to provide care to retirees and dependents on a space available basis. MTFs scoring between 0.75 and 0.50 are generally midsized and include MTFs from the Navy or Air Force located in the U.S.

4. This fourth MTF class roughly corresponds to the fourth MTF class in the structural analysis. It treats many of the same major medical problems that the third class of facility treats, but with surgery and procedures that are more technologically sophisticated and specialized.

5. The fifth MTF class provides a wide range of high technology medical and surgical services. For example, most chemotherapy and radiation therapy for cancer is performed in these MTFs, as is most surgery for tumors, major cardiovascular procedures, angioplasty, cardiac catherization, and major diagnostic procedures. A wide range of medical specialties is represented. This class is similar to the fifth MTF class (major teaching facilities) identified in the structural analysis.

   The identify of the MTFs with high scores in Class 5 confirms this observation. The nine largest medical centers in the military health care system all had high g_{ik} for this class. However, no MTF had a $g_{ik} > 0.75$. This means that no single MTF treated large numbers of cases in all of the wide variety of complex DRGs that characterize the fifth functionally defined class. This illustrates that a facility can fill the profile of services associated with more than one MTF class. For example, Hospital A is 75% like Class 5 *and* 16% like Class 1, i.e., MTF A primarily is a medical center, but also handles normal deliveries and provides primary medical services. MTF D treats a case load 40% like Class 5 and 35% like the third (retirees) class, and 17% like Class 1 (provision of basic or primary medical services).

The cases representative of this class include cerebrovascular problems, TIA, fractures, kidney stones, viral diseases, parasites, and COPD. High g_{ik} are found for larger Air Force and Army hospitals including larger overseas hospitals; most MTFs with $g_{ik} > 0.50$ in Class 5 are located in the continental U.S.

Though there is correspondence between several of the MTF classes in the structural and functional analyses there are also differences, i.e., structure and function are imperfectly correlated in the MTFs because of their necessary maintenance of standby capacity and services. A common factor is that acute medical problems are treated in all MTF classes (e.g., acute myocardial infarctions, heart attacks) because of the need for immediate medical response for these conditions.

In addition, in the functional analysis, the definition of certain MTF classes is dominated by high-frequency medical conditions. The third MTF class provides a broad range of medical specialties and services. The first two MTF classes provide care associated with small community hospitals in the civilian sector. Since location was not in the functional analysis, foreign MTFs do not emerge as a separately identifiable class as they did in the structural analysis.

The second parameter to examine from the fuzzy set analysis of function is the scale factor, i.e., the u_{ik}. In Table 4.4 both the scale adjustment factor $\Sigma_k u_{ik}$ (column 1) and the g_{ik} are presented to describe the extent to which the cases treated in a facility are described by the treatment role represented by each of the five classes for 15 selected MTFs.

This illustrates the interaction of case-mix and the treatment technologies actually used in a MTF. Large facilities (e.g., MTF A) have large u_{ik} values and have high scores for Class 5. The highest score was 0.75 (MTF A), implying that no MTF treats as many as the average number of cases for the DRGs associated with this class.

4.3. AN APPLICATION: QUADRATIC PROGRAMMING MODEL FOR OPTIMIZING RESOURCE ALLOCATION TO GOM CLASSES

As mentioned in Chapter 2, the GoM procedure produces summary scores (the g_{ik}) that capture individual variation. In that assessment we showed that, depending on how large the value of J was, either the scores (g_{ik}) themselves (Woodbury et al., 1993) or selected moments (μ_k) of the scores (Tolley and Manton, 1992, 1994) could be used in derivative risk calcula-

Table 4.4. Scale Adjustment Factors (u_{ik}) and the Grade of Membership Score (g_{ik}) for 15 selected MTFAs

		g_{ik} for Types				
Name	u_{ik}	1	2	3	4	5
Hospital A	3.69	0.16	0.07	0.00	0.02	0.75
Hospital B	3.39	0.14	0.03	0.04	0.12	0.67
Hospital C	4.06	0.06	0.19	0.22	0.04	0.48
Hospital D	2.68	0.17	0.06	0.35	0.01	0.40
Hospital E	0.24	0.00	0.00	0.00	0.86	0.14
Hospital F	0.81	0.27	0.01	0.00	0.73	0.00
Hospital G	0.39	0.00	0.89	0.05	0.06	0.00
Hospital H	2.20	0.00	0.42	0.03	0.00	0.54
Hospital I	2.84	0.43	0.07	0.39	0.02	0.08
Hospital J	0.87	0.01	0.32	0.57	0.07	0.03
Hospital K	0.20	0.26	0.56	0.00	0.14	0.05
Hospital L	5.19	0.41	0.01	0.35	0.04	0.20
Hospital M	0.14	0.69	0.00	0.18	0.13	0.00
Hospital N	0.41	0.62	0.06	0.23	0.08	0.00
Hospital O	0.21	0.00	0.06	0.00	0.94	0.00

tions for mixing prices for health care categories, etc. In this example, we use those scores in two ways to make optimal allocations of budgets across MTFs. Here, optimality is defined by minimizing a least-squares objective function, $\phi(x)$, representing the squared differences between the predicted budget and the observed budget under scale constraints to modify allocations for currently unperformed functions. We first present the quadratic programming model and then illustrate the use of both types of GoM models in implementing the quadratic programming model using data on the 167 MTFs and the patients they served in 1985.

4.3.1. A Quadratic Programming Model

The "optimally" allocated budget has three components. The first is resource allocation to MTFs using the case-mix relative weight estimates. In the objective function, weights are estimated for case-mix categories that produce the smallest mean-square discrepancy between predicted and current budgets.

Second, allocation weights are calculated with side conditions imposed for case-mix prices derived from nine other case-mix-based reimbursement systems, e.g., Medicare, CHAMPUS, Maryland Medicaid. These ensure that the allocation weights estimated from the model have a structure similar to other case-mix systems. This maintains consistency with other systems (even allowing that treatment costs for a medical condition can vary widely between types of hospitals, the prices in other systems represent a plausible starting point for the algorithm) and preserves the medical content of the case-mix weights; e.g., surgical cases tend to be more expensive than medical cases.

Since there are fewer MTFs (167) than categories (e.g., 473 DRGs), many different combinations of case-mix prices will minimize budget error. Side conditions imposed in the programming model are used to resolve this indeterminacy and select an unique solution (i.e., budget allocation).

Third, budget adjustments are calculated using multivariate indicators describing the hospital's role in providing services—for MTFs this includes war readiness, provision of emergency or disaster services, as well as more conventional roles, e.g., operation of tertiary care centers. This component is needed because case-mix weights will not represent those non–case-mix standby functions that are important to maintain in the MTF system.

As before we define certain terms. Each MTF is again designated by i ($i = 1, 2, \ldots, 167$) and each of 473 DRGs by j ($j = 1, 2, \ldots, 473$). l represents one of the other ($L = 9$) cost schedules. We also define the following:

$\{h_{ij}\}$ $= \mathbf{H}$ is a $I \times J$ matrix of number of patients in a MTF in a DRG.

$\{c_j\}$ $= \mathbf{c} =$ vector of cost allocation weights estimated for J case-mix categories.

$\{e_i\}$ $= \mathbf{e} =$ vector of 1s used to produce certain sums, e.g., $\mathbf{e}^{\mathrm{T}} h_{ij} = h_j^{\mathrm{T}}$ is the number of cases in each DRG category summed over MTFs (i).

$\{b_i\}$ $= \mathbf{b} =$ vector of budgets for the I MTFs; $\mathbf{B}$ is the total budget ($\mathbf{B} = \mathbf{e}^{\mathrm{T}}\mathbf{b}$).

$\{y_{ijl}\}$ $= Y = J$ structural features of MTFs (l represents levels of the Jth variable).

$\{g_{ik}\}$ $= \mathbf{G} =$ matrix of g_{ik} describing MTF structure calculated from realizations of the random vector ξ in R^K.

$\{a_{lj}\}$ $= \mathbf{A} =$ DRG weights from L external health systems; $\mathbf{A}$ is normalized, $\mathbf{e}^{\mathrm{T}}\mathbf{HA} = \mathbf{B}$.

$\{d_k\}$ $= \mathbf{d} =$ regression coefficients for K covariates (i.e., the 5 g_{ik} derived for the 167 MTFs from the structural analysis) predicting the budget portion independent of case mix variation over MTFs.

$\{w_l\}$ $= \mathbf{w} =$ is the vector $I \times L$ of weights for the $L = 9$ external sources.

Since $I = 167$ and $J = 473$, a procedure is needed to replace h_{ij}^* by its expectation. Since each element of $\mathbf{H}$ (the number of cases in each DRG in each MTF) is stochastic, this variation must be removed to estimate allocation weights to produce accurate facility specific estimates. If not removed, the heteroskedastic effects of the larger variance of rates in small facilities will produce estimates biased by the MTF size. Since the h_{ij} are discrete event counts, higher-order moments of the distribution of counts may contain significant information. Principal component analysis (an alternate rank reduction procedure) reproduces only the second-order moments of **H**. Instead we used the aggregate GoM model to identify $K-1$ classes that characterize the MTFs in terms of patient frequencies.

The procedure assumed that the frequencies h_{ij} are generated by hospital- and DRG-specific Poisson processes operating with rate parameters λ_{ij} (see Section 4.1). The basic aggregate GoM (equation (4.1)) has the u_{ik} constrained to represent the total number (167) of MTFs. The u_{ik} adjust for the size of the institutions. The v_{jk} the Poisson rate parameter for the jth DRG category for the kth type, are constrained to be positive, but are not bounded above by 1.

Estimation is accomplished by maximum likelihood using (4.5). Tests of the order of the model are based on likelihood ratio χ^2 for solutions with K and $K+1$ classes. Thus when the test for the $K+1$ class is no longer significant, we used the K-class representation of **H**, where the constrained expected counts h_{ij} were calculated as

$$\mathbf{E}(h_{ij}) = \lambda_{ij}^* = \sum_{k=1}^{K} u_{ik} v_{jk} \,. \tag{4.8}$$

Using $\mathbf{E}(h_{ij})$ from (4.8) ensures positivity in the quadratic programming solution.

Indicators of MTF characteristics are generated by estimating g_{ik} and λ_{kjl} from J observed MTF characteristics with L_j response levels (y_{ijl}, $l = 1, 2, \ldots, L_j$, $j = 1, 2, \ldots, J$). To produce these indicators, the standard form of GoM was used, where the unit of observation is the MTF. The indicator, g_{ik} is a multivariate score representing the kth profile of MTF structural characteristics used in the optimization model.

Needed ancillary relations are

$\mathbf{H}^*\mathbf{c}$ = expected income vector for MTFs produced by crediting the MTFs budget an amount c_j for each patient in a DRG; $\mathbf{H}^*$ is the fitted matrix with each element $\mathbf{E}(h_{ij})$ from the GoM structural analysis (4.8).

$\mathbf{Gd}$ = budgetary allocation to MTFs based upon operating and structural characteristics.

The budget is approximated by the expected income from treating specific classes of patients and a budgetary allocation to maintain the MTFs basic structural characteristics, i.e.,

$$\mathbf{b} \sim \mathbf{H}^*\mathbf{c} + \mathbf{Gd} \tag{4.9}$$

The objective function is generated by making (4.9) into a quadratic form and adding in necessary side conditions. This produces, for ϕ,

$$\phi = (\mathbf{H}^*\mathbf{c} + \mathbf{G}\mathbf{d} - \mathbf{b})^{\mathrm{T}}(\mathbf{H}^*\mathbf{c} + \mathbf{G}\mathbf{d} - \mathbf{b}) + \psi(\mathbf{c} - \mathbf{A}\mathbf{w})^{\mathrm{T}}(\mathbf{c} - \mathbf{A}\mathbf{w})$$
$$+ 2\lambda(e^{\mathrm{T}}\mathbf{H}^*\mathbf{c} + \mathbf{e}^{\mathrm{T}}\mathbf{G}\mathbf{d} - \mathbf{e}^{\mathrm{T}}\mathbf{b}) + 2\theta(\mathbf{Q} - \mathbf{e}^{\mathrm{T}}\mathbf{w}) , \tag{4.10}$$

where ψ is a scale factor that relates errors in the MTF budget b_i to errors in the DRG cost allocation. The terms involving λ and θ provide Lagrangian side constraints on the solution. ϕ is the term minimized in a least-squares sense. The weights c_j are estimated as

$$\mathbf{c} \sim \mathbf{A}\mathbf{w} . \tag{4.11}$$

In (4.11), **A** is normalized so that $\mathbf{e}^{\mathrm{T}}\mathbf{H}^*\mathbf{A} = \mathbf{b}\mathbf{e}^{\mathrm{T}}$ (i.e., the total budget or **B**), θ and λ are Lagrangian multipliers constraining the solution for the $L = 9$ sets of external DRG relative weights (**A**) and for MTF indicators of structure (**G**) respectively. To evaluate (4.11) we need a value for **Q**, i.e., the proportion of the budget determined by the external case-mix weights. Without indicators of MTF function, **Q** is 1, i.e., all weight is placed on the ancillary DRG weights. With the indicators (**G**), the contribution of the external case-mix weights (the **A**) to the total budget must be shared. The parameter **Q** in the model *with* structural indicators represents the portion of the budget associated *with* **G** through the regression function $\mathbf{e}^{\mathrm{T}}\mathbf{G}\mathbf{d}$. Consider the constraint

$$\mathbf{e}^{\mathrm{T}}(\mathbf{H}^*\mathbf{c} + \mathbf{G}\mathbf{d} - \mathbf{b}) = 0 . \tag{4.12}$$

If we insert $\mathbf{c}^* = \mathbf{A}\dot{\mathbf{w}}$ for **c** in (4.12), the constraint must still hold, i.e.,

$$\mathbf{e}^{\mathrm{T}}(\mathbf{H}^*\mathbf{A}\mathbf{w} + \mathbf{G}\mathbf{d} - \mathbf{b}) = 0 , \tag{4.13}$$

However, given $\mathbf{e}^{\mathrm{T}}\mathbf{H}^*\mathbf{A} = \mathbf{b}^*\mathbf{e}\mathbf{B}$ (i.e., the constraint that the frequency-weighted DRG weights from external sources reproduce the total budget), we have

$$\mathbf{b}^*\mathbf{e}^{\mathrm{T}}\mathbf{w} + (\mathbf{e}^{\mathrm{T}}\mathbf{G})\mathbf{d} - \mathbf{B} = 0 . \tag{4.14}$$

To obtain $\mathbf{e}^{\mathrm{T}}\mathbf{w} = \mathbf{Q}$ we solve (4.14) as

$$\mathbf{b}^*\mathbf{Q} + (\mathbf{e}^{\mathrm{T}}\mathbf{G})\mathbf{d} = \mathbf{b} \tag{4.15}$$

where

$$\mathbf{Q} = 1 - \frac{(\mathbf{e}^{\mathrm{T}}\mathbf{G})\mathbf{d}}{\mathbf{e}^{\mathrm{T}}\mathbf{b}} . \tag{4.16}$$

This produces the objective function (total mean-square budget discrepancy)

$$\phi = (\mathbf{H}^*\mathbf{c} + \mathbf{Gd} - \mathbf{b})^{\mathrm{T}}(\mathbf{H}^*\mathbf{c} + \mathbf{Gd} - \mathbf{b}) + \psi(\mathbf{c} - \mathbf{Aw})^{\mathrm{T}}(\mathbf{c} - \mathbf{Aw})$$

$$+ 2\lambda(\mathbf{e}^{\mathrm{T}}\mathbf{H}^*\mathbf{c} + \mathbf{e}^{\mathrm{T}}\mathbf{Gd} - \mathbf{e}^{\mathrm{T}}\mathbf{b}) + 2\theta\left(1 - \frac{(\mathbf{e}^{\mathrm{T}}\mathbf{G})\mathbf{d}}{\mathbf{e}^{\mathrm{T}}\mathbf{b}} - \mathbf{e}^{\mathrm{T}}\mathbf{w}\right), \tag{4.17}$$

where **Q** from (4.16) has been substituted to produce the decomposition to allow introduction of the indicators of hospital structure (**G**).

Since θ is an indeterminate Lagrange multiplier, by replacing θ with $\theta\mathbf{e}^{\mathrm{T}}\mathbf{b}$ the last term in (4.17) is simplified to

$$2\theta(\mathbf{e}^{\mathrm{T}}\mathbf{b} - (\mathbf{e}^{\mathrm{T}}\mathbf{G})\mathbf{d} - \mathbf{e}^{\mathrm{T}}\mathbf{w}\cdot\mathbf{e}^{\mathrm{T}}\mathbf{b}) \tag{4.18}$$

or

$$2\theta(\mathbf{B}^* - (\mathbf{e}^{\mathrm{T}}\mathbf{G})\mathbf{d} - \mathbf{B}^*\mathbf{e}^{\mathrm{T}}\mathbf{w})\,. \tag{4.19}$$

To minimize the objective function we need the partial derivatives of ϕ with respect to **c**, **d**, **w**, and the two Lagrangian parameters λ and θ. These may be written

$$\frac{1}{2}\frac{\partial\phi}{\partial\mathbf{c}} = \mathbf{H}^{*\mathrm{T}}(\mathbf{H}^*\mathbf{c} + \mathbf{Gd} - \mathbf{b}) + \psi(\mathbf{c} - \mathbf{Aw}) + \mathbf{H}^{*\mathrm{T}}\mathbf{e}\lambda$$

$$= (\mathbf{H}^{*\mathrm{T}}\mathbf{H}^* + \psi\mathbf{I})\mathbf{c} + \mathbf{H}^{*\mathrm{T}}\mathbf{Gd} - \mathbf{Aw}\psi - \mathbf{H}^{*\mathrm{T}}\mathbf{b} + \mathbf{H}^{*\mathrm{T}}\mathbf{e}\lambda \tag{4.20}$$

$$\frac{1}{2}\frac{\partial\phi}{\partial\mathbf{d}} = \mathbf{G}^{\mathrm{T}}(\mathbf{H}^*\mathbf{c} + \mathbf{Gd} - \mathbf{b}) + \mathbf{G}^{\mathrm{T}}\mathbf{e}(\lambda - \theta) \tag{4.21}$$

$$\frac{1}{2}\frac{\partial\phi}{\partial\mathbf{w}} = -\mathbf{A}^{\mathrm{T}}(\mathbf{c} - \mathbf{Ac})\psi - \mathbf{e}^{\mathrm{T}}\mathbf{b}^*\theta \tag{4.22}$$

$$\frac{1}{2}\frac{\partial\phi}{\partial\lambda} = (\mathbf{e}^{\mathrm{T}}\mathbf{H}^*\mathbf{c} + \mathbf{e}^{\mathrm{T}}\mathbf{Gd} - \mathbf{e}^{\mathrm{T}}\mathbf{b}) \tag{4.23}$$

$$\frac{1}{2}\frac{\partial\phi}{\partial\theta} = \mathbf{B} - (\mathbf{e}^{\mathrm{T}}\mathbf{G})\mathbf{d} - \mathbf{Be}^{\mathrm{T}}\mathbf{w}\,. \tag{4.24}$$

These equations are used in the computing tableau commonly used in the operations research literature as (see e.g., Hillier and Lieberman, 1986).

	$\mathbf{c}$	$\mathbf{d}$	$\mathbf{w}$	λ	θ	$\mathbf{P}$
$\mathbf{c}$	$(\mathbf{H}^{*\mathrm{T}}\mathbf{H}^* + \psi\mathbf{I})$	$\mathbf{H}^{*\mathrm{T}}\mathbf{G}$	$-\mathbf{A}\psi$	$\mathbf{H}^{*\mathrm{T}}\mathbf{e}$	0	$-\mathbf{H}^{*\mathrm{T}}\mathbf{b}$
$\mathbf{d}$	$\mathbf{G}^{\mathrm{T}}\mathbf{H}^*$	$\mathbf{G}^{\mathrm{T}}\mathbf{G}$	0	$\mathbf{G}^{\mathrm{T}}\mathbf{e}$	$-\mathbf{G}^{\mathrm{T}}\mathbf{e}$	$-\mathbf{G}^{\mathrm{T}}\mathbf{b}$
$\mathbf{w}$	$-\mathbf{A}^{\mathrm{T}}\psi$	0	$\mathbf{A}^{\mathrm{T}}\mathbf{A}\psi$	0	$\mathbf{eb}$	0
λ	$\mathbf{e}^{\mathrm{T}}\mathbf{H}^*$	$\mathbf{e}^{\mathrm{T}}\mathbf{G}$	0	0	0	$-\mathbf{e}^{\mathrm{T}}\mathbf{b}$
θ	0	$-(\mathbf{e}^{\mathrm{T}}\mathbf{G})$	$\mathbf{e}^{\mathrm{T}}\mathbf{b}$	0	0	$\mathbf{b}^{\mathrm{T}}\mathbf{e}$
$\mathbf{P}$	$-b^{\mathrm{T}}\mathbf{H}^*$	$-\mathbf{b}^{\mathrm{T}}\mathbf{G}$	0	$\mathbf{b}^{\mathrm{T}}\mathbf{e}$	$\mathbf{b}^{\mathrm{T}}\mathbf{e}$	$\mathbf{b}^{\mathrm{T}}\mathbf{b}$

We found the optimal solution using the exchange method with the numerical version of this tableau. The exchange method is a generalization of Gaussian elimination similar to the simplex method. The solution is calculated by "pivoting" on elements on the diagonal of the matrix corresponding to the partial derivatives defined in (4.20) to (4.24). At the solution, all derivatives and quantities in the **P** column must be nonnegative. Pivoting "exchanges" derivatives and variables. For a minimum to be achieved, the derivative must be 0 unless a constraint would be violated. In that case the variable must satisfy the constraint (in this case, be 0). The (λ, λ) and (θ, θ) elements cannot serve as pivots initially. One must pivot other elements until nonzero λ and θ elements are generated. There will be elements of **c** and **w** set to 0 since we prohibit negative values. Repeated simulations indicated that **Q** needs only to be nonnegative to select a sets of weights.

4.3.2. Results

The data come from several sources. The h_{ij} are the number of visits for each of 473 DRGs for 167 MTFs for the year 1985 (provided by DoD). In addition to the h_{ij} (and their reduced rank fitted representation h^*_{ij}), 1985 budget figures (b_i) are available for each MTF. Exogenous case-mix weights (a_{jl}) were the 473 DRG weights from the 1986 National Medicare System, the 1985 Maryland Medicare and Medicaid weights, Maryland Blue Cross-Blue Shield weights for 1985, private payment weights for Maryland, and CHAMPUS weights for 1987.

Facility structural characteristics were summarized (in **G**) in order to improve the fit of the model—especially for MTFs serving special medical care functions. The result of this analysis is the set of structural indicators g_{ik} derived from the analysis of 19 MTF operational and structural characteristics produced in Section 4.2. The five classes are described in Section 4.2. Clearly, though having relations to the classes from the functional analysis, they represent additional information for the optimization.

There are four outputs of the quadratic optimization solution. The predicted budgets b_i for each of the 167 MTFs, the 473 DRG costs c_j, the weights w_l on the $L = 9$ ancillary DRGs, cost structures, and ϕ, the objective function representing "unexplained" variance (in a mean-square error sense) for a preselected value of "ψ." The solution is controlled by ψ, which determines the relative contribution of the DRG weights c_j associated with the current budget and external weights **A**. ψ is a "policy" variable because it can be varied to emphasize the external weights at the expense of fitting the budgets (i.e., the decision to "disrupt" the current allocation to represent external evidence) or the budgets can be more closely matched at the expense of matching external weights (i.e., a decision to minimize disruption). Thus changes in the parameter reflect how much we allow the budget structure to be altered for any given budget cycle. The magnitude of ψ, since it relates facility budget to case-mix characteristics for the individual, is a scale factor representing the average annual number of discharges per facility (squared for the quadratic function). Thus we should explore ψ values in an empirically appropriate numerical range.

To illustrate, we present two solutions, one with $\psi = 1{,}000{,}000$ and one where $\psi = 2{,}000{,}000$. The MTF predicted budgets (i.e., b_i) are graphed against their actual budgets in Figures 4.1 and 4.2 for the two values of ψ.

There is a strong relation of predicted to actual budget for both values of ψ. In both cases 97% of the variance of the actual budget is explained. For $\psi = 1.0 \times 10^6$ there is slight nonlinearity. In both cases the fixed MTF budget is 1.9 to 2.0 million dollars, with an increase of 0.91 for $\psi = 2.0 \times 10^6$ and 0.95 for $\psi = 1.0 \times 10^6$ in the actual budget for each dollar in the estimated budget, i.e., the model redistributes between 9 and 5% of the actual budget with more redistribution produced by increasing the weight on the external case-mix prices (i.e., when $\psi = 2{,}000{,}000$ the external weights

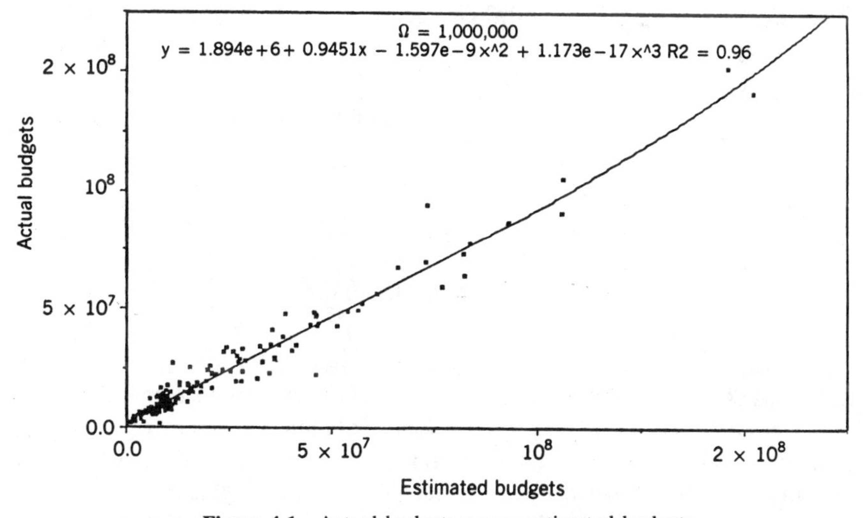

Figure 4.1 Actual budgets versus estimated budgets.

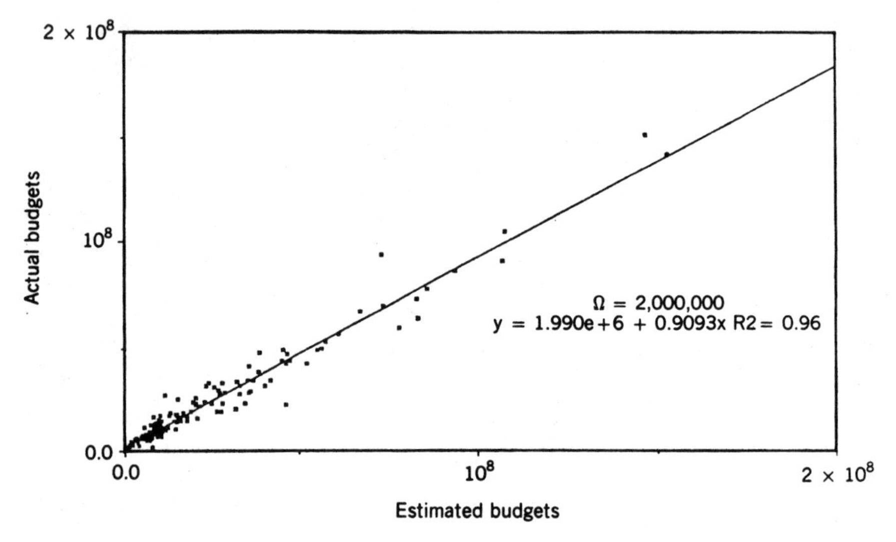

Figure 4.2 Actual budgets versus estimated budgets.

have greater influence). We restrict our attention below to the model with $\psi = 2.0 \times 10^6$.

One result are the weights assigned to the nine sets of external DRG weights. Only one of the nine DRG weights, those used by Medicare in 1986, contributed to the budget (i.e., had a nonzero weight). It explained 23.6% of each MTFs budget. Thus, 76.4% of the budget is determined by non–case-mix variation, i.e., cost variation due to facility structure. Since 23.6% of the budget is controlled by case mix, a large proportion of the fixed budget is due to volume and per-case costs in a MTF.

The variation of MTF structural characteristics is represented by the five g_{ik}. The coefficients **d** estimated for the five structural indicators are

Type	d (cost/bed)
1	\$15,170.00
2	\$13,515.00
3	\$17,062.00
4	\$16,395.00
5	\$16,805.00

The most costly MTFs to maintain on a per-bed basis are Class 3—overseas MTFs. Least expensive are those serving the highest proportion of retired personnel (Class 2) where additional payments are provided through CHAMPUS. The MTFs with the most serious case mix (Classes 4 and 5) are also expensive. The per-bed maintenance cost varies about 20%. To get the total budget for a MTF the **d** coefficients are multiplied by the number of beds in the MTF. Thus the structural indicator scores and bed size determines 76% of the budget variation with variation in case mix explaining the remainder. These factors explain 97% of the variance in budget over facilities.

To see how individual budgets are affected by the allocation, we present the plot of the log actual and estimated budgets. This plot (Figure 4.3 shows that the error of the prediction is less than proportional to the size of the budget. The explained variance is slightly less (92%) because of emphasis on smaller facilities. The relations for $\psi = 1.0 \times 10^6$ is similar.

Of interest are the prices estimated for the DRGs for the MTFs and their comparison to Medicare prices. These are presented on a log plot in Figure 4.4.

The linear relation for this relation is Medicare DRG weight $= -252.2 +$ $(0.98 \times$ estimated weight) with an R^2 of 92%. Thus the estimated DRG prices are, on average, \$252 more than the 1986 Medicare DRG prices, and for each dollar increase in estimated DRG costs, the Medicare price increases \$0.98. Thus the initial costs are higher for the MTFs, reflecting additional standby costs for MTFs and scale effects (i.e., that small MTF must have full services in isolated areas with a military population).

The optimization identified DRG resource allocation weights specific to the MTF system without using patient-specific charge information. An interesting result is that certain DRGs had zero weight (i.e., $c_j = 0$). This occurred because these weights are analogous to shadow prices and represent the (marginal) cost of treating an additional case in each DRG. This is different in concept from the average resource use concept implicit in the DRG weights from other payment systems. The majority of DRGs with

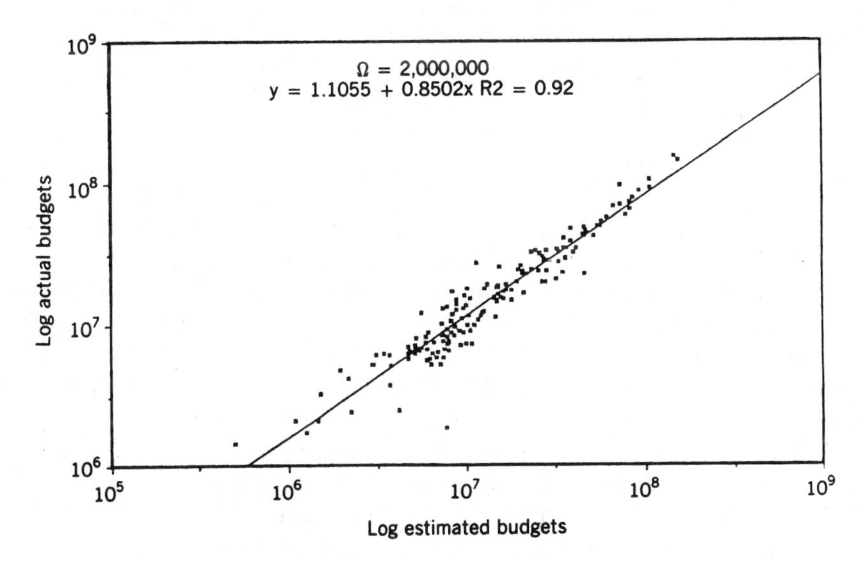

Figure 4.3 Log actual budgets versus log estimated budgets.

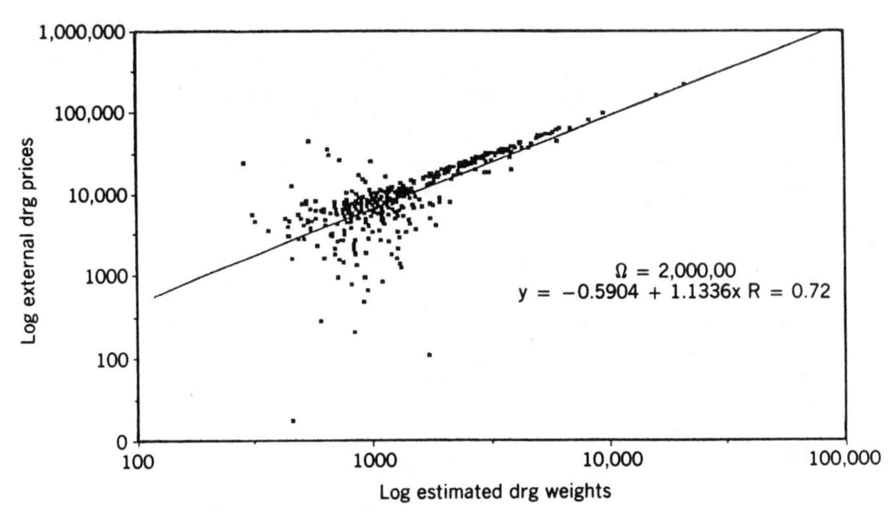

Figure 4.4 Log external DRG prices versus log estimated DRG weights.

zero weights were conditions prevalent in the CHAMPUS population, e.g., myringotomy, carpal tunnel syndrome, cesarean section delivery. Thus this finding implies that, since these cases are treated on a space-available basis and since CHAMPUS costs are not included in MTF budgets, they do not represent an additional (over long-run, fixed) cost to the MTF. Alternatively, a stable number of specific types of patients can be expected and their costs are treated as fixed. The implication is that an allocation formula must contain a term representing fixed expenses. Alternatively, if the DRG allocation model is used to differentially influence the rate of change of the facilities' budgets; the constant term is the baseline budget. Finally, MTFs have treatment patterns differing from hospitals in the civilian sector. This leads to cost differences for a few DRGs; e.g., a tooth extraction is often done as an inpatient service in MTFs. Thus the cost for this procedure is one-tenth of that for Medicare, where this would likely only be done in a civilian inpatient facility in extreme medical circumstances.

In addition to allocating budgets, the fact that the prices estimated by the model reflect marginal cost implies that similarly estimated prices could be used in other types of health care delivery systems. The marginal costs of a group of HMO operated hospitals (which, given that the model contains information on hospital attributes, need not be functionally homogeneous; e.g., some could specialize in maternal and pediatric care, others in treating cancer patients with chemotherapy, surgery and radiation) could be compared to a standard DRG prices. DRGs whose marginal cost to the HMO is less than the standard represent services where expansion could be profitable; DRGs whose marginal cost exceeds the standard represent services where contraction would be profitable.

4.4. SUMMARY

In this chapter we introduced a GoM model that produces fuzzy sets or classes by analyzing group or aggregate event rates using a Poisson modification of the GoM likelihood. We also showed that the g_{ik} from different types of fuzzy set analysis could be used to solve a quadratic optimization problem.

First, they were used (actually the v_{ik} from the aggregate form estimated for the functional model) to reduce the stochasticity in the h_{ij} elements from a 167×473 matrix **H**. The h_{ij}^* produced by this rank reduction retained higher-order moments of the frequency distribution. Second, GoM, in its standard form, was applied to MTF structural characteristics to define a small number of regression variables to control for MTF structural characteristics in the quadratic optimization procedure. The g_{ik} (i.e., the fuzzy set weights) performed well in both circumstances, characterizing the structural information for specific MTFs in a set of indicators and maintaining the positivity and rank constraints required in the optimization algorithm for the frequencies h_{ij}. Thus the fuzzy set procedures were useful in several ways in controlling for features of the MTFs in the budget optimization exercise.

The reasonableness of the results also suggested that the use of the individual g_{ik} scores seem to accurately represent the information in the original structural measures—even though the value of J in this analysis was moderate. This may be because, while J was moderate many of the variables had multiple answers (i.e., $L_j > 2$).

REFERENCES

Goldsmith, H. F., Jackson, D. J., Doenhofer, S., Johnson, W., Tweed, D. L., Stiles, D., Barbano, J. P., and Warheit, G. (1984). *The Health Demographic Profile System's Inventory of Small Area Social Indicators*. National Institute on Mental Health, Series BN, No. 4. DHHS Pub. No. (ADM) 84-1354. Washington, D.C., USGPO.

Hillier, F., and Lieberman, G. (1986). *Introduction to Operations Research*. Holden-Day, Oakland, California.

Morris, C. N. (1982). Natural exponential families with quadratic variance functions. *Annals of Statistics* **10**: 65–80.

Tolley, H. D., and Manton, K. G. (1992). Large sample properties of estimates of discrete Grade of Membership model. *Annals of Statistical Mathematics* **44**: 85–95.

Tolley, H. D., and Manton, K. G. (1994). Testing for the number of pure types of a fuzzy partition. In review at *Journal of the Royal Statistical Society, Series B*.

Vertrees, J. C., and Manton, K. G. (1986). A multivariate approach for classifying hospitals and computing blended payment rates. *Medical Care* **24**(4): 283–300.

Woodbury, M. A., Manton, K. G., and Vertrees, J. C. (1993). A model for allocating budgets in a closed system which simultaneously computes DRG allocation weights. *Operations Research* **41**(2): 298–302.

CHAPTER 5

Longitudinal and Event History Forms of the GoM Model

INTRODUCTION: INTRODUCING TIME INTO FUZZY SET STATE SPACES

In this chapter we generalize the GoM model to produce fuzzy classes or extreme profiles that span time (or "multiple" time) domain(s). This chapter covers both statistical estimation and mathematical issues in generalizing fuzzy state models to represent time. In Chapter 6 we discuss the use of parameter estimates from time-dependent GoM models in forecasting and simulating fuzzy state changes. This involves examining how well the estimated parameters of the model represent the "true" stochastic processes generating future realizations of complex multidimensional systems. In particular, the different stochastic propagation of a discrete versus fuzzy state model will be examined. Those forecasts will illustrate the realizations of some of the statistical assumptions not explicitly identified in cross-sectional or limited longitudinal data.

5.1. TEMPORAL GENERALIZATION OF THE GOM MODEL

The introduction of time into the GoM likelihood function is relatively straightforward once a stochastic process to model changes in time is specified. For example, we might posit that each individual is in one of a finite number of states and that each individual can "jump" between states at discrete points in time. The states occupied by an individual, coupled with his Grade of Membership scores, define the likelihood function for observed discrete responses. From these assumptions we might assume that the stochastic process for each individual was a Markov chain, a constrained random walk, etc. The range of stochastic processes that may be represented is broad. Therefore model specification requires that attention be paid to representing the temporal dependency in the model and the likelihood equations.

The crucial formula to generalize is the likelihood function (2.1), which needs to be extended over the appropriate time domains. This can be written in a general form as

$$L = \prod_i \prod_j \prod_l \prod_t \prod_e \left(\sum_k g_{ik \cdot te} \cdot \lambda_{kjl \cdot te} \right)^{y_{ijl \cdot te}} . \qquad (5.1)$$

In (5.1) we introduce the indices t and e. The first, t, indexes the times of "assessments" or measurement usually of a wide range of measures, made at fixed times t (i.e., $t = 1, 2, \ldots, T$). The second, e, represents "episodes" (i.e., $e = 1, 2, \ldots, E$) that begin and end at "random times" within an interval bounded between t and $t + 1$.

Implicit in (5.1) is the assumption of independence conditional on the g_{ik} (Suppes and Zanotti, 1981) over both assessment times and episodes. In this case, the assumption means that two individuals with the same g_{ik} at a point in time produce responses identically distributed to responses of a single individual over time provided that the individuals' g_{ik} do not change.

To fix ideas about these additional terms and how they are defined by the observational plan we provide the cross-temporal observational plan for the 1982, 1984, 1989, and 1994 (planned) National Long Term Care Surveys (NLTCS). This plan and its elements are contained in Figure 5.1.

In Figure 5.1, we see that detailed assessments of chronically disabled elderly (65 years and older) Medicare beneficiaries were conducted in four-month field periods in 1982, 1984, 1989, and planned for 1994. At these times ($t = 1, 2, 3, 4$), detailed questionnaires are delivered to community and institutional resident Medicare-eligible elderly persons who were identified as chronically (90+ days) impaired in ADL or IADL functions on a screening interview, either at the time of the current survey or from some prior survey (i.e., all previously disabled *and* institutionalized persons are continuously followed; survivors are automatically eligible for a detailed interview at the next survey). These detailed survey assessments are "partly" fixed assessments in that they occur within a *fixed* field time interval that is short relative to the intersurvey time interval. Though there is a time dimension to the field period, so long as individual assessments are randomly distributed during the field period (an assumption examined later), we can treat assessments in a short field period as if they occurred at a fixed time t. In this case we treat time t as an "average" time over the distribution of assessment times in the field period. As we can see (and will discuss in the context of nonresponse bias adjustments in Section 5.4) in Figure 5.2 the field period, which usually covered four months, was long enough to allow some individuals to change state during data collection, especially since mortality and disabling events occur at high rates among very elderly and highly impaired persons (Manton and Suzman, 1992).

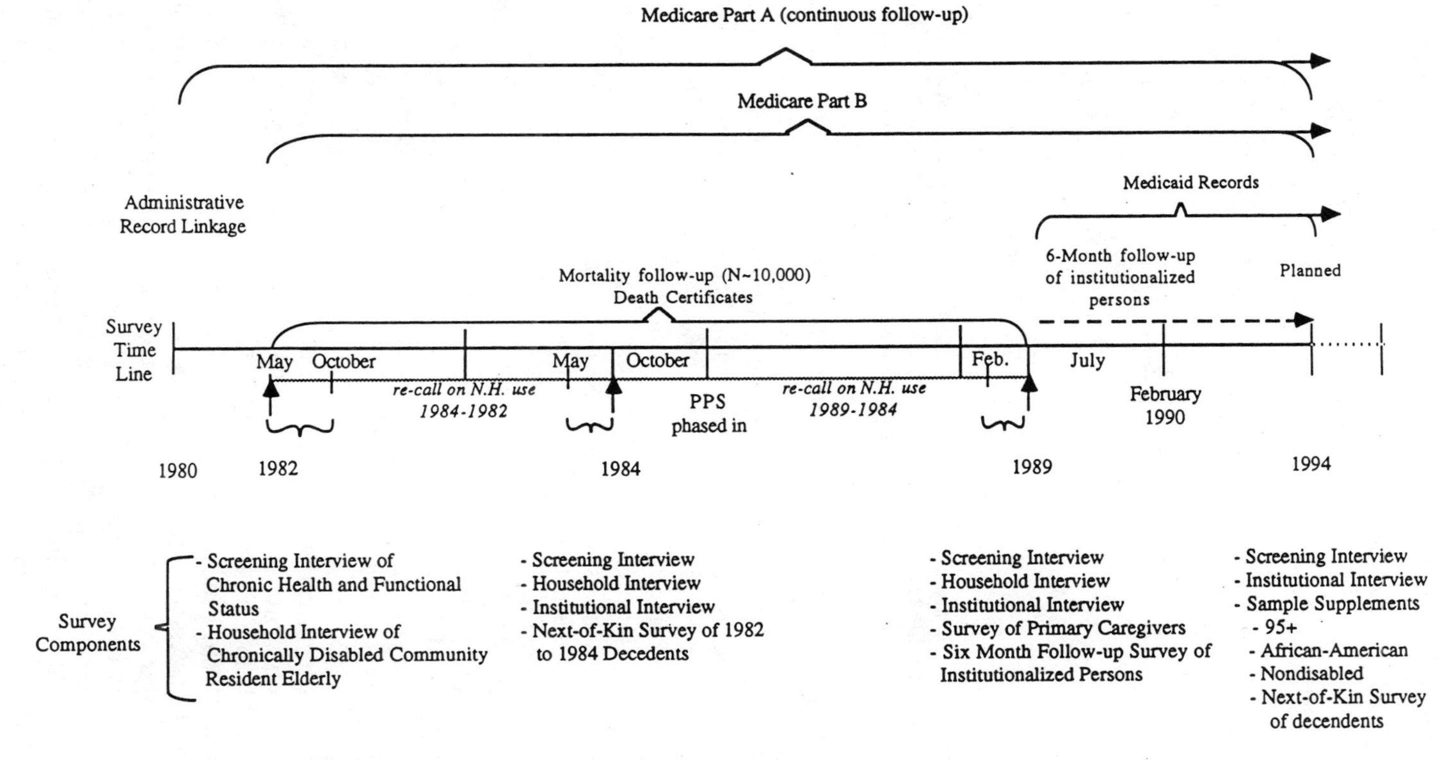

Figure 5.1 Overview of observational plan for the 1982, 1984, 1989 and 1992 NLTCS.

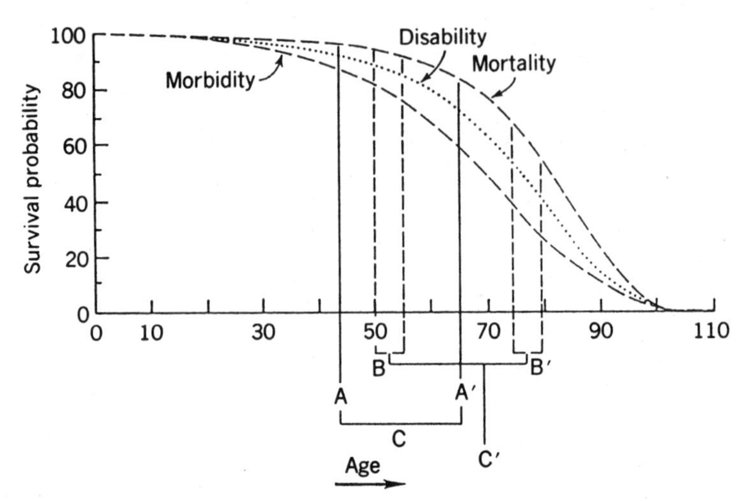

Figure 5.2 *A* Time at first evaluation. *A′* Time at follow-up. *B* Time interval over which initial evaluation is made. *B′* Time interval over which follow-up evaluation is made. *C* Follow-up interval. *C′* Average time between initial (*B*) and follow-up (*B′*) assessment interval.

In addition to the survey data collected in 1982, 1984, 1989, and 1994 (planned) ($t = 1, 2, 3, 4$), the survey records of each person are linked to his continuous Medicare Part A and B service use record from 1982 onward. Since the original NLTCS sample was a "list" sample, drawn from Medicare administrative records, such persons could be tracked continuously and completely (i.e., there was 100% of individuals follow-up from 1982 to 1992 in the Medicare records system). From the Medicare records, the beginning and end dates and costs of any Part A (i.e., acute hospital, skilled nursing facility, or Home Health) or Part B (outpatient and physician) services used by the individual can be determined. It is the information from the Medicare continuous record of dates of service from which one can define "episodes" of service used to define e in equation (5.1).

In addition to data on Medicare Part A and B use, the Medicare records also provide information on the exact dates of death for all persons in the file (e.g., for the period 1982 to 1991 there were about 11,000 deaths recorded). Thus, the temporal information in this type of file is complete and, for such an extremely elderly U.S. population, very accurate. For example, the recording of age has followed strict evidentiary requirements since 1937 before a person is accepted as qualified for Social Security and Medicare (1966) benefits (Bayo and Faber, 1985; Kestenbaum, 1992).

With the assessments ($t = 1, \ldots, T$) and episodes ($e = 1, \ldots, E$) defined, we see that, to conduct an analysis, the cross-temporal observational record for a person must be reorganized into components based upon the temporal structure of measurement. Thus, at each assessment at t, a new g_{ik} vector (i.e., $g_{ik \cdot t}$) will be specified in the likelihood and estimated from health and functional status variables measured at that time (actually within the short field work period). The set of $g_{ik \cdot t}$ define a "state" at time t. Thus, the $K-1$ dimensions defined by the λ_{kjl} represent the state space for individuals. This state space definition (i.e., the λ_{kjl} profiles) is held constant over time so changes in $g_{ik \cdot t}$ are comparable.

Episodes are characterized by one type of event, its duration (which is measured from T to the event e or from the end of the prior event $e-1$ to the end of the current episode e), and the mode of termination of the episode. These episodes are described by sets of transition variables containing the parameters necessary to generate multiple life tables for each episode type. More generally, if there are multiple episodes between two assessment times t to $t+1$, this represents the event history space. The event history space is assumed to be estimated conditionally on the state of the person (i.e., his $g_{ik \cdot t}$). For an event history space where there are multiple episodes occurring between two assessments (i.e., within t to $t+1$), the $g_{ik \cdot t}$ from the prior time point t is assumed to be invariant at the beginning of the second and subsequent episodes contained in the interval, until a new assesment is made at $t+1$. At $t+1$ there is a jump in information, and a new vector of scores $g_{ik \cdot t+1}$ is calculated. This implies that all information on state space changes occurs in discrete jumps at the time of assessment. In contrast, the event space is allowed to evolve continuously between $t+1$ given conditioning on $g_{ik \cdot t}$. If the g_{ik} truly describe the states, conditioning on their values should make events independent of type and duration of all other events in the survey period.

This last assumption is important to the organization of the GoM analysis and bears more evaluation. Explicitly, if the information regarding the individual's response distribution is contained in the individual's g_{ik}, then, conditional on the g_{ik} values for an individual, the responses are independent over time. Since this is a basic assumption for temporal analyses using GoM, we formally state the assumption as

Assumption 6. *Conditional on the g_{ik} scores of an individual over time, repeated observations of the same individual over time are independent.*

Assumption 6 is actually a special case of Assumption 3 (Chapter 1), where the specialization is for indexes of j over time.

Other assumptions could be made about changes of $g_{ik \cdot t}$ to $g_{ik \cdot t+1}$ over the intersurvey assessment interval. One could, for example, interpolate the $g_{ik \cdot t}$ between t and $t+1$ using, say, a linear function. Thus, at the midpoint of the interval the value of $g_{ik \cdot t+1/2}$ would be the average of the state at t and $t+1$. Each assumption about the change in $g_{ik \cdot t}$ within an interval implies a different model of the continuous time process generating changes in $g_{ik \cdot t}$. The assumption we have adopted for most analyses is the simplest; i.e., information on the individual's state jumps at every t, with the uncertainty about $g_{ik \cdot t}$ increasing as a function of dt within the interval. At a reassessment. say $t+1$, the $g_{ik \cdot t}$ changes to $g_{ik \cdot t+1}$ and the variance or uncertainty about the state value for persons instantaneously drops to zero. Clearly, the more densely distributed assessments are on the time line, the better the empirical approximation of the state trajectory of the $g_{ik \cdot t}$ and the more closely the underlying continuous time model will be approximated. Other assumptions regarding changes in $g_{ik \cdot t}$ (e.g., a linear increase between t and $t + 1$) still have increasing uncertainty with the time since the prior assessment and, in addition, require the justification of a specific model of change over the interval.

Of course the validity of this decomposition of the observational record into its temporal elements involves other assumptions that are derived from the type of stochastic process being modeled. The primary one is that the $g_{ik \cdot t}$ vector adequately describes the "state" of the individual. In this context, the state assumption means that the duration and mode of termination of each of the episodes occurring at t to $t + 1$ are conditionally independent given the $g_{ik \cdot t}$ (Suppes and Zanotti, 1981). The validity of this assumption increases as the information content of the $g_{ik \cdot t}$ is increased—either by increasing J (where $J + 1$ is not perfectly correlated with the J other variables) or decreasing the time of t to $t + 1$ (i.e., shortening the length of the interval).

The "state" assumption is testable. To test if the episodes within the interval t to $t + 1$ are conditionally independent one needs to estimate two models. The first is just as we have described with the temporal component in (5.1) and an implicit "first-order" Markov assumption. The second increases the information available to define each vector of $g_{ik \cdot t}$ by, say, adding in the identity and duration of the immediately preceding episode. This can be viewed as either assuming that the process describing changes in the state is semi-Markovian or by retaining the assumption of Markovity by expanding the definition of the state space to include episodes or events occurring in a given prior span of time. In the limit, of course, the entire event history between $t - 1$ to t can be included in the definition of the state $g_{ik \cdot t}$ by introducing indicators of the event history $t - 1$ to t in the set of J variables used to estimate the $g_{ik \cdot t}$.

If the new $g^*_{ik \cdot t}$ formed by, say, including information from episodes between t and $t - 1$, produce a model with a significantly better fit then a first-order assumption, then temporal dependence among the episodes must be assumed to exist. Such temporal dependence must be included in the model to retain the factorability (of i, j, and l) of the likelihood. The state space itself must be made more complex and include the necessary temporal information. A related approach to the problem would be to assume that events are dependent due to their sampling from a common distribution. The likelihood becomes factorable in this case by conditioning on the hyperparameters of the distribution of events by type, duration, and mode of exit. This is similar to the approach used in Chapter 3 to deal with distributional contamination in the case of cluster samples (i.e., contamination due to PSU intraclass correlations) or in the case of genetic or other determinants of event characteristics. Of course, ancillary information must be available to specify the form of the prior distribution to use the so-called "unconditional" GoM model. In some cases this may be natural. For example, in the case of "point-in-time" surveys of event durations, the distribution of event durations suffers from length-biased sampling. In this case, relatively simple temporal relations that describe the effects of length-biased sampling can be used to transform the observed cross-sectional event duration distribution to its unbiased form.

In prior analyses of the 1982 and 1984 NLTCS, we found that the 56 health and functioning variables selected from the detailed community survey provided an adequate description of a person's state; i.e., the inclusion of variables describing the prior episode in the variables in J to generate an enhanced state space $g^*_{ik\cdot t}$ did not significantly improve the ability of the model to prospectively describe changes in Medicare Part A service use or mortality in time windows before, and after, the introduction of the Prospective Payment System for case-mix reimbursement of acute hospital stays (Manton and Liu, 1990; Manton et al., 1993).

To be specific, suppose that between t and $t+1$ there are four episodes for a person. Each episode generates a term in the likelihood. The J variables measured at t are assumed to remain the same for the four episodes so that the $g_{ik\cdot t}$ are assumed to be the same for each of the four episodes. In the first-order temporal dependence model a $J+1$ and $J+2$ variable can be added to represent the identity (e.g., a hospital stay) and duration (e.g., 1–2 days) of the episode $e-1$. Then a second model is estimated from a modification of (5.1) to include the temporal dependence, and the fits of the two models are tested with a likelihood ratio test.

In addition to specifying the temporal dependence of the g_{ik} we need to consider how to constrain λ_{kjl} over time, e.g.,

$$L = \prod_i \prod_j \prod_l \prod_t \prod_e \left(\sum_{k=1}^{K} g_{ik\cdot t} \cdot \lambda_{kjl} \right)^{y_{ijl\cdot te}} . \tag{5.2}$$

The likelihood in (5.2) is estimated without episode dependence in either $g_{ik\cdot t}$ or λ_{kjl}. The $g_{ik\cdot t}$ in a model *with* episode or event dependence could be written $g_{ik\cdot t(e-1)}$. In (5.2) the λ_{kjl} are constrained to be equal over all episodes e and times of assessment t. This is because the λ_{kjl} define the state simplex, which must span all heterogeneity of persons over time. This means that any $g_{ik\cdot t}$ (or $g_{ik\cdot t(e-1)}$) can be given a common interpretation on the fuzzy states at any time. If there is systematic heterogeneity in characteristics over time, this may require a large number of fuzzy classes K to be estimated, with some classes decreasing in prevalence early in the follow-up and others increasing in prevalence later, but with the space spanned by the λ_{kjl} defined broadly enough that all individual tracks of $g_{ik\cdot t}$ over time are spanned. Since the $g_{ik\cdot t}$ need not be sampled from a specific distribution of trajectories, the fact that the density along certain of the $K-1$ dimensions of the state space may change with time is not a problem if J is large enough that the Jth-order moments are sufficient to describe the $g_{ik \cdot t}$ over the state space and time.

Of course temporal dependency may be increased beyond one prior episode to as many prior episodes as necessary. Thus, if the J health and functioning measures available are inadequate to describe the trajectory of episodes that occur between t to $t-1$, then inclusion of the identity of the prior episode (or episodes) represents proxy variables for unobserved health measures that determine the time of episodes between t and $t+1$ beyond the J variables observed. This also provides a way of describing "tracking" (persistence) of individual state variables.

It is appropriate to expand further on the concept of the $g_{ik \cdot t}$ as a temporally changing state variable. One point that often causes confusion in the evaluation of processes is that, in hazard analysis, the hazard is itself a proxy measure for unobserved variables that cause the hazard to change with age and/or time. Covariates in this type of model only stratify the hazard function (Logue and Wing, 1986); i.e., they are not time varying. In contrast, the fuzzy set state variables can be time varying. We will present models below that specifically examine these changes in states.

To illustrate the differences in a hazard model and a time-variable state space model, we present Figure 5.3.. In the figure there are two panels. Figure 5.3(a) describes a hazard function that increases with time. Curve A might represent a Gompertz hazard function for human mortality as observed in an elderly human population (e.g., Spiegelman, 1968). Curve B represents what happens when the hazard function is estimated conditionally on informative, non–time-varying covariates. In this case the shape parameter increases, indicating that the time to death for a given individual has become more "precise" due to stratification of the hazard (i.e., some variables affecting the time to death have been measured). Curve C represents what happens when *all* informative covariates are entered into the hazard function; i.e., the age at death, conditional upon the covariates, is known exactly, and the hazard (and the likelihood function itself) degenerates to a collection of infinite spikes. This just shows that a hazard function is *not* a temporal model of a process but a model of an error distribution that is person homogenous (at least on the shape parameter) and time *in*homogeneous (Manton and Tolley, 1991).

This is what happens in a fixed or static covariate hazard model. A more general, and realistic, model describes such a "failure" process by a dynamic state variable process linked with a jump process representing death or, more generally, other discrete exchanges in state. We have discussed such models in the context of multivariate Gaussian processes where the state variables evolving with time are observed chronic disease risk factors (e.g., Manton, Stallard, and Singer, 1994).

In Figure 5.3(b) we represent such a process in a simple (one-dimensional) fuzzy state space. As in Figure 5.3(a) movement to the right indicates the

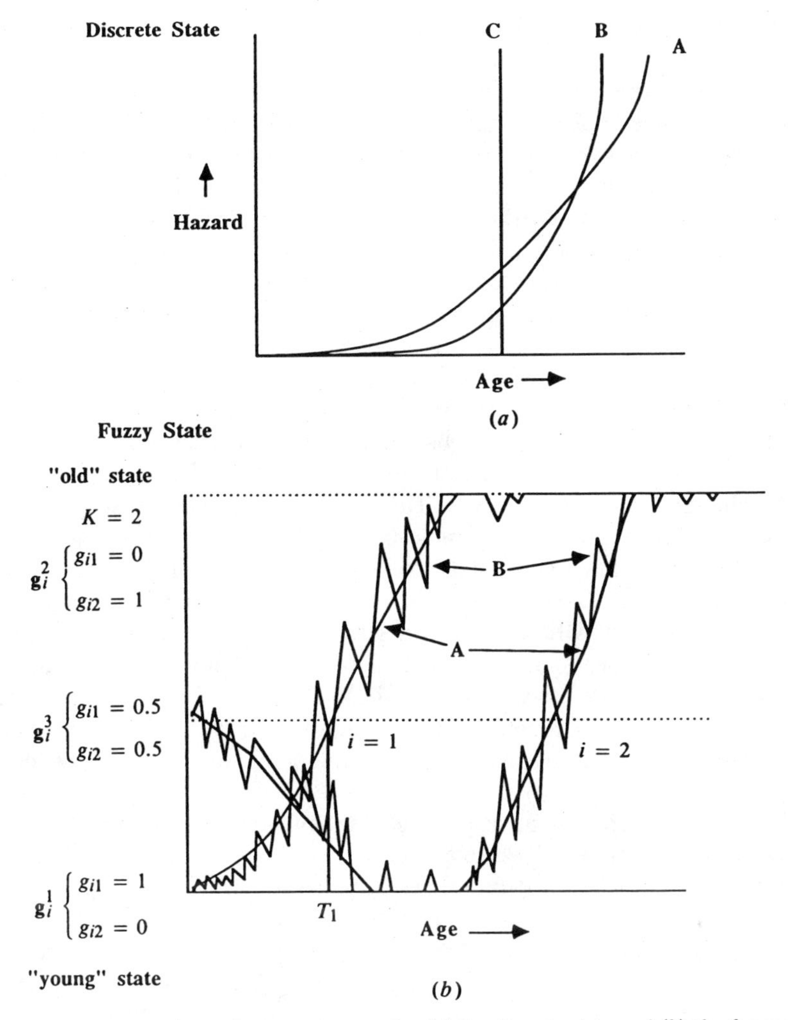

Figure 5.3 A comparison of event processes for (*a*) the discrete state and (*b*) the fuzzy state.

passage of time (or aging). The space is unidimensional and defined by two fuzzy sets. The first state represents a set of λ_{kjl} for J characteristics that are associated with "old" age. The second state represents the λ_{kjl} for J variables that represent the physiological state of being "young." In the figure the state trajectories of two persons are plotted. One starts at $g_{i1} = g_{i2} = 0.5$ and the other at $g_{i1} = 0$ and $g_{i2} = 1$. The person starting in a mixed health state initially "improves" (i.e., increases his g_{i1} and decreases his g_{i2}) before progressing to an old disabled state. The other has a regular temporal progression, moving from the young to old state (i.e., as g_{i1} declines and g_{i2} increases).

The trajectories of these states express certain substantial properties of interest. First we note that the letters A and B are linked to the curves in Figure 5.3(*b*). These represent temporal phenomena for the fuzzy states analogous to those reflected by A and C in the first panel. The trajectory of curve B is jagged. This is due to the fact that J is not large enough to represent all perturbations of the state trajectory so that the trajectory is "uncertain." This is somewhat like curve B and A in Figure 5.3(*a*), where there is uncertainty about the state of the person at time t. As J increases, however, the trajectories change from being jagged (B) to smooth (A). This is the true state trajectory and reflects the same condition of knowledge about state that defined C in the first panel. However, instead of being a spike or discontinuity, A is a set of smooth curves. This means that the state-dependent risk of death will change as the $g_{ik \cdot t}$ changes and not be explicitly dependent on time. In effect, the $g_{ik \cdot t}$ trajectories represent "aging" while "aging" is an unobserved or latent process in Figure 5.3(*a*). This also illustrates an important substantive point about the study of human health changes as it relates to age. That is, that aging mechanisms cannot be studied by the simple hazard function in Figure 5.3(*a*). This is due to the problem of identifiability of state changes from a simple hazard curve fit only to "outcome" data. The hazard curve describing the increase in the rate of, say, death with age can be generated by a number of different underlying processes—some with, and others without, time-varying elements. The GoM process will describe both fixed and time-varying elements in $g_{ik \cdot t}$ trajectories. If it describes all temporal variation in the risk of an event like death, then the event death is independent of time given the $g_{ik \cdot t}$ (Manton et al., 1991).

A second insight on measurement can be gotten from Figure 5.3(*b*). Specifically, in Figure 5.2 we saw what could happen with an observational plan for an elderly frail population with even a four-month field period; i.e., the process of measurement could interact with substantively meaningful events within the time period. One can see the same problems emerge at the individual level in Figure 5.3(*b*). Specifically, if measurements are taken far apart (i.e., t to $t+1$ is a long time), then, depending on their placement on the time axis relative to the shape of the trajectory, they could give very different impressions of the state trajectory for person 2 if the measurement of the trajectory is as for the jagged trajectory B (Manton and Suzman, 1992). As time resolution increases (i.e., t to $t+1$ decreases in size), then we have more finely detailed observations on the trajectory of $g_{ik \cdot t}$, but, when they become narrow enough, the perturbations of the state trajectory (i.e., a single spike) may dominate the measurement. There has to be a balance between the field period and the entire measurement period that allows signals to be isolated from noise. The best ratio of the length of these periods for resolving various types of short- and long-term health changes depends, of course, on the relative strength of the signal and the degree of noise. This will be covered explicitly in our discussion of forecasting where we use a parameter θ to explicitly represent the effect at (a) unobserved variables whose effects are correlated with age and (b) information lost (unobserved) on intermediate transition.

What actually happens in a longitudinal survey is that both effects become confounded. That is, the field period is short enough that a person listed in a sample on April 1 could experience a change in status by June 1 (i.e., changing within the field period). Likewise, the time between surveys is long (e.g., 2 to 5 years), so that some systematic changes that are rapid may be missed. Of course one hopes that averaging states over the population will help resolve certain issues of confounding, but that clearly requires that superpopulation sampling models (e.g., Cassel et al., 1977) be employed where individual processes with different parameter values are represented as a heterogeneous mixture. In finite population sample theory, where inferences are made only about population parameters, the confounding of individual process parameters is total.

5.2. "EXTERNAL" VARIABLES: ANCILLARY LIKELIHOODS

With the conceptual development of Section 5.1 we can now consider specific implementations and examples of the model for cross-temporal analysis. The first example, one which we have employed extensively, is the use of "external" variables. These involve using the MIP algorithm to maximize only the λ_{kjl} for a subset of variables (say, a set of J^*) that are *not* used to calculate the g_{ik}. Thus, only the score function for λ_{kjl} equation (3.4) is optimized, while the g_{ik} scores are held fixed [(i.e., equation (3.3)]. In practice the g_{ik} values used are those estimated previously using different measurement variables measured on the same individuals. Usually the first J variables are used to estimate λ_{kjl} and g_{ik} jointly, and the subsequent J^* variables are used to form the ancillary likelihood. This defines an ancillary likelihood function

$$L_{\text{ext}} = \prod_{i} \prod_{j^*} \prod_{l} \left(\sum_{k=1}^{K} g_{ik}^* \cdot \lambda_{kj^*l} \right)^{y_{ij^*}} \tag{5.3}$$

estimated conditionally on the g_{ik}—note j^* indexes the $J+1$ to $J+J^*$ variables not used to estimate $g_{ik \cdot t}$. The $g_{ik \cdot t}^*$ variables are the score estimates using *only* the first J variables. This gives us a conditional distribution of λ_{kjl} estimates for a set of J^* variables conditional on the $g_{ik \cdot t}$ estimated in an earlier step. This procedure has a number of applications. First, it provides a "regression" model of external variables on the g_{ik}—but one estimated using the maximum likelihood machinery with the appropriate boundary constraints imposed for the fuzzy state space. It also provides a way to "validate" the substance of the K fuzzy classes on independent measures.

To illustrate, suppose the $(J+1)$th variable is the cost of a hospital stay. Let us suppose it is divided into 10 categories and that those 10 categories are defined from the discrete limits of the observed deciles of hospital cost. Let us also assume that we have, for each of the $l_{j+1} = 10$ categories, the within-category mean cost, i.e., $\bar{x}_{(J+1)l_{J+1}}$. From these mean costs we can calculate the expected cost for one of the K classes by

$$\bar{x}_{K(J+1)} = \sum_{l=1}^{10} \bar{x}_{(J+1)l} \cdot \lambda_{k(J+1)l} \; . \tag{5.4}$$

The $\lambda_{k(J+1)l_{J+1}}$ give the proportion of the kth class in the l_{J+1}th cost interval, and the $\bar{x}_{(j+1)l_{J+1}}$ provide the overall sample means for the cost categories. A sum of the products produces a total average cost for the kth class. This may then be used in subsequent actuarial formulations to predict costs overall (assuming the $\bar{g}_k$ estimated from the sample; Tolley and Manton, 1991). For example, the expected cost of hospitalization for a person is

$$x_{i(J+1)} = \sum_{k=1}^{K} g_{ik} \cdot \bar{x}_{k(J+1)} \; . \tag{5.5}$$

Thus, if a person is partly in both Classes 1 and 2 (say, 0.5 and 0.5), then his expected hospital cost will be $0.5(\bar{x}_{1(J+1)}) + 0.5(\bar{x}_{2(J+1)})$. This provides a useful alternative to case-mix reimbursement systems where prices or costs are estimated for *discrete* categories using a recursive partitioning algorithm (Mills et al., 1976) where costs are the dependent measure used to define partitions. In such a discrete category reimbursement system (e.g., DRGs) there is often instability in category price estimates due to the large number of categories (e.g., 473) and to the effects of category misclassification. In addition, since the discrete categories are defined on the ability to predict *costs*, the system, when used to estimate reimbursement, is tautological. That is, the categories, though related to health measures J, are not established to explain health, but to explain service costs. Thus, if economic resources historically were poorly allocated to persons with specific health needs (e.g., services in nursing homes for demented residents; Manton et al., 1990), then the cost categories will be necessarily be poorly related to health and may provide inappropriate incentives to the distribution of resources.

Furthermore, the partitions will *only* be meaningful in the particular reimbursement system for which they were formulated and calibrated (e.g., DRGs for Medicare hospital stay). The g_{ik} determined only from the J health measures could be applied in the context of any service system or market since they are not constructed to be a function of reimbursement within a specific payment system. Such tautologies exist for other reimbursement systems, e.g., the RUGS-II categories (Manton et al., 1990).

Any variable can be used externally. For example, the sample could have been divided on sex and two sets of external variables formed—one for males and one for females. If a third set of external variables is created (for the total sample), a likelihood ratio test of the significance of K specific cost functions could be generated (i.e., to see if the pair of sex-divided cost variables explain significantly more variation than the single total-cost variable). This is because, from the MIP algorithm, the individual likelihood elements may be calculated and extracted, and multivariate hypotheses can be formed with those likelihood elements. Indeed, the estimation of the χ^2 for these variables is less problematic than for internal variables, since they are estimated conditionally upon the g_{ik} (Tolley and Manton, 1987).

Thus, one could estimate a χ^2 for the total variable and the χ^2 for the male and female variables, and generate a χ^2 test of the gender difference in hospital costs (say) conditional upon the health measures represented in the g_{ik}.

5.2.1. Tests for Unobserved Variables

The tests for external variables provide a way of dealing with the effects of unobserved variables. There have been several strategies proposed for eliminating the effects of unobserved variables. These involve either assuming some specific mixing distribution (e.g., the gamma or inverse Gaussian) for the unobservable variables (e.g., Manton et al., 1986a) or having a proxy measure that is correlated with the missing variable available (Heckman et al., 1985).

In GoM, the external variable strategy provides a semiparametric approach specifically designed for testing for the effects of unobserved factors (e.g., Gill, 1992). What is done is first to take the variable of interest and treat it as an external variable. All heterogeneity predictable in that variable from the g_{ik} will be represented in the $\lambda_{k(j+1)l}$ that are estimated conditional upon the g_{ik} created solely from internal variables. Then one increases the set of internal variables to $J+1$ to include the variable of interest. One can determine the new likelihood value for the variable when it is allowed to effect the g_{ik}. The change in the likelihood for that variable represents heterogeneity that (a) is not predictable from the g_{ik} based only on the J variables (b) is systematic (if the χ^2 is significant), and (c) is forced to be averaged in and covary with the original sets of measures. Thus, a test of unobservable variable effects and the introduction of systematic variation into the $\lambda_{k(j+1)l}$ elements can be done by this procedure (Manton et al., 1986).

5.2.2. "External" Life Table Calculations

An important use of the "external" variable computation is in calculating life tables to represent the timing, or history, of discrete events (i.e., as in Section 5.1). This can be done by constructing a two-dimensional variable (computationally the two dimensions can be treated as a vector), where in place of l one has a set of intervals representing duration of the episode of interest (i.e., l_t) and a set of intervals indicating how the episode would terminate (i.e., l_m). In this case we also need to identify the initial condition of the person. This can be simply characterized by the particular variable, $J + M$, used in the life table calculation.

Thus, suppose we designate a person who is currently in a hospital as, after completion of his stay, being in a hospital episode of duration l_t, measured in discrete time units that ended by a transition to one of the other states l_m. Thus, we may estimate an external variable (i.e., with g_{ik} fixed) with response probabilities for each of the K fuzzy classes designated $\lambda_{kJ^*(l_t \times l_m)}$. The J^* indicates the starting state (i.e., hospital) and the $l_t \times l_m$ indicates that a multinomial variable has been created with $l_t \times l_m$ probabilities. These have the same characteristics as the probabilities for multiple decrement life tables—say, $d_{x \cdot A}$, where A is a particular force of decrement.

Thus, if there are, say, six ways to terminate a specific episode: (1) end of study, (2) death, (3) being remeasured at t (4–6) three types of moves to other service episodes (e.g., hospital to SNF, HHA, or back home), then $l_m = 6$ for any type of episode (from the example of the NLTCS these can be hospitals, SNFs, HHAs, or in the community). Depending upon the type of episode (described by the initial state) the duration intervals l_t will likely differ in number and length (e.g., for hospital stay duration it would be in days; SNF would be in weeks or months). After choices are made about (a) the nature of starting points (type of episode), (b) coding of duration (i.e., setting the number and category boundaries for l_t; if the multinomial likelihood is used, these should be short enough that the probabilities are small to well approximate the underlying hazards) and (c) determining the modes of termination (l_m), then transition variables $\lambda_{k(J^*)(l_t \times l_m)}$ can be calculated conditional upon k, g_{ik} scores. From those variables, life tables can be calculated. There will be a set of such life tables for each type of episode. For each type of episode, K sets of life tables will be generated where there will be M different life tables—one for each mode of terminating the episode. From these life tables one can calculate duration ("life expectancy") measures and survival curves (l_x) and make competing risk calculations for the censoring effect of other modes of termination (e.g., the life table for death could be adjusted to remove the right censoring due to end of study, e.g., Chiang, 1968).

It is important to realize that these life tables actually deal with *three* types of competing risk effects. The first, as mentioned above, is for simple censoring among the M modes of termination (Chiang, 1968).

Of greater importance is the adjustment for *dependent* competing risks. That is, each set of M tables is (for a type of episode) made dependent on the g_{ik} scores for one of the fuzzy classes. Thus, it represents the risk and time to one of m transitions, for the kth fuzzy class. If one of the life tables is for mortality and that life table is altered by reducing the transition rates, then the average $\bar{g}_k$, the weighted prevalence of that class in the population, will increase due to slower mortality selection. This would change the $\bar{g}_k$ for all K types (because of the convexity constraint on the g_{ik}) and affect the mortality trajectory for the total population, which is a weighted function of the $\bar{g}_k$ and their associated transition rates or schedules.

There is a third type of dependent competing risk effect if the g_{ik} are time variable, i.e., $g_{ik \cdot t}$. The $g_{ik \cdot t}$ can be influenced by multiple factors. One is the history of the process. This can be included as prior episode information in the J measures used to define $g_{ik \cdot t}$. As $g_{ik \cdot t}$ changes, so will $\bar{g}_{k \cdot t}$, which changes the mix of fuzzy set specific life tables in the population. Thus, life tables are not only dependent on $g_{ik \cdot t}$ but also on factors influencing the processes changing $g_{ik \cdot t}$.

The complex dependency of the competing risks on $g_{ik \cdot t}$ is determined without distributional assumptions (except being bounded by 0 and 100) about the fuzzy state variables generating the dependency. Furthermore, given that transition variables are estimated externally, the transition variables (i.e., the $\lambda_{K(J^*)(l_t \times l_m)}$) can be made internal to determine if the life tables express unobservable variable effects not explained by the K sets of $g_{ik \cdot t}$ or their trajectories.

These external variables can be related back to Figure 5.3 because, for each state of the system (i.e., $g_{ik \cdot t}$), there is a set of $K \times M$ life tables describing exits or discrete jumps from the system. That is, Figure 5.3 can be generalized both for $K > 2$ and for multiple-jump processes. If the $g_{ik \cdot t}$ do a good job in explaining age changes, then one would expect that the mortality parameters would not increase over time within episode (this is explained below).

5.2.3. Implications for Calculation of Models of Aging and Human Mortality

The models presented above have implications for the conceptualization of human mortality and aging processes. Human survival curves are often examined to see if they show indications of becoming more rectangular suggesting that a population is nearing a genetically determined life span (Fries, 1980, 1983). It can be shown that, under special conditions, one can test a hypothesis of curve squaring using the shape parameters of the hazard function at advanced ages (Manton and Tolley, 1991). This is because the trajectory of mortality risks can be produced, indeed is likely to be produced, by time-varying covariates. In the Gompertz hazard function, for example, all of these covariates are assumed to be both (a) unobserved and (b) temporally fixed.

Thus, we need a bridge between the unknown effects that are correlated with age in the Gompertz function and the fuzzy state transitions represented by the GoM model. This can be represented by generating the definition of state in a two-component stochastic process model (Strehler, 1975). The first component represents change in $g_{ik \cdot t}$ over time. This, thus, is a extension of the basic GoM likelihood estimation problem since it requires inferences to be made about state transitions between times of assessment. This equation is written as a K-dimensional autoregressive process,

$$\mathbf{g}_{i(t+1)} = \mathbf{C}_t \mathbf{g}_{it} + \mathbf{e}_{i(t+1)} , \tag{5.6}$$

where

$$\mathbf{C}_t = (\mathbf{D}_t^{-1} \cdot \mathbf{G}_t)^{\mathrm{T}} .$$

In equation (5.6), $\mathbf{g}_{it}$ is the K element vector of g_{ikt}; $\mathbf{D}_t$ is a $K \times K$ diagonal matrix of age specific means of g_{ikt}; $\mathbf{G}_t$ is a $K \times K$ matrix of the average cross products of g_{ikt} and $g_{ik(t+1)}$. Hence, $\mathrm{diag}(\mathbf{D}_t) = E(\mathbf{g}_{it} \mid T > t + 1)$ and $\mathbf{G}_t = E(\mathbf{g}_{it} \cdot \mathbf{g}_{i(t+1)} \mid T > t + 1$ where conditioning on $T > t + 1$ indicates that only survivors to $t + 1$ are used to estimate $\mathbf{C}_t$. The form of $\mathbf{C}_t$ involves replacing the $K \times K$ matrix $\mathbf{H}_t = E(\mathbf{g}_{it} \cdot \mathbf{g}_{it} \mid T > t + 1)$ in the least squares estimation by $\mathbf{D}_t$. $\mathbf{D}_t$ yields a $\mathbf{C}_t$ which is stochastic. $\mathbf{H}_t$ is not. $\mathbf{C}_t$ represents the flow of condition weighted prevalence between K fuzzy states (i.e., Δg_{ik}) and describes the temporal change in the condition weighted burden of disability in the population. This generalizes the transition probabilities in a Markov chain model with discrete states to the parameters of a fuzzy state process. Other strategies to estimate such a matrix are discussed in the section on structural equations (Section 5.3.2.2).

The matrix $\mathbf{C}_t$ represents aging changes in the model of human aging and mortality. One issue is how those dynamics explicate the average effects of unobserved processes represented in the Gompertz hazard. This is represented by using an age dependent quadratic hazard function,

$$\mu(\mathbf{g}_{it}) = \{\mathbf{g}_{it}^{\mathrm{T}} \mathbf{Q} \mathbf{g}_{it}\} \exp(\theta t) , \tag{5.7}$$

where $\mathbf{Q}$ is a real positive definite matrix of hazard coefficients. The current (i.e., at t) values of the $g_{ik \cdot t}$ from (5.6) are substituted in the mortality function (5.7) for each time t. The $\exp(\theta t)$ is a Gompertz multiplier representing time or age effects not captured by the $g_{ik \cdot t}$ as they evolve according to (5.6). If the $g_{ik \cdot t}$ are related to the process, then θ, which is the per annum aging rate, will decrease. This type of multidimensional survival analysis with time varying covariates is illustrated in Chapter 6, where its dual role in representing types of missing information (i.e., on factors correlated with age affecting health and on observed changes in state) is presented.

5.2.4. Continuous Time Representation

Continuous time can be represented in the episode component of (5.2). There is a complication, however, when the measurement process is determined in part by the course of the process. That is, it may occur that health assessment is done not at fixed times t but on a random schedule according to the underlying rate of change of basic health processes. To handle this problem (i.e., to generate the g_{ik}) one has to translate the data structure so that episodes can be encoded in the T, or fixed assessment, space. This was done for several long-term care demonstration studies (e.g., of the Social/Health Maintenance Organization; Manton et al., 1994) by defining the lowest common denominator of the duration metric. The three year follow-up was, thus, divided into 36 monthly time periods where the changes in g_{ikt} from t to $t+1$ are held fixed from month to month until a new assessment (at a time determined by monitoring of the population) is made to assess health and functioning. The results of that translation can be reflected in a series of life tables estimated for each group. These are shown in Table 5.1. We can see that there was a 42.5-day difference in the mean time to assessment between the control group and the intervention group. The results of the full analysis are in Chapter 6.

5.3. MODELS OF STRUCTURAL RELATIONS

With cross-temporal data one can directly estimate process parameters. However, often one may not have full information on the temporal ordering of changes so that substantive theory, and prior information must be used to construct the parametric structure to reflect hypothesized causal relations. This is done frequently in econometric analyses where variables in economic systems are divided into predetermined (exogenous) sets and endogenous sets. Social scientists have imposed parametric constraints in maximum likelihood factor analysis to estimate such equation systems (Jöreskög, 1969; Jöreskög and Sörbom, 1979). We developed an alternative scheme to provide parameter estimates for structural equation systems using the more general semiparametric form of the fuzzy set model. This is discussed in steps.

5.3.1. The Measurement Model

The initial element of the general structural equation system estimation algorithm currently available (Jöreskög and Sörbom, 1986a,b, 1988; i.e., LISREL) is dealing with "errors in variables" using a measurement model. The measurement model, in those systems, is usually based on a maximum likelihood factor analysis. This is a natural interpretation of GoM parameters as well, where the discrete response variable y_{ijl} are explained by K sets of underlying fuzzy set scores (i.e., the g_{ik}) with the relation of observed to latent variables represented by λ_{kjl}.

Table 5.1. Transitions For S/HMO, FFS, and HMO Episodes Updated with Health Data from CAF

A. S/HMO

	Destination									
	HMO		Death		FFS		CAF		Retention Rate	
Class	Observed	CAF Risk Eliminated	Observed	CAF Risk Eliminated	Observed	CAF Risk Eliminated	Observed	CAF Risk Eliminated	Observed	CAF Risk Eliminated
1	0.058	0.060	0.030	0.031	0.107	0.108	0.036	—	0.324	0.361
(Healthy)	(597.8)	(604.9)	(561.8)	(567.2)	(426.2)	(432.2)	(567.8)		(1060.4)	(1061.4)
2	0.018	0.029	0.221	0.273	0.272	0.294	0.655	—	0.065	0.195
(Acutely ill)	(395.6)	(539.3)	(227.8)	(290.6)	(214.4)	(315.8)	(253.0)		(840.6)	(926.1)
3	0.034	0.051	0.064	0.094	0.115	0.154	0.754	—	0.080	0.296
(Chronically impaired)	(443.6)	(607.1)	(360.0)	(474.0)	(292.2)	(391.4)	(235.6)		(998.2)	(1026.6)
4	0.021	0.017	0.074	0.080	0.049	0.060	0.809	—	0.325	0.724
(Pulmonary and cancer)	(275.8)	(506.4)	(205.0)	(378.5)	(192.0)	(333.1)	(219.8)		(320.8)	(516.2)
5	0.041	0.053	0.086	0.108	0.088	0.108	0.401	—	0.129	0.288
(Cardiopulmonary)	(535.5)	(609.6)	(450.5)	(509.1)	(423.5)	(478.3)	(369.5)		(1043.8)	(1053.2)
6	0.006	0.013	0.517	0.195	0.034	0.071	0.788	—	0.378	0.537
(Frail)	(262.8)	(452.5)	(110.7)	(229.9)	(215.0)	(405.1)	(198.8)		(180.4)	(409.1)
Case Mix	0.039	0.050	0.074	0.083	0.094	0.109	0.380	—	0.217	0.367
Weighted	(569.3)	(612.9)	(347.6)	(392.1)	(378.6)	(420.6)	(272.1)		(946.0)	(975.0)
Unweighted	0.0380	0.0490	0.0760	0.0830	0.0900	0.1050	03850	—	0.2210	0.3720
	(570.9)	(614.0)	(342.2)	(387.9)	(382.4)	(425.0)	(269.7)		(936.1)	(967.4)

B. FFS

	Destination									
	S/HMO		HMO		Death		CAF		Retention Rate	
Class	Observed	CAF Risk Eliminated	Observed	CAF Risk Eliminated	Observed	CAF Risk Eliminated	Observed	CAF Risk Eliminated	Observed	CAF Risk Eliminated
1	0.021	0.012	0.207	0.208	0.037	0.037	0.012	—	0.351	0.369
(Healthy)	(386.0)	(388.5)	(354.4)	(356.8)	(518.2)	(520.8)	(730.6)		(1016.4)	(1018.0)
2	0.016	0.016	0.144	0.145	0.144	0.145	0.010	0.288	0.967	
(Acutely ill)	(336.1)	(336.6)	(281.7)	(282.2)	(456.1)	(456.2)	(260.3)		(1041.8)	(104.9)
3	0.007	0.009	0.112	0.143	0.095	0.131	0.761	—	0.186	0.319
(Chronically impaired)	(455.6)	(468.6)	(301.3)	(322.0)	(425.8)	(440.4)	(94.7)		(1029.3)	(1032.0)

Class	S/HMO Observed	S/HMO CAF Risk Eliminated	HMO Observed	HMO CAF Risk Eliminated	Death Observed	Death CAF Risk Eliminated	FFS Observed	FFS CAF Risk Eliminated	CAF Observed	CAF CAF Risk Eliminated	Retention Rate Observed	Retention Rate CAF Risk Eliminated
4	0.003	0.004	0.066	0.071	0.121	0.169	0.569	—			0.204	0.378
(Pulmonary and cancer)	(424.4)	(497.3)	(270.2)	(338.6)	(396.5)	(458.9)	(210.2)				(774.1)	(852.3)
5	0.007	0.007	0.164	0.166	0.224	0.235	0.076	—			0.182	0.215
(Cardiopulmonary)	(321.8)	(322.3)	(255.5)	(263.1)	(458.2)	(468.0)	(425.5)				(1033.9)	(1037.5)
6	0.003	0.004	0.032	0.032	0.282	0.387	0.702	—			0.165	0.285
(Frail)	(347.5)	(465.1)	(202.4)	(311.5)	(282.4)	(389.8)	(180.7)				(578.8)	(766.7)
Case Mix	0.009	0.010	0.152	0.162	0.097	0.108	0.218	—			0.263	0.338
Weighted	(385.0)	(399.9)	(333.6)	(350.7)	(427.3)	(442.9)	(233.9)				(980.0)	(988.2)
Unweighted	0.009	0.010	0.148	0.166	0.104	0.115	0.237	—			0.256	0.335
	(376.3)	(391.2)	(330.4)	(348.9)	(423.7)	(440.6)	(227.2)				(976.5)	(985.6)

C. HMO

	Destination											
	S/HMO		HMO		Death		FFS		CAF		Retention Rate	
Class	Observed	CAF Risk Eliminated	Observed	CAF Risk Eliminated	Observed	CAF Risk Eliminated	Observed	CAF Risk Eliminated	Observed	CAF Risk Eliminated	Observed	CAF Risk Eliminated
6	0.002	0.002	0.368	0.367	0.030	0.020	0.098	0.097	0.005	—	0.432	0.437
(Healthy)	(301.2)	(302.4)	(221.3)	(221.9)	(392.1)	(393.6)	(300.2)	(301.2)	(537.2)		(647.6)	(650.1)
2	0.009	0.009	0.346	0.348	0.123	0.124	0.165	0.165	0.012	—	0.300	0.304
(Acutely ill)	(416.1)	(416.4)	(243.9)	(244.1)	(307.7)	(308.7)	(282.0)	(282.4)	(285.0)		(706.5)	(708.0)
3	0.002	0.002	0.437	0.460	0.090	0.094	0.108	0.113	0.182	—	0.282	0.310
(Chronically impaired)	(466.6)	(466.9)	(231.9)	(234.5)	(306.6)	(312.5)	(298.8)	(302.9)	(134.0)		(667.5)	(673.9)
4	0.024	0.024	0.212	0.228	0.095	0.114	0.126	0.144	0.467	—	0.281	0.416
(Pulmonary and cancer)	(45.0)	(45.0)	(187.0)	(219.3)	(304.5)	(357.7)	(301.2)	(368.4)	(222.1)		(525.8)	(613.0)
5	0.003	0.003	0.328	0.330	0.109	0.113	0.131	0.134	0.069	—	0.316	0.341
(Cardiopulmonary)	(400.5)	(400.6)	(209.5)	(214.9)	(396.3)	(404.2)	(349.6)	(359.7)	(306.7)		(649.1)	(662.1)
6	0.008	0.008	0.052	0.0073	0.303	0.277	0.095	0.091	0.789	—	0.266	0.459
(Frail)	(45.0)	(45.0)	(199.4)	(319.7)	(149.2)	(293.1)	(168.7)	(316.5)	(187.4)		(321.8)	(585.5)
Case Mix	0.004	0.003	0.319	0.329	0.078	0.081	0.106	0.113	0.184	—	0.350	0.407
Weighted	(300.9)	(317.3)	(220.4)	(231.1)	(312.0)	(328.2)	(303.3)	(315.4)	(215.7)		(626.5)	(644.4)
Unweighted	0.004	0.003	0.346	0.351	0.061	0.062	0.107	0.108	0.098	—	0.375	0.410
	(310.0)	(318.4)	(221.4)	(226.9)	(337.0)	(345.0)	(299.8)	(311.4)	(229.4)		(640.1)	(649.6)

5.3.2. The Structural Equation

To translate the measurement model into a structural equation system, constraints must be imposed to generate a conditional relation of endogenous sets (actually, latent factors describing those sets) of variables and exogenous sets (again their latent factors) of variables. The standard equation (e.g., Jöreskög and Sörbom, 1979) relating systems of endogenous (η) and exogenous variables (ξ) is,

$$\boldsymbol{\beta}\eta = \boldsymbol{\Gamma}\xi + \zeta \tag{5.8}$$

where $\boldsymbol{\beta}$ is a $(N \times N)$ and $\boldsymbol{\Gamma}$ is a $(N \times P)$ matrix describing, respectively, the interrelation of N endogenous variables (the diagonal of $\boldsymbol{\beta}$ is an identity matrix) and the dependency of N endogenous variables on P exogenous variables. η is assumed to be a set of factors estimated from y observed variables. The ξ are factors estimated from x exogenous variables and ζ represents random error in the estimate of η. The measurement factor relations are

$$x = \Lambda_x \xi + \delta \tag{5.9}$$

$$y = \Lambda_y \eta + \epsilon \tag{5.10}$$

where Λ_x and Λ_y are analogous to the λ_{kjl} in GoM.

Estimation in LISREL uses the second-order moments matrix of the joint distribution function for x and y. If we designate this covariance matrix as $\mathbf{S}_T$ with subcomponents $\mathbf{S}_{xx}$, $\mathbf{S}_{xy}$, and $\mathbf{S}_{yy}$ we can write the estimation equations in matrix form as

$$\mathbf{S}_T = \begin{pmatrix} \mathbf{S}_{yy} & \mathbf{S}_{xy} \\ \mathbf{S}_{yx} & \mathbf{S}_{xx} \end{pmatrix} = \begin{pmatrix} \Lambda_y \boldsymbol{\beta}^{-1}(\boldsymbol{\Gamma}\boldsymbol{\phi}\boldsymbol{\Gamma}' + \boldsymbol{\psi})[\boldsymbol{\beta}^{-1}]'\Lambda_y' + \theta_\epsilon & \Lambda_y \boldsymbol{\beta}^{-1}\boldsymbol{\Gamma}\boldsymbol{\phi}\Lambda_x' \\ \Lambda_x \boldsymbol{\phi}\boldsymbol{\Gamma}'[\boldsymbol{\beta}^{-1}]'\Lambda_y' & \Lambda_x \boldsymbol{\phi}\Lambda_x' + \theta\delta \end{pmatrix}. \tag{5.11}$$

In (5.11) $\mathbf{S}_{xx}$ (i.e., exogenous inputs) is simply a factor analysis with correlated ($\boldsymbol{\phi}$) factors. The $\mathbf{S}_{yx}$ and $\mathbf{S}_{xy}$ submatrices represent the regression of dependent and independent variables reflected through the measurement equation parameters (i.e., 5.9 and 5.10). The $\mathbf{S}_{yy}$ matrix represents (a) a variance component of the y's constructed in the factor space of the exogenous variables, (b) a variance component that represents the interrelation of the y's with the x's conditioned out, and (c) stochasticity.

In generalizing (5.11) to the GoM structure, all moments to order J must be represented. To do this equation (2.1) is modified. Define M sets of variables y_{ijl}^m such that y_{ijl}^1 is causally prior to $y_{ij_2l}^2, y_{ij_3l}^3, \ldots, y_{ij_ml}^m$ where the matrix of these variables each containing J_m variables can be written $\mathbf{Y}_1, \mathbf{Y}_2, \ldots, \mathbf{Y}_m$. For $m = 1$ we have the standard GoM model where, using matrix notation (i.e., $\mathbf{G}_1^{(1)}$ equals the g_{ik} for set 1, and Λ_{11} equals the λ_{kjl} for set 1), we can write

$$\Pi_1 = E(\mathbf{Y}) = \mathbf{G}_1^{(1)}\Lambda_{11} . \tag{5.12}$$

In scalar form this is

$$\Pi_{ijl}^{(1)} = E(y_{ijl}^{(1)}) = \sum_{k=1}^{K} g_{ik}^{(1,1)}\lambda_{kjl}^{(1,1)} . \tag{5.13}$$

The elements of (5.13) may be related to the terms in (5.8)–(5.10), i.e., the $g_{ik}^{(1,1)}$ correspond to ξ, and the $\lambda_{kjl}^{(1,1)}$ correspond to Λ_x elements. Since neither exogenous $(y_{ijl}^{(1)})$ nor state variables $(g_{ik}^{(1,1)})$ are normally distributed, $e_{ijj}^1 = y_{ijl}^{(1)} - \Pi_{ijl}^{(11)}$ corresponds to δ not being normally distributed. The problem is that the (2.1) has to be generalized so that a second set of endogenous variables may be added with the normalization $\Sigma_k\, g_{ik} = 1$ maintained at *both* stages. This requires use of the Poisson likelihood form of GoM, i.e.,

$$L = \prod_i \prod_j \prod_l \left(\frac{\Sigma_{k=1}^{K} g_{ik} \cdot \lambda_{kjl}}{\Sigma_k\, (g_{ik}\, \Sigma_l\, \lambda_{jkl})} \right)^{y_{ijl}} \tag{5.14}$$

We discuss the properties of this likelihood in greater detail in Chapter 6. Note that the constraint $\Sigma_l\, \lambda_{kjl} = 1$ is not needed in (5.14). This allows us to normalize g_{ik} for sets of factors extracted conditionally upon prior sets of factors—where the order of conditioning represents substantive theory about the order of causal effects. If (5.12) is rewritten,

$$E(\mathbf{Y}) = \mathbf{W} \cdot \mathbf{Z} \tag{5.15}$$

or, in scalar terms,

$$E(y_{ijl}) = \sum_{k=1}^{K} w_{ik} \cdot z_{kjl} , \tag{5.16}$$

the rotation of $\mathbf{W}$ and $\mathbf{Z}$ to $\mathbf{G}$ and $\boldsymbol{\Lambda}$ is produced by the normalization,

$$\mathbf{W} = \mathbf{G}\mathbf{D}^{-1} \quad \text{and} \quad \mathbf{Z} = \mathbf{D}\mathbf{A} \tag{5.17}$$

where $\mathbf{D}$ is a diagonal matrix such that

$$d_{kk} = \sum_{1} \frac{g_{ik}}{I} = \bar{g}_k . \tag{5.18}$$

Thus $\mathbf{W}$ and $\mathbf{Z}$ vary from $\mathbf{G}$ and $\boldsymbol{\Lambda}$ only by $\bar{g}_k$. To estimate w and z, since

$\Sigma_l \lambda_{kjl} = 1$ is no longer enforced, a modification of (5.14) is used,

$$L = \prod_i \prod_j \exp\left\{-y_{ij+} \sum_{k=1}^{K} w_{ik} \sum_l z_{kjl}\right\} \prod_l \left(\sum_{k=1}^{K} w_{ik} z_{kjl}\right)^{y_{ijl}} \Big/ y_{ijl}^{(1)} , \quad (5.19)$$

which is maximized with the constraint,

$$\sum w_{ik} = I , \quad (5.20)$$

where y_{ij+} is an indicator that a response was obtained to y_{ijl} (to represent missing data). For w and z we use (5.20) to obtain

$$\mathbf{G}^* = \mathbf{WC} \quad \text{and} \quad \boldsymbol{\Lambda}^* = \mathbf{C}^{-1}\mathbf{Z} \quad (5.21)$$

where $\mathbf{C}$ is a diagonal matrix (akin to (5.6)) where

$$C_{kk} = \sum_{(jl)} Z_{k(jl)} / J . \quad (5.22)$$

The asterisks on $\mathbf{G}$ and $\boldsymbol{\Lambda}$ in (5.21) indicate the normalization

$$g_{ik}^* = g_{ik} \Big/ \sum_m g_{im} \quad (5.23)$$

and

$$\lambda_{kjl}^* = \lambda_{kjl} \Big/ \sum_m \lambda_{kjm} . \quad (5.24)$$

Thus, (5.12) is replaced by

$$E(\mathbf{Y}) = \mathbf{G}^* \boldsymbol{\Lambda}^* , \quad (5.25)$$

which differs by the normalization in (5.21). In the calculations, we reparameterize stage one as

$$E(\mathbf{Y}_1) = w_1 z_{11} . \quad (5.26)$$

The second stage is

$$E(\mathbf{Y}_2) = w_i z_{21} + w_2 z_{22} . \quad (5.27)$$

The third stage is

$$EE(\mathbf{Y}_3) = w z_{31} + w_2 z_{32} + w_3 z_{33} , \quad (5.28)$$

etc.

Computationally we use the scalar form of the equation of a given order to solve for the maximum likelihood estimates using (5.19). If K_m is the number of classes introduced in the mth stage, and $K^{(m)}$ is the total number at all stages, then w_m is of dimension $I \times K_m$. Also $N_m = \Sigma_j L_j^{(m)}$ is the number of responses to the jth variable in Y_m and $z_{mn} = K_m \times N_n$. Once w_m and z_{mn} are estimated, the renormalization yields

$$E(\mathbf{Y}_1) = \mathbf{G}_1^{(1)} + \mathbf{\Lambda}_{11} \tag{5.29}$$

$$E(\mathbf{Y}_2) = \mathbf{G}_1^{(2)}\mathbf{\Lambda}_{21} + \mathbf{G}_2^{(2)}\mathbf{\Lambda}_{22} \tag{5.30}$$

$$E(\mathbf{Y}_3) = \mathbf{G}_1^{(3)}\mathbf{\Lambda}_{31} + \mathbf{G}_2^{(3)}\mathbf{\Lambda}_{32} + \mathbf{G}_3^{(3)}\mathbf{\Lambda}_{33} \tag{5.31}$$

with stage specific GoM scores,

$$\mathbf{G}^{(1)} = \mathbf{G}_1^{(1)} \tag{5.32}$$

$$\mathbf{G}^{(2)} = [\mathbf{G}_1^{(2)} \vdots \mathbf{G}_2^{(2)}] \tag{5.33}$$

$$\mathbf{G}^{(3)} = [\mathbf{G}_1^{(3)} \vdots \mathbf{G}_2^{(3)} \vdots \mathbf{G}_3^{(3)}] \tag{5.34}$$

or,

$$\mathbf{g}^{(m)} = [\mathbf{G}_1^{(m)} \vdots \dots \vdots \mathbf{G}_m^{(m)}] \tag{5.35}$$

In GoM, scores are obtained simultaneously with structural coefficients as parameters. Thus, they contain all order-moment dependence (up to J) between the various stages. The structure of **G**, including all moment relations, is complex, so that we may restrict the information considered to certain order moments.

5.3.2.1. State Variable Changes

In this section we discuss how the flow from class k to l is estimated. This is represented as $T^{(1,2)} = \{t_{kl}\}$ such that,

$$\mathbf{G}^{(2)} = \mathbf{G}^{(1)}\mathbf{T}^{(1,2)} + \mathbf{E}^{(1,2)} \tag{5.36}$$

or,

$$\mathbf{G}_1^2 = \mathbf{G}_1^{(1)}\mathbf{T}_1^{(1,2)} + \mathbf{E}_1^{(1,2)} \tag{5.37a}$$

and

$$\mathbf{G}_1^2 = \mathbf{G}_1^{(1)}\mathbf{T}_2^{(1,2)} + \mathbf{E}_2^{(1,2)} \tag{5.37b}$$

The $K_1 \times K_2$ matrix $\mathbf{T}_2^{(1,2)}$ describes flows from K_1 classes in stage 1 to K_2 classes in stage 2, while $\mathbf{T}_1^{(1,2)}$ is the flow between exogenous state variables. These equations can be generalized to m stages.

Estimates of **T** can be generated in several ways. **T** is not a conditional probability matrix since the g_{ik} are in a closed interval $[0, 1]$ rather than being a binary set $\{0, 1\}$. The elements of **T** might be estimated by least squares. This, however, will not guarantee that constraints on the individual g will be satisfied (i.e., range and summation constraints). Note that for fuzzy states the t_{kl} must satisfy certain conditions as the fuzzy states approximate discrete states, i.e.,

$$t_{kl} = \Pr[(k_i^{(1)} = k) \cap (k_i^{(2)} = l)] / \Pr(k_i^{(1)} = k) \tag{5.38}$$

$$= E(g_{ik}^{(1)} \cdot g_{ik}^{(2)}) / E(g_{ik}^{(1)}) \tag{5.39}$$

$$= E(g_{ikl}^{(1)}) / E(g_{ik}^{(1)}) \tag{5.40}$$

where $g_{ik}^{(1)} \in \{0, 1\}$, $g_l^{(2)} \in \{0, 1\}$, and $g_{ikl}^{(1,2)}$ is the joint membership of i in state k and l. Dividing by $E(g_{ik}^{(1)})$ ensures that t_{kl} reflects changes for persons starting in state k. For fuzzy state transitions, t_{kl} is

$$t_{kl} = E(g_{ikl}^{(1,2)}) / E(g_{ik}^{(1)}) \tag{5.41}$$

but (5.39) is not required to hold.

There are two methods to provide estimates of T which satisfy the condition:

$$g_{ikl}^{(1,2)} = g_{ik}^{(1)} g_{il}^{(2)} \tag{5.42}$$

and

$$g_{ikl}^{(1,2)} = \min(g_{ik}^{(1)}, g_{ik}^{(2)}) \tag{5.43}$$

The first estimation scheme (5.42) reflects joint membership of the products of individual membership. This satisfies (5.39) but will not produce an identity matrix for **T** in $g_i^{(1)} = g_i^{(2)}$. The second alternative (Kaufman and Gupta, 1985; p. 14) generates a maximum information solution under marginal constraints and with the necessary properties (Manton et al., 1992). Thus, these equations describe changes in state over time or between exogenous and endogenous variables.

5.3.2.2. GoM Structural Equations

The GoM structural equations can be written

$$\mathbf{G} = \mathbf{G}\mathbf{B} + \mathbf{E}$$

where **B** is block upper triangular with an identity in the first block and zeros on the remainder of the main diagonal. Thus

$$\mathbf{G}_2 = \mathbf{G}_1\mathbf{B}_{12} + \mathbf{G}_2\mathbf{B}_{22} + \mathbf{E}_2 \tag{5.44}$$

or

$$\mathbf{G}_2(\mathbf{I} - \mathbf{B}_{22}) = \mathbf{G}_1\mathbf{B}_{12} + \mathbf{E}_2 , \tag{5.45}$$

which is the same form as (5.8) where $\boldsymbol{\beta} = (\mathbf{I} - \mathbf{B}_{22})$ and $\boldsymbol{\Gamma} = \mathbf{B}_{12}$. For $\mathbf{G}_3$ we have

$$\mathbf{G}_3(\mathbf{I} - \mathbf{B}_{33}) = \mathbf{G}_1\mathbf{B}_{13} + \mathbf{G}_2\mathbf{B}_{23} + \mathbf{E}_3 \tag{5.46}$$

where $m_{ikl} \neq g_{ikl}$ since $m_{ikl} \geq 0$ for $k = l$ for $b \leq l$. Thus, we can use GoM to represent complex structural relations for simultaneous equation systems without strong multivariate distributional assumptions.

5.4. ADJUSTMENT FOR SURVEY NONRESPONSE

In any of the estimation schemes there is a problem that nonrespondents may cause bias if they are systematically missing. In standard survey practice, if censoring is *not* informative about the process, then one simply can renormalize the sample weights to population values. This is not reasonable in studies of health and mortality at advanced ages where morbidity and death rates are high and likely to be correlated with the probability of nonresponse. In this case we need to adjust for the distribution of nonrespondents. This can be done with a survey like the NLTCS which is based on a list sample which provides ancillary information on nonrespondent characteristics (i.e., mortality and Medicare service use) from linked administrative record sources.

First, we present the λ_{jkl} generated from a combined (1982 and 1984) analysis of the NLTCS.

In Table 5.2, six fuzzy classes are defined from 27 functional measures. Individuals in Class 1 are relatively intact; those in Class 6 are the most highly impaired. The intermediate classes represent different types or patterns of disability (e.g., those in Class 2 are cognitively impaired but have few significant physical impairments). Data from this analysis is used in the example below.

First, define the probability of response (μ_{0i}) or nonresponse (μ_{1i}) as a function of the GoM parameters, i.e.,

$$\mu_{0i} = \sum_k g_{ik}\lambda_{0k} \quad \text{and} \quad \mu_{1i} = \sum_k g_{ik}\lambda_{1k} . \tag{5.47}$$

Let $\bar{\mu}_0 = E(\mu_{0i})$ and $\bar{\mu}_1 = E(\mu_{1i})$ be the expected probabilities. These can be functions of as yet unknown parameters

$$\mu_0 = \sum_k \bar{g}_k\lambda_{0k} \quad \text{and} \quad \bar{\mu}_1 = \sum_k \bar{g}_k\lambda_{1k} . \tag{5.48}$$

Table 5.2. Estimates of Response Profile Probabilities ($\lambda_{kjl} \times 100$) for the Combined 1982 and 1984 NLTCS Sample, 11,535 Complete Detailed Interviews

Variable	Observed Frequency	Profiles 1	2	3	4	5	6
ADL—Needs Help							
Eating	6.1	0.0	0.0	0.0	0.0	0.0	46.2
Getting in/out of bed	26.3	0.0	0.0	0.0	0.0	76.7	100.0
Getting around inside	40.6	0.0	0.0	0.0	0.0	100.0	100.0
Dressing	19.8	0.0	0.0	0.0	0.0	0.0	100.0
Bathing	44.0	0.0	0.0	0.0	42.0	100.0	100.0
Using toilet	21.3	0.0	0.0	0.0	0.0	41.5	100.0
Bedfast	0.8	0.0	0.0	0.0	0.0	0.0	5.3
No inside activity	1.4	0.0	0.0	0.0	0.0	0.0	9.8
Wheel chair fast	3.4	0.0	0.0	0.0	0.0	0.0	23.0
IADL—Needs Help							
With heavy work	76.8	24.1	100.0	100.0	100.0	100.0	100.0
With light work	24.2	0.0	0.0	0.0	0.0	0.0	100.0
With laundry	46.1	0.0	100.0	18.2	100.0	45.3	100.0
With cooking	33.0	0.0	100.0	0.0	0.0	0.0	100.0
With grocery shopping	63.3	0.0	100.0	100.0	100.0	100.0	100.0
Getting about outside	63.5	0.0	52.7	55.1	100.0	100.0	100.0
Traveling	61.6	0.0	100.0	100.0	100.0	100.0	100.0
Managing money	29.7	0.0	100.0	0.0	0.0	0.0	100.0
Taking medicine	24.6	0.0	93.3	0.0	0.0	0.0	100.0
Making telephone calls	17.5	0.0	83.0	0.0	0.0	0.0	96.0
Function Limitations							
How much difficulty do you have:							
Climbing (1 flight of stairs)							
none	15.8	45.3	20.3	0.0	0.0	0.0	0.0
some	28.9	54.7	79.7	22.1	0.0	0.0	0.0
very difficult	33.0	0.0	0.0	77.9	53.1	67.1	7.0
cannot at all	22.3	0.0	0.0	0.0	46.9	32.9	93.0
Bending, e.g., putting on socks							
none	41.4	100.0	100.0	0.0	0.0	100.0	0.0
some	28.5	0.0	0.0	100.0	0.0	0.0	0.0
very difficult	19.0	0.0	0.0	0.0	100.0	0.0	0.0
cannot at all	11.1	0.0	0.0	0.0	0.0	0.0	100.0
Holding a 10 lb. package							
none	26.4	77.4	45.2	0.0	0.0	0.0	0.0
some	17.8	22.6	54.8	21.0	0.0	22.9	0.0
very difficult	16.7	0.0	0.0	79.0	0.0	0.0	0.0
cannot at all	39.0	0.0	0.0	0.0	100.0	77.1	100.0
Reaching overhead							
none	54.0	100.0	100.0	0.0	0.0	100.0	0.0
some	21.8	0.0	0.0	100.0	0.0	0.0	37.6
very difficult	14.7	0.0	0.0	0.0	77.5	0.0	0.0
cannot at all	9.5	0.0	0.0	0.0	22.5	0.0	62.4
Combing hair							
none	69.8	100.0	100.0	0.0	0.0	100.0	0.0
some	17.1	0.0	0.0	100.0	35.1	0.0	37.4
very difficult	7.6	0.0	0.0	0.0	65.0	0.0	0.0
cannot at all	5.5	0.0	0.0	0.0	0.0	0.0	62.6

Table 5.2. (*Continued*)

Variable	Observed Frequency	Profiles 1	2	3	4	5	6
Washing hair							
none	53.4	100.0	100.0	0.0	0.0	100.0	0.0
some	15.2	0.0	0.0	100.0	0.0	0.0	0.0
very difficult	10.0	0.0	0.0	0.0	100.0	0.0	0.0
cannot at all	21.4	0.0	0.0	0.0	0.0	0.0	100.0
Grasping an object							
none	64.8	100.0	100.0	0.0	0.0	100.0	27.3
some	20.8	0.0	0.0	100.0	0.0	0.0	33.6
very difficult	10.5	0.0	0.0	0.0	94.4	0.0	10.8
cannot at all	3.9	0.0	0.0	0.0	5.6	0.0	28.3
Can you see well enough to read a newspaper?							
yes	73.1	100.0	0.0	100.0	71.3	100.0	46.4
Mean Scores ($\bar{g}_k \times 100$)		30.7	11.9	13.5	9.0	16.5	15.2

We initially assume a self-weighting sample (i.e., all $W_i = 1$). Let $\hat{g}_{0k}$ be the observed mean of the g_{ik} for R_0 respondents, i.e.,

$$\hat{g}_{0k} = \sum_{i \in R_0} g_{ik} / n_0 \tag{5.49}$$

where n_0 is the number of respondents. Assume $\bar{\mu}_0 = \hat{\mu}_0$, where $\hat{\mu}_0 = N_0 / I$, and $\bar{g}_k$ is related to $\hat{g}_{0k}$ as

$$\hat{g}_{0k} = \bar{g}_k \lambda_{0k} \Big/ \sum_l \bar{g}_l \lambda_{0l} \,. \tag{5.50}$$

Solving for $\bar{g}_k$ in (5.50) and substituting in (5.48), the estimates of $\hat{\mu}_1$ are

$$\hat{\mu}_1 = \sum_k \hat{g}_{0k} \frac{\lambda_{1k}}{\lambda_{0k}} \left(\sum_l \bar{g}_l \lambda_{0l} \right), \tag{5.51}$$

$$= \sum_k \hat{g}_{0k} \frac{\lambda_{1k}}{\lambda_{0k}} \hat{\mu}_0 \tag{5.52}$$

$$= \sum_k \hat{g}_{0k} b_k \hat{\mu}_0 \tag{5.53}$$

which involves $\hat{g}_{0k}$ and $\hat{\mu}_0$, which are observed, and b_k, the proportionality factor which is the odds of nonresponse for k_i. Both probabilities can be represented as

$$\lambda_{0k} = \frac{1}{1 + b_k} \tag{5.54}$$

and

$$\lambda_{1k} = \frac{b_k}{1 + b_k} . \tag{5.55}$$

The estimator of $\hat{g}_{1k}$ (i.e., the $\bar{g}_k$ for nonrespondents) can be calculated from $\hat{g}_{0k}$ and b_k from (5.50) as

$$\hat{g}_{ik} = \bar{g}_k \lambda_{1k} \Big/ \sum_l \bar{g}_l \lambda_{1l} \tag{5.56}$$

$$= \bar{g}_k \lambda_{0k} b_k \Big/ \sum_l \bar{g}_l \lambda_{0l} b_l \tag{5.57}$$

$$= \hat{g}_{0k} b_k \Big/ \sum_l \hat{g}_{0l} b_l . \tag{5.58}$$

With (5.58) the $\hat{g}_{1k}$ for nonresponders can be calculated with the b_k. The b_k are estimated from the probabilities of nonresponse and mortality for observed cases in two steps. First, short-term mortality coefficients are estimated from the g_{ik} and mortality for responders. We use short-term (e.g., 6 months) mortality because the fuzzy state specific mortality rate in the frailest states changes rapidly; i.e., there are significant changes in the rate over a two-year period. Second, mortality risks are assumed to be fixed for each profile. This information can then be used to parameterize the joint likelihood of nonresponse and mortality to estimate b_k.

The likelihood can be constructed if we first recognize the sets of (a) respondents (R_0), (b) nonrespondents (R_1), (c) survivors (S_0), and (d) decedents (S_1) for a short time period (e.g., 6 months or a year). Mortality is predicted as a linear hazard with an exponential age increase; i.e.,

$$\mu(g_{ik}) = \sum_k g_{ik} q_k \exp(\theta t_i) \tag{5.59}$$

where parameters were estimated using

$$L = \prod_{i \in R_0 \cap S_0} \exp[-\mu_t(g_{ik})] \prod_{i \in R_0 \cap S_1} \{1 - \exp[-\mu_t(g_{ik})]\} . \tag{5.60}$$

The short-term probability of death for class K is

$$q^*_{k_t} = 1 - \exp[-\hat{q}_k \exp(\hat{\theta} t)] \tag{5.61}$$

where $\hat{g}_k$ and $\hat{\theta}$ are MLEs from (5.60). With these coefficients (assumed to apply on age, sex, and survey year basis) we have

$$\hat{\mu}_{10t_i} = \Pr\{i \in R_1 \cap S_0\} \tag{5.62}$$

$$= \sum_k \hat{g}_{0kt_i} \hat{\mu}_{0t_i} (1 - g^*_{kt_i}) b_k \tag{5.63}$$

and

$$\hat{\mu}_{11t_i} = \Pr(i \in R_i \cap S_i\} \tag{5.64}$$

$$= \sum_k \hat{g}_{0kt_i} \hat{\mu}_{0ti} g^*_{kt_i} b_k \tag{5.65}$$

where the probability of nonresponse for S_0 and S_i satisfies $\hat{\mu}_{10t} + \hat{\mu}_{1k} = \hat{\mu}_{1t}$. With equations (5.63) and (5.65) and the $q^*_{k_t}$ estimated from (5.61) the b_k were estimated using

$$L = \prod_{i \in R_0} (1 - \hat{\mu}_{1t_i}) \prod_{i \in R_1 \cap S_0} \hat{\mu}_{10t_i} \prod_{i \in R_1 \cap S_1} \hat{\mu}_{11t_i} \tag{5.66}$$

In effect, if nonresponse is not informative, the $\bar{g}_k$ apply to all nonrespondents (on an age and sex specific basis), so that the b_k would be constant over K. In this case the g_{ik} distribution (on an age and sex specific basis) would be proportionally allocated over K to reproduce the count. If the total mortality for nonrespondents is different than expected if the nonrespondent had the same $\bar{g}_k$ as respondents, then that information can be used to proportionally allocate persons according to b_k. The calculation assumes that each person is in R_0 or R_1 prior to determination of survival (S_0 and S_i; Hoem, 1985, 1989), i.e., that follow-up begins when respondent status is determined.

5.4.1. An Example

The techniques described above for nonresponse adjustments were applied to $g_{ik \cdot t}$ $(t = 1, 2)$ estimates for the 1982 and 1984 NLTCS where age was divided into 30 single year-of-age groups (65 to 94+) for $K = 7$ classes, separately for survey year and sex (see Table 5.2). The mortality status was linked to 25,320 persons in the longitudinal sample with 5,473 deaths identified in the two years following each survey (i.e., from 1982 to 1986). The results for short-term mortality and nonresponse rates are shown in Table 5.3.

The two stages of nonresponse occur because, after list sample selection, there was a screening interview before the detailed interview. We see that nonresponse differs by stage and type. For Stage 2, Classes 4 and 6 have the largest adjustments. For Stage 1 (prescreening) the adjustment is strongest for Classes 4 for females and 5 for males.

Table 5.3. Profile Specific Probabilities (×100) of Two Types of Nonresponse and Short-Term Mortality Rates

Profile	*Stage* 1 ($\lambda_{1k} \times 100$) (Nonresponse Between the Sample Dates and the Screen				*Stage* 2 ($\lambda_{1k} \times 100$) (Nonresponse Between the Screen and Detailed Interview)				$q_k \times \exp(75\theta) \times 100$ Short-Term Mortality Rates at Age 75			
									One Year Follow-up		Six-Month Follow-up	
	Males		Females		Males		Females		Males	Females	Males	Females
k	1982	1984	1982	1984	1982	1984	1982	1984	($\theta = 0.0395$)	($\theta = 0.0415$)	($\theta = 0.0335$)	($\theta = 0.0446$)
1	1.19	1.42	0.73	1.57	2.47	0.41	0.0	5.75	3.54	1.84	3.12	1.67
2	0.00	0.00	0.00	17.34	0.00	4.10	21.74	0.00	10.38	7.59	11.25	5.59
3	0.00	0.00	0.00	3.97	0.00	0.00	2.96	0.00	16.27	2.96	16.80	3.55
4	0.00	0.00	37.36	9.12	14.83	24.74	0.00	1.41	3.46	3.43	3.12	2.77
5	28.21	23.11	0.00	0.00	0.00	0.00	0.00	8.29	13.38	5.18	10.48	4.47
6	0.00	0.00	1.10	0.00	9.84	10.88	1.92	3.83	39.72	22.78	48.99	25.15
7	1.50	0.00	2.44	0.00	—	—	—	—	22.00	15.56	23.59	17.27
$\hat{\mu}_0$	1.88	1.89	2.00	2.02	3.73	4.25	3.81	3.97	5.99	4.58	5.86	4.80
χ^2	17.66*	12.22*	38.52*	15.74*	3.79	4.45	9.95*	2.01	1414.10*	2489.31*	869.79*	1586.81*
Total χ^2				84.14*				20.20*		3903.41*		2456.60*

*$p < 0.05$.

Note: $\lambda_{1k} = b_k/(1 + b_k)$.

Also in the table are short-term mortality rates. We see (parallel to the patterns of disability shown on the 27 disability measures used to generate the profile (Table 5.2); a seventh profile was created for persons in institutions) that mortality is high for the most frail groups with the mortality risks being higher in the short than long term.

5.5. SUMMARY

In this chapter, we showed that, by appropriate parameterization of the likelihood, GoM can be used for much more than classification or structuring the parameter space. We showed how systematic nonresponse adjustments can be made for longitudinal surveys and how stochastic processes and systems of structural equations can be estimated. We also discussed how temporal models with fuzzy state representations differed from standard hazard models and how they could be applied to a wide range of observation plans.

These modifications were made possible by recognizing that the fuzzy state definition could be generalized to include time or duration of several types. The information used to determine an individual's status (i.e., his $g_{ik \cdot t}$) and the structure of conditioning in estimating $g_{ik \cdot t}$ and λ_{kjl} can be used to define the state and conditional event space. More general processes could be represented by including event information in an expanded fuzzy state definition. This could be done without making specific assumptions about either the form of the individual mixing parameters distributions or of the time trajectory of state change or event occurrence. All of this required that the states are defined in a way that the factorizations in the likelihood function hold, i.e., that there is no correlation (or, more generally, contamination) in the likelihood of independent product terms associated with time.

REFERENCES

Bayo, F. R., and Faber, J. F. (1985). Mortality rates around age one hundred. *Transactions of the Society of Actuaries* **35**: 37–59.

Cassel, C.-M., Sarndal, C.-E., and Wretman, J. H. (1977). *Foundations of Inference in Survey Sampling*. Wiley, New York.

Chiang, C. L. (1968). *Introduction to Stochastic Processes in Biostatistics*. Wiley, New York.

Fries, J. F. (1980). Aging, natural death, and the compression of morbidity. *New England Journal of Medicine* **303**: 130–135.

Fries, J. F. (1983). The compression of morbidity. *Milbank Memorial Fund Quarterly* **61**: 397–419.

Gill, R. D. (1992). Multistate life-tables and regression. *Mathematical Population Studies* **3**(4): 259–276.

Heckman, J., Hotz, J., Walker, J. (1985). New evidence on the timing and spacing of births. *American Economic Review* **75**: 179–184.

Hoem, J. M. (1989). The issue of weights in panel surveys of individual behavior. In *Panel Surveys* (D. Kasprzyk, G. Duncan, G. Kalton and M. P. Singh, Eds.). Wiley, New York, pp. 539–559.,

Hoem, J. (1985) Weighting, misclassification and other issues in the analysis of survey samples of life histories. In *Longitudinal Analysis of Labor Market Data* (J. Heckman, and B. Singer, Eds.). Cambridge University Press, Cambridge, Massachusetts, pp. 249–293.

Jöreskög, K. G. (1969) A general approach to confirmatory maximum likelihood factor analysis. *Psychometrika* **34**: 183–202.

Jöreskög, K. G. and Sörbon, D. (1979). *Advances in Factor Analysis and Structural Equation Models*. ABT, Cambridge, MA, 1979.

Jöreskög, K. G., and Sörbon, D. (1986a). *LISREL VI: Analysis of Linear Structural Relationships by Maximum Likelihood and Least Square Methods*. Scientific Software, Mooresville, Indiana.

Jöreskög, K. G., and Sörbon, D. (1986b). *PRELIS: A Preprocessor for LISREL*. Scientific Software, Mooresville, Indiana.

Jöreskög, K. G., and Sörbon, D. (1988). *LISREL 7: A Guide to the Program and Applications*. SPSS, Chicago.

Kaufmann, A., and Gupta, M. M. (1985). *Introduction to Fuzzy Arithmetic: Theory and Applications*. Van Nostrand Reinhold, New York.

Kestenbaum, B. (1992) A description of the extreme aged population based on improved Medicare enrollment data. *Demography* **29**(4): 565–580.

Logue, E. E., and Wing, S. (1986). Life-table methods for detecting age-risk factor interactions in long-term follow-up studies. *Journal of Chronic Disease* **39**: 709–717.

Manton, K. G., and Liu, K. (1990). Recent changes in service use patterns of disabled Medicare beneficiaries. *HCF Review* **11**(3): 51–66.

Manton, K. G., and Suzman, R. M. (1992). Conceptual issues in the design and analysis of longitudinal surveys of the health and functioning of the oldest old. Chapter 5 in *The Oldest-Old* (R. M. Suzman, D. P. Willis, and K. G. Manton, Eds.). Oxford University Press, pp. 89–122.

Manton, K. G., and Tolley, H. D. (1991). Rectangularization of the survival curve: Implications of an ill-posed question. *Journal of Aging and Health* **3**(2): 172–193.

Manton, K. G., Stallard, E., and Vaupel, J. W. (1986a). Alternative models for the heterogeneity of mortality risks among the aged. *Journal of the American Statistical Association* **81**: 635–644.

Manton, K. G., Stallard, E., Woodbury, M. A., and Yashin, A. I. (1986). Applications of the Grade of Membership technique to event history analysis: Extensions to multivariate unobserved heterogeneity. *Mathematical Modelling* **7**: 1375–1391.

Manton, K. G., Vertrees, J. C., and Woodbury, M. A. (1990). Functionally and medically defined subgroups of nursing home populations. *HCF Review* **12**: 47–62.

Manton, K. G., Stallard, E., and Woodbury, M. A. (1991) A multivariate event history model based upon fuzzy states: Estimation from longitudinal surveys with informative nonresponse. *Journal of Official Statistics* **7**: 261–293 (published by Statistics Sweden, Stockholm).

Manton, K. G., Woodbury, M. A., Corder, L. S., and E. Stallard (1992). The use of Grade of Membership techniques to estimate regression relationships. In *Sociological Methodology, 1992* (P. Marsden, Ed.). Basil Blackwell, Oxford, England, pp. 321–381.

Manton, K. G., Woodbury, M. A., Vertrees, J. C., and Stallard, E. (1993). Use of medicare services before and after introduction of the Prospective Payment System. *Health Services Research* **28**(3): 269–292.

Manton, K. G., Newcomer, R., Lowrimore, G., Vertrees, J. C., Harrington, C. (1994). Differences in health outcomes over time between S/HMO and fee for service delivery systems. *Health Care Financing Review*, Special Issue on Managed Care, in press.

Manton, K. G., Stallard, E., and Singer, B. H. (1994). Projecting the future size and health status of the U.S. elderly population. In *The Economics of Aging* (D. Wise, Ed.). National Bureau of Economic Research, University of Chicago Press.

Mills, R., Fetter, R., Reidel, J., Aderill, R. (1976). AUTOGRP: An interactive computer system for the analysis of health care data. *Medical Care* **14**: 603–615.

Spiegelman, M. (1968). *Introduction to Demography*. Harvard University Press, Cambridge, Massachusetts.

Strehler, B. L. (1975) Implications of aging research for society. *Proceedings 58th Annual Meeting of the Federation of American Societies for Experimental Biology, 1974*, **34**: 5–8.

Suppes, P., and Zanotti, M. (1981). When are probabilistic explanations possible? *Synthese* **48**: 191–199.

Tolley, H. D., and Manton, K. G. (1987) A Grade of Membership approach to event history data. *Proceedings of the 1987 Public Health Conference on Records and Statistics*. DHSS Pub. No. (PHS) 88-1214, USGPO, Hyattsville, MD, pp. 75–78.

Tolley, H. D., and Manton, K. G. (1991). A Grade of Membership method for partitioning heterogeneity in a collective. *SCOR Notes*, International Prize in Acturial Science, April, 1991, pp. 121–151.

CHAPTER 6

Empirical Bayesian Generalizations of the GoM Model

INTRODUCTION: EMPIRICAL BAYESIAN PRINCIPLES

So far we have concentrated on the basic GoM model and multinomial and Poisson likelihood estimation schemes presented in Chapters 2, 3, and 4. In this chapter we extend the basic model to show how fuzzy set models can embody empirical Bayesian features.

The work of Robbins (1955) on empirical Bayes methods has spawned a series of research efforts to estimate components of the prior distribution from observed data. The philosophy of Bayesian methods is that the parameters underlying the distribution of the observations are, themselves, random variables. The distribution of these random variables, referred to as the prior distribution, determines much of the structure of the observable variables. Unfortunately, since the realizations of these variables are never observed, the only way to analyse the distribution of parameters is to use marginal information. Robbins described the framework of the problem in which one tries to estimate the prior distribution using density estimation methods (see, e.g., Robbins 1964, 1980, 1983). When the data to be analyzed come from experiments that are exact replicates of each other, the density estimation step is relatively straightforward. However, when the nature of the experiments vary, the problem becomes more complex. Since little is assumed about the prior distribution, forming density estimates from the marginal observations of different experiments is often problematic, entailing the specification of a mathematical relation (see, e.g., Rencher and Krutchkoff, 1975).

Efron and Morris (1972, 1973, 1975) extend the method of empirical Bayes by making assumptions about the form of the prior distribution. By assuming that the prior distribution is of a specific form, with unknown parameters, or by assuming that it is a conjugate prior, relative to the distribution of the observations, these researchers provide a general method for forming posterior estimates of the parameters of the observed distribution (see also Morris, 1982). With additional assumptions, empirical Bayes methods can more readily be applied to estimation problems where exact experimental replication is not available. The need for this occurs frequently in population health studies. An example of this is given in estimation of the variation of site specific spatial cancer mortality rates over U.S. counties (Manton et al., 1989).

In this chapter we focus on the types of models proposed by Efron and Morris to develop an empirical Bayes model for GoM. By estimating the "hyper" parameter, or parameters of the distribution of the parameters of the observable outcomes, empirical Bayes methods allows us to "borrow strength" or information across the whole statistical ensemble of outcomes to improve estimates for individual outcomes (see e.g., the Stein effect; Efron and Morris, 1973).

6.1. THE MODEL

To generalize GoM to the empirical Bayes case we modify (2.1), i.e.,

$$p_{ijl} = \Pr[y_{ijl} = 1] = \sum_k g_{ik} \cdot \lambda_{kjl} \tag{6.1}$$

to be

$$E(p_{ijl} \mid \theta) = \sum_k g_{ik}(\lambda_{kjl} \mid \theta) \tag{6.2}$$

where, for each i and j, $(p_{ijl}, p_{ij2}, \ldots, p_{ijL_j})$ is a random vector distributed according to a specific distribution. The parameter of this distribution is denoted θ and is the "hyperparameter." Equation (6.2) says that the expectation of an individual outcome, given the parameter θ, is a combination of the grade of membership functions for the individual. Each function is peculiar to the individual and is dependent on both the profiles λ and the parameter θ.

An alternative formulation might be

$$E(p_{ijl} \mid \theta) = \sum_k g_{ik}(\theta)\lambda_{kjl} \tag{6.3}$$

where θ is a prior distribution for the g_{ik}, e.g., a multivariate Dirichlet distribution. In some sense, the specification of p_{ik} to be distributed as a member of an equivalence class of distributions with moments identifiable to the Jth order is a general empirical Bayesian constraint to smooth the g_{ik} to get consistency if the number of variables (J) is small. Since the Jth-order moment distribution is not of a specified form, it is more general than the usual empirical Bayesian formulation.

We use equation (6.2) as the basic empirical Bayesian form. The heterogeneity of the p_{ijl} is hypothesized to result not only from the g_{ik} and λ_{kjl} parameters, but also from a third heterogeneity component representing the heterogeneity of individual responses to the λ_{kjl}. This third heterogeneity component serves to weight the contribution of each variable to each of K types. To implement the parameterization in equation (6.2) we must select a distribution for random variables over a simplex. One suitably rich distribution is the Dirichlet distribution (see Johnson and Kotz, 1972). Besides being able to assume a variety of shapes over different parameter values, the Dirichlet distribution is the conjugate prior distribution for the multinomial. As a result, certain optimality properties of the empirical Bayesian estimates (i.e., minimax properties) are guaranteed by using a conjugate distribution if no other information is available (see Morris, 1982; Cox and Hinkley, 1974).

In this chapter we assume that the profiles λ_{kjl} are distributed across the population as the Dirichlet distribution with parameter θ_t. As a result, the optimal forms of the estimates follow the development given by Morris (1982). However, if the Dirichlet distribution does not apply, the linear combination of individual and overall parameter estimates is still appropriate using methods proposed by Buhlmann (1970). Note, however, that in this case we do not get the same optimality results as we do with the use of conjugate priors in empirical Bayesian estimation.

6.2. THE DIRICHLET DISTRIBUTION

Because of the use of the properties of the Dirichlet distribution in deriving the fuzzy empirical Bayesian estimates, we first review its general properties. The Dirichlet distribution of response profiles p_{ijl}, $l = 1, 2, \ldots, L_j$, over the simplex with L_j vertices is defined with

$$\sum_l p_{ijl} = 1 \qquad (p_{ijl} \geq 0, l = 1, 2, \ldots, L_j) . \tag{6.4}$$

The probability density for the Dirichlet distribution for the set of realizations $\{p_{ijl}\}$ is, for each i and j, given by

$$f(p_{ijl} \mid \alpha_{ijl}) = \Gamma(\alpha_{ij+}) \prod_{l=1}^{L_j} [p_{ijl}^{\alpha_{ijl}-1} / \Gamma(\alpha_{ijl})] , \tag{6.5}$$

where $\alpha_{ij+} = \Sigma_l \alpha_{ijl}$ and where $\Gamma(\alpha)$ represents the Gamma function (see, e.g., Johnson and Kotz, 1972).

The first and second moments of the Dirichlet distribution are

$$E(p_{ijl} \mid \boldsymbol{\alpha}_{ij}) = \frac{\alpha_{ijl}}{\alpha_{ij+}} = \bar{p}_{ijl} \tag{6.6}$$

$$\mathrm{Var}(p_{ijl}) = \bar{p}_{ijl}(1 - \bar{p}_{ijl})/(\alpha_{ij+} + 1) \tag{6.7}$$

$$\mathrm{Cov}(p_{ijl}, p_{ijr}) = -\bar{p}_{ijl} \cdot \bar{p}_{ijr}/(\alpha_{ij+} + 1)\,, \qquad \text{for} \quad l \neq r\,. \tag{6.8}$$

The realizations y_{ijl} of a multinomial with cell probabilities generated as realizations of Dirichlet distributions across values of i and j is

$$L(p_{ijl}, y_{ijl} \mid \boldsymbol{\alpha}_{ij}) = \prod_i \prod_j f(p_{ijl} \mid \boldsymbol{\alpha}_{ij}) y_{ij+}! \prod_l p_{ijl}^{y_{ijl}}/y_{ijl}!\,, \tag{6.9}$$

which, when the Dirichlet probability density in (6.5) is substituted for the functional f, provides

$$L(p_{ijl}, y_{ijl} \mid \boldsymbol{\alpha}_{ij}) = \prod_i \prod_j \Gamma(\alpha_{ij+}) y_{ij+}! \prod_l (p_{ijl}^{y_{ijl}+\alpha_{ijl}-1}/(y_{ijl}!\Gamma(\alpha_{ijl}))\,. \tag{6.10}$$

In these expressions, $y_{ij+} = \Sigma_l\, y_{ijl}$ is a constant integer, say n_{ij}, and the p_{ijl} are constrained to sum to 1 across l,

$$\left(\left(\sum_{l=1}^{L_j} p_{ijl} = 1, (p_{ijl} \geq 0, l = 1, 2, \ldots, L_j), j = 1, 2, \ldots, J\right), i = 1, 2, \ldots, \Gamma\right). \tag{6.11}$$

The likelihood in (6.10) cannot be evaluated directly because of the gamma-distributed $\{p_{ijl}\}$, which are not observed (i.e., the p_{ijl} are gamma distributed with hyperparameters that must be estimated). To integrate out the gamma-distributed p_{ijl} to get the marginal likelihood for y_{ijl} we rewrite (6.10) as

$$L(p_{ijl}, y_{ijl} \mid \boldsymbol{\alpha}_{ij}) = \prod_{i=1}^{I} \prod_{j=1}^{J} \left\{ \left[\frac{y_{ij+}!\Gamma(\alpha_{ij+})}{\Gamma(y_{ij+} + \alpha_{ij+})} \cdot \prod_{l=1}^{L_j} \frac{\Gamma(y_{ijl} + \alpha_{ijl})}{y_{ijl}!\Gamma(\alpha_{ijl})} \right] \cdot \left[\frac{1}{\Gamma(y_{ij+} + \alpha_{ij+})} \cdot \prod_{l=1}^{L_j} \frac{p_{ijl}^{y_{ijl}+\alpha_{ijl}-1}}{\Gamma(y_{ijl} + \alpha_{ijl})} \right] \right\}. \tag{6.12}$$

This expression is derived by normalizing factors in the second pair of brackets to make the total probability evaluate to 1; and its reciprocal in the first pair of brackets to produce a discrete hypergeometric distribution. The expression in the second pair of brackets is the conditional posterior Dirichelet distribution of the p_{ijl} random variables given the y_{ijl}. The first bracketed term in (6.12) is the marginal distribution of the y_{ijl}. The conditional expectation

$$E\left(\prod_{l=1}^{L_j} p_{ijl}^{rl} \mid (y_{ijl}, l = 1, 2, \ldots, L_j), A\right), \tag{6.13}$$

where $A = \{\alpha_{ijl}\}$ has the value

$$\frac{\Gamma(\alpha_{ij+} + y_{ij+})}{\Gamma(\alpha_{ij+} + y_{ij+} + r_+)} \prod_{l=1}^{L_j} \frac{\Gamma(\alpha_{ijl} + y_{ijl} + r_l)}{\Gamma(\alpha_{ij+} + y_{ij+})}, \tag{6.14}$$

where $r_+ = \Sigma_{l=1}^{L_j} r$.

This result is produced with the correct normalizing factor. The first and second moments of the posterior Dirichelet distribution are sufficient to determine the distribution. The moments for the posterior distribution of the L_j multinomials are

$$\bar{p}_{ijl} = \nu_l = E(p_{ijl} \mid y_{ij+}, A) = \alpha_{ijl}/\alpha_{ij+}, \tag{6.15}$$

$$\nu_{lm} = E(p_{ijl} \cdot p_{ijm} \mid y_{ij+}, A), \tag{6.16}$$

$$\mathrm{Cov}(p^*_{ijl} p^*_{ijm}) = \nu_{lm} - \nu_l \nu_m, \tag{6.17}$$

$$\sigma^2 = \mu m,$$

producing

$$\sigma^2(p_{ijl}) = \bar{p}_{ijl}(1 - \bar{p}_{ijl})/(y_{ij+} + \alpha_{ij+} + 1), \tag{6.18}$$

$$\mathrm{Cov}(p_{ijl}, p_{ijm}) = -\bar{p}_{ijl}\bar{p}_{ijm}/(y_{ij+} + \alpha_{ij+} + 1). \tag{6.19}$$

When the data for y_{ij+} are missing in (6.15), $\bar{p}_{ijl}$ reduces to

$$\bar{p}_{ijl} = E(p_{ijl} \mid A) = \frac{\alpha_{ijl}}{\alpha_{ij+}} \quad (l = 1, 2, \ldots, L_j). \tag{6.20}$$

This can be seen from (6.12) since $y_{ijl}!$ and $y_{ij+}!$ are both 1. The marginal likelihood (except for factorial terms) is

$$L(y_{ijl} \mid A) = \prod_{i=1}^{I} \prod_{j=1}^{J} \prod_{l=1}^{L_j} \left(\frac{\alpha_{ijl}}{\alpha_{ij+}}\right)^{y_{ijl}}. \tag{6.21}$$

The value of $E(p_{ijl} \mid \alpha_{ij+})$ is unchanged if α_{ijl} is multiplied by a positive constant; i.e., only the relative degree of heterogeneity is determined. The ratio $\alpha_{ijl}/\alpha_{ij+}$ is undefined if $\alpha_{ij+} = 0$. In this case, a limiting form for the term in the likelihood (as $\alpha_{ij+} \to 0$) is

$$\Gamma(\alpha_{ij+}) \prod_{l=1}^{L_j} p_{ijl}^{y_{ijl}+\alpha_{ijl}} / \Gamma(\alpha_{ijl}) = \frac{1}{\alpha_{ij+}} \prod_{l=1}^{L_j} p_{ijl}^{y_{ijl}-1} \alpha_{ijl} + 0(\alpha_{ij+})$$

$$\to \prod_{l=1}^{L_j} E(p_{ijl}) p_{ijl}^{y_{ijl}-1} . \tag{6.22}$$

As $\alpha_{ij+} \to 0$, $\alpha_{ij+} \geq 0$, the gamma distribution of $\{p_{ijl}\}$ becomes concentrated at the extreme points (vertices) of the simplex. The mass at each vertex l (characterized by $p_{ijl} = 1$ for one and only one l in L_j and p_{ijl} for all others) is $E(p_{ijl})$. This represents a condition of maximum heterogeneity for y_{ij+}. In this case, in (6.21), we replace $\alpha_{ijl}/\alpha_{ij+}$ by $\bar{p}_{ijl} = E(p_{ijl})$.

To make explicit the relation of the Dirichelet distribution to the GoM model, the Dirichlet parameter α_{ijl} is assumed to be related to the basic GoM parameters as

$$\alpha_{ijl} = \sum_{k=1}^{K} g_{ik} \cdot \lambda_{kjl} \, . \tag{6.23}$$

Therefore,

$$E(p_{ijl}) = \frac{\alpha_{ijl}}{\sum_{l=1}^{L_j} \alpha_{ijl}} \tag{6.24}$$

$$= \frac{\sum_{k=1}^{K} g_{ik} \cdot \lambda_{kjl}}{\sum_{k=1}^{K} \left(g_{ik} \cdot \sum_{l=1}^{L_j} \lambda_{kjl} \right)} , \tag{6.25}$$

which is the same expression as the Poisson form of the likelihood function in Chapter 2. Thus, the Poisson formulation has a direct interpretation in the empirical Bayesian context with the normalization factor being interpretable as representing heterogeneity of the p_{ijl}.

The denominator in (6.25) is actually related to the heterogeneity of the p_{ijl}, i.e., the smaller its value, the *greater* the heterogeneity of responses to the question and the less important is the *jth* variable in defining the kth extreme type. When $\Sigma_{l=1}^{L_j} \lambda_{kjl}$, for all values of k and j, is 1, the standard (i.e., multinomial or Poisson with all variables with equal weights) and empirical Bayesian forms of the models are *identical*. Thus, this term can be referred to as a "relevance factor" for question j, which indicates the relative importance (or weight) of the jth question in defining the K fuzzy classes. The question relevance factor is important when the explanatory power of the individual variables differs significantly and can be used in "item" analyses to select the most meaningful variables.

In empirical Bayesian GoM the g_{ik} are still state variables (i.e., functionally a convex mixing coefficient) and not probabilities. The mathematical implication of being a state variable was described in Chapter 2 and illustrated in Table 2.1. The g_{ik} are, however, applied to $\lambda_{kjl}/\Sigma_l \lambda_{kjl}$, which is the expectation of response l to variable j for the kth type, i.e.,

$$E(p_{ijl}) = \bar{p}_{ijl} = \frac{\alpha_{ijl}}{\Sigma_{l=1}^{L_j} \alpha_{ijl}}. \tag{6.26}$$

The likelihood for this case, is identical to the "Poisson" likelihood in Chapter 2 and is

$$L = \prod_i \prod_j \prod_l \left(\frac{\alpha_{ijl}}{\alpha_{ij+}}\right)^{y_{ijl}} = \prod_i \prod_j \prod_l \left(\frac{\sum_k g_{ik} \cdot \lambda_{kjl}}{\sum_k g_{ik} \cdot \sum_{l=1}^{L_j} \lambda_{kjl}}\right)^{y_{ijl}}. \tag{6.27}$$

The arguments about statistical inference and properties of the parameters presented in Chapter 2 can be extended to the parameters in (6.27). Specifically, the p_{ijl} in the "simple" model are bilinear functions of g_{ik} and λ_{kjl}. In (6.27) they are *fractional* bilinear functions of g_{ik} and λ_{kjl}. In the "simple" model, the λ_{kjl} sum to 1 over l. In the empirical Bayesian model there is an additional parameter α_{ij+} of the Dirichelet mixing distribution A. Thus, each data element (in a given sample) must be closer on average to $\bar{p}_{ijl}$ than to $\hat{p}_{ijl}$ for a given K-class solution. This will be illustrated in our example. The larger heterogeneity of p_{ijl}, the more the functionals can become curved. This nonlinearity may help reduce the dimensionality of the space.

6.3. EXAMPLES

We provide two examples. One is based on the combination of the Duke and Johns Hopkins ECA data sets ($N = 7184$), which were examined in Chapter 2 where we compared the relative performance of GoM and LCM. In the first analysis we use the Poisson form of the likelihood and report the degree of heterogeneity associated with each of the 33 psychiatric symptoms. The second example is based on data from a study of nursing home reimbursement and quality of care (Manton et al., 1994). In this analysis 4525 nursing home residents over age 65 in 6 states were evaluated on 111 clinical and behavioral measures using (a) LCM, (b) the simple GoM model, (c) the empirical Bayesian form of the GoM model with question relevance relevance factors that approach 0, and (d) the empirical Bayesian model with an approximation of the λ_{kjl} when the question relevant factors approach 0.

6.3.1. Psychiatric Symptoms in Two Communities

In the first analysis 7184 persons in Baltimore, Maryland, and Durham, North Carolina, were assessed using a common psychiatric review instrument (i.e., the Diagnostic Interview Survey; Robins, 1981). From that instrument 33 symptoms were selected that characterized six distinct psychiatric disorders, i.e., chronic anxiety, panic, depression, simple phobia, social phobia, and somatization associated with depression. This analysis differs from that reported in Chapter 2 because we analyzed all subjects in both the Hopkins and Duke data sets.

In Table 6.1, for $K = 6$, we present estimates of $\lambda_{kjl} = (\alpha_{kjl}/\alpha_{kj+})$ and of the relative heterogeneity of individual responses to each question. In the table the first column describes the 33 symptomatic variables, the second describes the proportion of the sample with the attribute, and the next six columns describe the λ_{kjl} for the six fuzzy classes. Underneath the λ_{kjl}, in parentheses, are the question relevance factors. These are $\Sigma_{l=1}^{L_i} \lambda_{kjl}$, which arc inversely related to the relative heterogeneity of J for each of the K classes. Thus, higher values (greater than 1.0) indicate *less* heterogeneity and hence a *greater* influence of the variable on the fuzzy class. As values approach 0, heterogeneity reaches a *maximum* (i.e., entropy is maximized) and the variable has little information to contribute to identifying the fuzzy class.

Thus, for the first variable (Sad for 2 weeks) the positive response for persons like Class V has 39% less heterogeneity (QRF = 1.39) than average for all variables in the analysis. In contrast, in Class II the variable is 15% more heterogeneous (QRF = 0.85) than the average variable. In Class I the heterogeneity parameter is 0.0 indicating that the heterogeneity is *maximum* (i.e., entropy is maximized; information is minimized) for this variable. The other three parameters are near 1.0 indicating an average degree of heterogeneity. Note that coefficients can be large but still only have a moderate degree of predictive influence. On the "Phobia: Crowds" question, Class 1 has a coefficient (24.6%) 9.3 times the marginal (2.7%), but the heterogeneity parameter suggests that this variable is 27% less influential than the average variable.

The effect of the heterogeneity parameter also changes the structure of the simplex in which parameters are estimated (if it deviates from 1.0). This can be seen by examining the equation relating the λ_{kjl} to the marginal proportions. The simple and empirical Bayesian GoM equations for the relation are compared below.

$$p_{ijl} = \sum_{k=1}^{K} g_{ik} \cdot \lambda_{kjl}$$

Table 6.1. Probability in % (λ_{kjl}) and Relative Heterogeneity (α) Parameter Sets for an Empirical Bayesian GoM Analysis of 33 Psychiatric Symptoms from the Duke and Johns Hopkins ECA Projects

Variable	Sample Proportion (%)	Pure Type Coefficients in % (Λ) and (A)					
		I	II	III	IV	V	VI
Sad for 2 weeks	3.83	0	0	0	0	100.00	0
		(0.0)	(0.85)	(1.00)	(0.99)	(1.39)	(1.00)
Sad for 2 years	1.09	0	0	0	0	57.47	0
		(1.04)	(0.99)	(1.01)	(1.03)	(0.69)	(1.01)
Fainting (chronic anxiety)	0.54	0	50.38	0	0	0	0
		(1.01)	(0.96)	(1.00)	(1.04)	(0.98)	(1.00)
Shortness of breath (chronic anxiety)	2.33	0	100.00	0	0	0	0
		(1.01)	(2.08)	(0.98)	(0.95)	(0.77)	(1.00)
Palpitations (chronic anxiety)	2.77	0	100.00	0	0	0	0
		(0.91)	(2.47)	(0.98)	(0.91)	(0.78)	(1.00)
Felt dizzy (chronic anxiety)	3.05	0	100.00	0	0	0	0
		(0.83)	(2.72)	(0.99)	(0.91)	(0.66)	(1.00)
Feel weak	2.19	0	100.00	0	0	0	0
		(0.89)	(1.96)	(0.99)	(0.93)	(0.80)	(1.00)
Nervous person	24.72	100.00	100.00	100.00	0	0	0
		(1.34)	(0.32)	(1.12)	(0.00)	(0.00)	(1.05)
Fright attack (panic)	1.51	0	100.00	0	0	0	0
		(0.73)	(1.35)	(1.00)	(0.88)	(0.89)	(1.01)
Phobias: Eating in public	0.68	15.45	0	0	17.53	0	0
		(0.88)	(0.96)	(1.01)	(0.97)	(0.99)	(1.00)
Phobias: Speaking in small group	1.58	0	0	0	49.61	0	0
		(1.04)	(0.84)	(1.00)	(1.04)	(0.95)	(1.00)
Phobias: Speaking to strangers	1.28	0	0	0	43.91	0	0
		(0.70)	(1.01)	(1.01)	(0.95)	(1.00)	(1.00)
Phobaias: Alone	1.20	69.04	3.28	0	15.19	0	0
		(0.89)	(0.94)	(1.00)	(0.90)	(1.04)	(1.00)
Phobias: Tunnels and bridges	2.76	0	0	0	100.00	0	0
		(0.80)	(1.06)	(1.01)	(0.90)	(1.04)	(1.00)
Phobias: Crowds	2.73	25.46	33.93	0	93.55	0	0
		(0.63)	(0.68)	(1.00)	(0.79)	(1.12)	(1.01)

Phobias: Public transportation	3.52	0	0	0	100.00	0	0
		(0.80)	(0.90)	(1.00)	(1.14)	(0.97)	(1.00)
Phobias: Outside house alone	1.02	27.09	0	0	25.82	0	0
		(0.78)	(0.98)	(1.00)	(0.96)	(1.05)	(1.00)
Phobias: Heights	6.23	0	0	0	100.00	0	0
		(0.78)	(0.81)	(0.96)	(2.02)	(0.91)	(1.00)
Phobias: Closed place	2.67	0	0	0	97.92	0	0
		(0.87)	(0.85)	(1.00)	(0.88)	(1.01)	(1.01)
Phobias: Storms	3.87	0	0	0	100.00	0	0
		(0.91)	(0.84)	(1.00)	(1.25)	(0.90)	(1.00)
Phobias: Water	4.14	0	0	0	100.00	0	0
		(1.06)	(0.88)	(1.00)	(1.35)	(0.89)	(0.99)
Phobias: Bugs	8.09	0	0	0	100.00	0	0
		(0.82)	(0.87)	(0.97)	(2.62)	(0.56)	(0.96)
Phobias: Animals	1.44	0	0	0	47.98	0	0
		(0.84)	(1.09)	(1.00)	(0.97)	(1.01)	(1.00)
Crying spells	4.36	0	0	0	0	100.00	0
		(1.06)	(0.64)	(0.96)	(0.87)	(1.58)	(1.00)
Felt hopeless (somatization)	3.95	0	0	0	0	100.00	0
		(0.23)	(0.69)	(0.99)	(1.02)	(1.43)	(1.00)
Change in weight or appetite	4.77	0	0	0	0	100.00	0
		(0.45)	(0.82)	(0.99)	(0.81)	(1.74)	(0.99)
Sleeping more or less	10.32	0	0	0	0	100.00	0
		(0.00)	(0.15)	(0.92)	(0.57)	(3.75)	(0.97)
Talking or moving slower	4.85	100.00	0	0	0	0	0
		(4.22)	(0.48)	(1.01)	(0.68)	(0.00)	(1.01)
Decreased interest in sex	1.78	100.00	0	0	0	0	0
		(1.54)	(0.92)	(0.99)	(0.89)	(1.09)	(1.00)
More tired	6.93	100.00	0	0	0	100.00	0
		(0.02)	(0.14)	(0.97)	(0.85)	(2.51)	(0.99)
Worthless, sinful, or guilty	2.32	100.00	0	0	0	0	0
		(2.00)	(0.95)	(1.01)	(1.00)	(0.42)	(1.00)
Difficulty concentrating or thinking	4.34	100.00	0	0	0	0	0
		(3.70)	(0.59)	(1.01)	(0.86)	(0.00)	(1.00)
Thoughts of death or suicide	7.94	0	0	0	0	100.00	0
		(0.00)	(0.18)	(0.98)	(0.62)	(2.89)	(0.98)

Note: Relative heterogeneity measures are in parentheses.

so the mean for the simple, or standard, (QRF = 1.0) GoM case is

$$\bar{x}_j = \sum_{k=1}^{K} \bar{g}_k \sum_{l=1}^{L_j} (\lambda_{kjl} \cdot \bar{x}_{jl}) , \tag{6.28}$$

and the mean for the empirical Bayesian case is

$$\bar{x}_j = \left[\sum_{k=1}^{K} \bar{g}_k \sum_{l=1}^{L_j} (\lambda_{kjl} \cdot \bar{x}_{jl}) \right] \sum_{l=1}^{L_j} \lambda_{kjl} . \tag{6.29}$$

That is, in reweighting the K classes to reproduce the marginal frequencies (or of a predicted mean for a continuous variable) the value has to be adjusted by the size of the heterogeneity parameter. Given also that the sum $\Sigma_{l=1}^{L_j} \lambda_{kjl}$ is nonlinear, this parameter introduces nonlinearity in the response. That is, all boundary constraints (e.g., $0 \leq g_{ik} \leq 1$; $\Sigma_{l=1}^{L_j} \lambda_{kjl} = 1$) must hold, but the state space in the simplex can be curved if the $\Sigma_{l=1}^{L_j} \lambda_{kjl}$ is different from 1. This nonlinearity is again not of a specified parametric form, but is a function of the degree of heterogeneity for a specific variable.

6.3.2. Nursing Home Residents

In the second example we analyzed 111 characteristics for 4525 nursing home residents in six states. The tables are extensive, so we present only portions of them to illustrate specific points. The complete tables and the substantive interpretations are presented in Manton et al. (1994).

The first analysis used the general (empirical Bayesian) form of the GoM model. Interestingly, this solution, which has $\Sigma_{j=1}^{J} L_j = 407$, which is large enough to produce consistent estimates for the g_{ik}, identified 11 classes of patients. These are summarized (on all 111 measures) from the λ_{kjl} in Table 6.2 for the standard empirical Bayesian model.

The description and labels for these 11 classes are presented below (Manton et al., 1994).

1. *Major Stroke*. This class is characterized by medically significant strokes. The class has a 100% chance of having a "stroke" and manifesting aphasia, hemiplegia, swallowing, and chewing problems, loss of voluntary arm and leg use, high incidence of diabetes, hypertension and atherosclerosis, problems with neuromuscular coordination (i.e., swallowing and chewing), seizures, and communication difficulties (all physical or neurological consequences of stroke). This class also has contractures, uses signs and motions to make needs known (because of aphasia), is seldom coherent, has edema, and, though stable, needs monitoring. This group has "sad" affect and extensive ADL impairment. It also has no characteristics associated with psychological well-being (e.g., social relations).

Though this profile describes residents affected by a "catastrophic" or

Table 6.2. λ_{kjl} Values for Measures of Psychological Well-being, Medical Conditions, Problems, Sign and Symptoms, Skin Conditions, Cognitive Patterns, Communication, Vision, Mood and Behavior Patterns, and Limitations in Activities of Daily Living

	Frequency	1 Major Stroke	2 Depressed	3 CVD Diabetes Dementia	4 Early Dementia	5 Late Stage Dementias	6 Alzheimer's Disease	7 Comatose and Pulmonary	8 Multiple Chronic Disease & Hip Fracture	9 CVD	10 Post-Acute CVD	11 Rehab.
Psychological Well-being												
Sense of initiative involvement												
At ease interacting with others	49.7	0.0	0.0	100.0	18.8	0.0	0.0	0.0	0.0	100.0	100.0	100.0
At ease doing planned/structured activities	30.5	0.0	0.0	100.0	0.0	0.0	0.0	0.0	0.0	12.7	100.0	73.4
Establishes own goals	24.1	0.0	40.6	100.0	0.0	0.0	0.0	0.0	0.0	0.0	0.0	66.0
Pursues involvement in life of facility	13.2	0.0	0.0	100.0	0.0	0.0	0.0	0.0	0.0	0.0	66.7	26.5
Accepts invitations into most group activities	33.6	0.0	0.0	100.0	0.7	0.0	0.0	0.0	0.0	87.9	100.0	27.6
Relationships/attitudes												
Covert/open conflict with and/or repeated criticism of staff	9.9	0.0	100.0	0.0	0.0	0.0	9.9	0.0	0.0	0.0	0.0	0.0
Unhappy with roommate	4.6	0.0	58.5	0.0	0.0	0.0	0.0	0.0	0.0	0.0	0.0	0.0
Verbal expressions of grief	11.1	0.0	100.0	0.0	0.0	0.0	0.0	0.0	0.0	0.0	0.0	0.0
Indicates pervasive concern with health status	12.6	0.0	100.0	0.0	0.0	0.0	0.0	0.0	0.0	0.0	0.0	0.0
Does not adjust easily to change in routine	29.2	0.0	100.0	0.0	0.0	0.0	50.2	0.0	0.0	0.0	0.0	0.0
Avoids interaction with others	14.1	0.0	100.0	0.0	0.0	0.0	0.0	0.0	0.0	0.0	0.0	0.0
Past Roles												
Strong identification with past roles and life status	24.1	0.0	100.0	100.0	0.0	0.0	0.0	0.0	0.0	0.0	100.0	30.8

Table 6.2. (*Continued*)

	Frequency	1 Major Stroke	2 Depressed	3 CVD Diabetes Dementia	4 Early Dementia	5 Late Stage Dementias	6 Alzheimer's Disease	7 Comatose and Pulmonary	8 Multiple Chronic Disease & Hip Fracture	9 CVD	10 Post-Acute CVD	11 Rehab.
Expresses sadness/anger/empty feeling over lost roles/status	17.7	0.0	100.0	0.0	0.0	0.0	0.0	0.0	0.0	0.0	0.0	0.0
Medical Conditions												
Arthritis	30.0	0.0	0.0	0.0	0.0	0.0	0.0	0.0	100.0	0.0	0.0	0.0
Anemia	11.1	0.6	0.0	0.0	9.0	11.0	0.0	0.0	100.0	0.0	0.0	10.1
Anxiety disorder	4.6	0.0	55.7	0.0	0.0	0.0	0.0	0.0	0.0	0.0	0.0	0.0
Alzheimer's disease	13.1	0.0	0.0	0.0	0.0	23.8	100.0	14.4	0.0	0.0	0.0	0.0
Other dementia	35.7	0.0	3.8	100.0	71.9	72.0	52.1	40.9	0.0	0.0	0.0	0.0
Aphasia	4.4	100.0	0.0	0.0	0.0	0.0	0.0	5.0	0.0	0.0	0.0	0.0
Cancer	6.4	0.2	0.0	0.0	0.3	0.0	9.5	14.3	67.7	0.0	0.0	11.8
Cataracts	6.3	0.0	0.0	0.0	0.0	0.0	0.0	0.0	100.0	0.0	0.0	0.0
Cerebrovascular accident (stroke)	22.3	100.0	0.0	0.0	0.0	0.0	0.0	0.0	0.0	0.0	0.0	0.0
Congestive heart failure	24.0	0.0	4.2	82.3	0.0	16.9	0.0	2.8	100.0	100.0	100.0	0.0
ASHD	21.1	0.0	6.0	98.0	0.0	20.2	0.0	0.0	100.0	100.0	100.0	0.0
Peripheral vascular disease	8.0	23.4	0.0	0.0	0.0	0.0	0.0	0.0	100.0	0.0	0.0	0.0
Other cardiovascular disease	20.7	30.9	0.0	100.0	0.0	10.6	0.0	0.0	20.4	100.0	100.0	0.0
Depression	9.1	0.0	100.0	0.0	0.0	0.0	0.0	0.0	0.0	0.0	0.0	0.0
Diabetes mellitus	17.6	40.7	0.0	96.5	0.0	0.0	0.0	0.0	0.0	100.0	100.0	0.0
Emphysema/asthma/COPD	10.7	0.0	0.0	0.0	0.0	0.0	0.0	90.9	0.0	0.0	0.0	0.0
Explicit terminal prognosis	0.8	0.0	0.0	0.0	0.0	0.0	0.0	24.8	0.0	0.0	0.0	0.0
Hypertension	26.9	86.8	0.0	100.0	0.0	0.0	0.0	0.0	41.4	100.0	100.0	0.0
Hypothyrodism	7.1	0.0	0.0	57.9	0.0	4.1	0.0	0.0	0.0	100.0	100.0	0.0
Osteoporosis	8.5	0.0	0.0	0.0	0.0	0.0	0.0	0.0	100.0	0.0	0.0	0.0
Parkinson's disease	6.4	9.0	0.0	0.0	9.2	9.4	1.2	0.0	5.0	14.7	7.2	7.2
Pneumonia	1.8	0.0	0.0	0.0	0.0	0.0	0.0	44.5	0.0	0.0	0.0	0.0

Septicemia	0.3	2.1	0.0	0.0	0.0	0.0	0.0	0.0	0.0	4.6	0.0	0.0
Urinary tract infection in last 30 days	7.3	19.3	0.0	0.0	0.0	2.2	0.0	7.5	43.9	100.0	57.7	0.0
Problems/Signs/Symptoms												
Intake/oral status												
Chewing problem	23.7	100.0	29.9	0.0	0.0	42.5	28.9	0.0	24.5	51.3	0.0	0.0
Dehydrated	3.4	0.0	0.0	0.0	0.0	6.0	0.0	6.1	82.5	0.0	0.0	0.0
Lung aspirations	1.0	0.0	0.0	0.0	0.0	0.0	0.0	29.9	0.0	0.0	0.0	0.0
Broken, loose, careous or missing teeth	19.0	100.0	22.8	21.7	10.9	23.8	24.4	21.3	97.6	0.0	0.0	0.0
Oral abscesses, reddened or bleeding gums	1.6	18.9	0.0	0.0	0.0	0.0	3.4	7.9	0.0	0.0	0.0	0.0
Dentures (or partial plates) fit improperly or are not used	43.9	0.0	5.8	100.0	0.0	0.0	0.0	0.0	100.0	100.0	100.0	0.0
Regular complaint of hunger	3.1	2.0	19.9	12.4	0.0	0.0	6.7	0.0	0.0	0.0	0.0	0.0
Regularly leaves 25% + food uneaten	20.6	0.0	92.1	0.0	18.9	12.2	21.6	0.0	100.0	49.2	0.0	0.0
Regularly complains about taste of many foods	6.9	0.0	100.0	0.0	0.0	0.0	0.0	0.0	0.0	0.0	0.0	0.0
Mouth pain associated with eating	1.1	7.3	5.8	0.0	0.0	0.0	0.0	0.0	4.3	0.0	3.1	0.0
Swallowing problem	11.9	100.0	0.0	0.0	0.0	0.0	0.0	7.1	0.0	0.0	0.0	0.0
Weight loss	10.5	4.7	17.7	0.0	0.0	8.0	11.3	0.0	100.0	41.2	24.8	0.0
Body control												
Fell in past month	9.1	0.0	0.0	0.0	0.0	0.0	30.1	0.0	100.0	0.0	47.0	0.0
Fell in past 2–6 months	16.6	0.0	6.3	24.4	0.0	0.0	48.6	0.0	100.0	0.0	100.0	0.0
Amputation	2.9	7.3	2.8	0.0	0.0	0.0	0.0	0.0	0.0	43.7	0.0	2.5
Hemiplegia	8.5	100.0	0.0	0.0	0.0	0.0	0.0	0.0	0.0	0.0	0.0	0.0
Quadriplegia	0.4	3.0	0.0	0.0	0.0	0.0	0.0	3.0	0.0	4.3	0.0	0.0
Partial or total loss of voluntary leg movement	28.6	100.0	0.0	0.0	0.0	0.0	0.0	0.0	0.0	0.0	0.0	0.0
Partial or total loss of voluntary arm movement	19.2	100.0	0.0	0.0	0.0	0.0	0.0	0.0	0.0	0.0	0.0	0.0

Table 6.2. (*Continued*)

	Frequency	1 Major Stroke	2 Depressed	3 CVD Diabetes Dementia	4 Early Dementia	5 Late Stage Dementias	6 Alzheimer's Disease	7 Comatose and Pulmonary	8 Multiple Chronic Disease & Hip Fracture	9 CVD	10 Post-Acute CVD	11 Rehab.
Unsteady gait	39.8	0.0	100.0	0.0	49.6	9.9	89.6	0.0	100.0	0.0	100.0	28.4
Bedfast all or most of the time	17.1	100.0	0.0	0.0	0.0	46.0	0.0	86.1	0.0	71.4	4.2	0.0
Other problems												
Cuts, bruises	8.6	0.0	0.0	0.0	0.0	0.0	1.7	13.3	100.0	0.0	0.0	0.0
Burns	0.0	0.0	0.0	0.0	0.0	0.0	0.0	0.0	0.0	1.0	0.0	0.0
Allergies	22.3	9.6	29.1	65.0	0.0	11.8	0.0	1.4	100.0	100.0	84.1	3.3
Chest pain	2.3	0.0	7.7	0.0	0.0	0.0	0.0	0.0	28.0	0.0	13.0	1.2
Constipation	29.4	38.5	43.6	40.4	0.0	35.2	13.9	3.0	100.0	100.0	100.0	0.0
Diarrhea	3.2	0.0	4.9	5.2	0.0	0.0	0.0	2.5	35.9	18.8	0.0	0.0
Fever	2.0	0.0	0.0	0.0	0.0	0.0	0.0	50.8	0.0	0.0	0.0	0.0
Hallucinations/delusions	5.2	0.0	0.0	0.0	0.0	0.0	48.6	0.0	0.0	0.0	0.0	0.0
Fracture of femur in last 6 months	2.4	0.0	0.0	0.0	0.0	0.0	0.0	0.0	45.6	0.0	16.2	0.0
Other fractures in last 6 months	3.5	0.0	0.0	0.0	0.0	0.0	0.0	5.2	77.6	0.0	0.0	0.0
Internal bleeding	0.8	5.0	0.0	0.0	0.0	0.0	0.0	1.3	11.8	0.0	0.0	0.0
Joint pain	16.5	0.0	0.0	0.0	0.0	0.0	0.0	0.0	100.0	0.0	0.0	0.0
Resident complains of pain daily or almost daily	16.7	0.0	90.0	0.0	0.0	0.0	0.0	0.0	100.0	0.0	0.0	0.0
Respiratory infection	3.5	0.0	0.0	0.0	0.0	0.0	0.0	67.9	0.0	0.0	0.0	0.0
Seizures	1.8	28.2	0.0	0.0	0.0	0.0	0.0	4.6	0.0	0.0	0.0	0.0
Shortness of breath	8.6	0.0	0.0	0.0	0.0	0.0	0.0	88.1	0.0	0.0	0.0	0.0
Vomiting	2.0	0.0	2.1	0.0	0.0	1.7	0.0	11.3	0.0	17.1	18.4	0.0
Diagnoses/Condition/Body Control												
Stability of conditions during prior 7 days												
Conditions/diseases make residents cognitive, ADL, or behavior status unstable-fluctuating, precarious, or deteriorating	18.0	21.9	84.9	0.0	0.0	20.9	89.3	34.1	75.7	0.0	0.0	0.0
Resident's status is stable, but conditions/diseases demand monitoring due to need to detect sudden change	57.9	78.2	0.0	100.0	12.8	52.6	0.0	0.0	0.0	100.0	100.0	0.0

Resident experiencing an acute episode or a flare-up of a chronic condition	2.9	0.0	15.1	0.0	0.0	0.0	0.0	26.4	24.3	0.0	0.0	0.0
None of the above	20.5	0.0	0.0	0.0	87.2	26.6	10.7	11.6	0.0	0.0	0.0	100.0
Edema												
General	3.4	31.3	0.0	0.0	0.0	0.0	0.0	30.3	13.4	0.0	0.0	0.0
Localized, not pitting	15.0	0.0	0.0	68.1	0.0	0.0	0.0	0.0	0.0	100.0	93.4	0.0
Pitting	8.0	0.0	0.0	0.0	0.0	0.0	0.0	0.0	86.6	0.0	0.0	0.0
None	73.1	68.7	100.0	32.0	100.0	100.0	100.0	61.0	0.0	0.0	0.0	100.0
Contractures												
Face/neck	2.1	4.6	0.0	0.0	0.0	8.2	0.0	20.3	3.2	0.0	0.0	0.0
Shoulder/elbow	8.9	49.4	0.0	0.0	0.0	12.6	0.0	22.0	0.0	0.0	0.0	0.0
Hands/wrist	6.0	42.3	0.0	0.0	0.0	0.0	0.0	0.0	0.0	0.0	0.0	0.0
Hips/knee	6.3	0.0	0.0	0.0	0.0	17.2	0.0	0.8	59.9	87.8	0.0	0.0
Feet/angle	1.3	3.7	0.0	0.0	0.0	0.0	0.0	2.3	10.6	12.3	0.0	0.0
None	75.5	0.0	100.0	100.0	100.0	62.1	100.0	54.6	26.3	0.0	100.0	100.0
Skin Condition												
Pressure scores (number of sites)												
Grade 1: Persistent area of skin redness												
None	91.6	79.9	100.0	100.0	100.0	96.7	100.0	70.1	7.7	33.0	74.4	100.0
1	6.8	20.1	0.0	0.0	0.0	3.3	0.0	9.6	60.5	67.0	25.6	0.0
2+	1.6	0.0	0.0	0.0	0.0	0.0	0.0	20.3	31.8	0.0	0.0	0.0
Grade 2: Partial thickness of loss of skin layers												
None	93.6	93.6	100.0	100.0	100.0	90.1	100.0	85.4	21.3	64.1	100.0	100.0
1	4.7	1.1	0.0	0.0	0.0	6.8	0.0	7.5	53.0	35.9	0.0	0.0
2+	1.7	5.4	0.0	0.0	0.0	3.2	0.0	7.1	25.7	0.0	0.0	0.0
Grade 3: Full thickness of loss of skin layers												
None	97.8	100.0	100.0	100.0	100.0	96.4	100.0	92.9	87.3	76.6	100.0	100.0
1	1.6	0.0	0.0	0.0	0.0	3.3	0.0	4.6	8.0	15.9	0.0	0.0
2+	0.6	0.0	0.0	0.0	0.0	0.4	0.0	2.5	4.7	7.5	0.0	0.0
Grade 4: Full thickness of skin and subcutaneous tissue loss exposing muscle/bone												
None	99.0	89.4	100.0	100.0	100.0	98.7	100.0	97.3	96.5	100.0	100.0	100.0
1	0.6	10.6	0.0	0.0	0.0	0.0	0.0	0.7	0.0	0.0	0.0	0.0
2+	0.4	0.0	0.0	0.0	0.0	1.3	0.0	2.0	3.5	0.0	0.0	0.0
Stasis ulcers	2.9	0.0	0.0	0.0	0.0	0.0	0.0	0.0	94.1	0.0	0.0	0.0
Other wounds/lesions	13.8	49.0	0.0	0.0	0.0	6.8	16.8	10.0	100.0	0.0	0.0	0.0

Table 6.2. (*Continued*)

	Frequency	1 Major Stroke	2 Depressed	3 CVD Diabetes Dementia	4 Early Dementia	5 Late Stage Dementias	6 Alzheimer's Disease	7 Comatose and Pulmonary	8 Multiple Chronic Disease & Hip Fracture	9 CVD	10 Post-Acute CVD	11 Rehab.
Cognitive Patterns												
Comatose	2.1	0.0	0.0	0.0	0.0	0.0	0.0	90.7	0.0	0.0	0.0	0.0
Cognitive skills for daily decision making												
Independent decisions consistent/reasonable	20.0	0.0	0.0	0.0	0.0	0.0	0.0	0.0	0.0	0.0	0.0	59.6
Modified independent—difficulty in new situations only	18.1	0.0	68.9	100.0	0.0	0.0	0.0	0.0	0.0	0.0	0.0	40.4
Moderately impaired—decisions poor; cues/supervision required	26.6	0.0	31.1	0.0	100.0	0.0	0.0	0.0	0.0	0.0	0.0	0.0
Severely impaired—never/rarely made decisions	34.3	0.0	0.0	0.0	0.0	100.0	100.0	0.0	0.0	0.0	0.0	0.0
Short term memory OK	44.7	0.0	100.0	0.0	0.0	0.0	0.0	0.0	0.0	0.0	0.0	100.0
Long term memory OK	51.3	0.0	100.0	0.0	18.0	0.0	0.0	0.0	0.0	0.0	0.0	100.0
Recall ability—able to recall events[1)]												
No recall	31.8	0.0	0.0	0.0	0.0	100.0	72.2	100.0	0.0	0.0	0.0	0.0
Can recall one event	9.9	0.0	0.0	0.0	27.0	0.0	27.8	0.0	0.0	0.0	0.0	0.0
Can recall two events	10.6	0.0	0.0	0.0	40.3	0.0	0.0	0.0	0.0	0.0	0.0	0.0
Can recall three events	10.0	0.0	16.8	0.0	32.7	0.0	0.0	0.0	0.0	0.0	0.0	0.0
Can recall four events	37.8	0.0	83.2	0.0	0.0	0.0	0.0	0.0	0.0	0.0	0.0	100.0
Resident's memory problem requires that some or all ADL activities be broken into a series of subtasks so residents can perform them	23.5	0.0	0.0	0.0	92.8	0.0	56.0	0.0	0.0	0.0	0.0	0.0
Periodic discordant thinking/awareness												
Recent onset of fluctuating disturbances of consciousness	2.2	0.0	1.2	0.0	0.0	0.0	7.7	0.0	63.5	27.9	0.0	0.0

Changing awareness of environment	5.6	7.5	0.0	0.0	12.1	4.6	19.7	0.0	0.0	0.0	0.0	0.0
Episodes of incoherent speech	7.1	56.8	0.0	0.0	0.0	5.4	42.3	0.0	0.0	0.0	0.0	0.0
Cognitive ability varies over course of day	13.2	0.0	18.2	0.0	37.1	0.0	27.9	0.0	0.0	72.1	87.6	0.0
Appears distracted or less alert than usual	1.4	15.3	0.2	0.0	0.0	0.0	2.4	8.2	36.5	0.0	0.0	0.0
None of the above	68.1	20.4	80.5	100.0	50.8	90.0	0.0	0.0	0.0	0.0	12.4	100.0
Communication/Hearing Patterns												
Resident has sound production/articulation difficulties that make speech unclear	25.3	100.0	0.0	50.7	11.3	37.9	25.2	0.0	0.0	0.0	0.0	0.0
Hearing												
Adequate	52.9	95.2	84.6	0.0	37.9	32.4	98.9	0.0	0.0	0.0	0.0	100.0
Minimal loss—difficulty when not in quiet listening conditions	30.1	0.0	0.0	76.9	38.3	35.2	0.0	0.0	54.4	100.0	94.6	0.0
Special situation only—speaker had to adjust tone quality and speak distinctly	10.0	4.8	13.3	14.6	17.7	19.9	0.0	0.0	25.5	0.0	5.4	0.0
Highly impaired/deaf	5.4	0.0	2.1	8.5	6.1	12.5	1.1	0.0	20.1	0.0	0.0	0.0
Making self understood	51.2	0.0	100.0	100.0	17.8	0.0	0.0	0.0	0.0	0.0	0.0	100.0
Usually understood	21.3	63.3	0.0	0.0	82.2	0.0	0.0	0.0	0.0	0.0	0.0	0.0
Sometimes understood	14.3	36.7	0.0	0.0	0.0	34.5	100.0	0.0	0.0	0.0	0.0	0.0
Rarely or never understood	12.1	0.0	0.0	0.0	0.0	65.5	0.0	0.0	0.0	0.0	0.0	0.0
Modes of expression resident uses to make needs known												
Speech	86.7	0.0	100.0	100.0	100.0	37.6	100.0	0.0	99.2	100.0	100.0	100.0
Writing	0.3	14.8	0.0	0.0	0.0	0.0	0.0	0.0	0.0	0.0	0.0	0.0
Signs/sounds	4.2	84.2	0.0	0.0	0.0	16.6	0.0	0.0	0.0	0.0	0.0	0.0
Speaks only in language nondominant at facility	0.3	0.0	0.0	0.0	0.0	1.4	0.0	0.0	0.8	0.0	0.0	0.0

Table 6.2. (*Continued*)

	Frequency	1 Major Stroke	2 Depressed	3 CVD Diabetes Dementia	4 Early Dementia	5 Late Stage Dementias	6 Alzheimer's Disease	7 Comatose and Pulmonary	8 Multiple Chronic Disease & Hip Fracture	9 CVD	10 Post-Acute CVD	11 Rehab.
Communication board	0.0	1.1	0.0	0.0	0.0	0.0	0.0	0.0	0.0	0.0	0.0	0.0
None of the above	6.7	0.0	0.0	0.0	0.0	44.4	0.0	0.0	0.0	0.0	0.0	0.0
Ability to understand others												
Understands	43.1	0.0	100.0	0.0	0.0	0.0	0.0	0.0	0.0	0.0	0.0	100.0
Usually understands	26.6	0.0	0.0	0.0	100.0	0.0	0.0	0.0	0.0	0.0	0.0	0.0
Sometimes understands	20.0	0.0	0.0	0.0	0.0	52.9	100.0	0.0	0.0	0.0	0.0	0.0
Rarely or never understands	9.1	0.0	0.0	0.0	0.0	47.1	0.0	0.0	0.0	0.0	0.0	0.0
Visual Patterns												
Perceptual difficulties	10.7	74.4	0.0	0.0	0.0	11.6	30.4	0.0	86.2	0.0	0.0	0.0
Vision												
Adequate	38.3	0.0	95.7	31.6	0.0	0.0	83.6	0.0	0.0	0.0	0.0	100.0
Impaired	35.9	0.0	0.0	68.4	80.6	23.8	9.4	0.0	0.0	100.0	97.6	0.0
Highly impaired	14.6	100.0	0.0	0.0	16.7	26.0	7.0	0.0	100.0	0.0	0.0	0.0
No vision	2.0	0.0	4.4	0.0	2.7	7.9	0.0	0.0	0.0	0.0	2.4	0.0
Unknown	9.2	0.0	0.0	0.0	0.0	42.4	0.0	100.0	0.0	0.0	0.0	0.0
Mood and Behavior Patterns												
Sad or anxious mood												
Verbal expression of distress	19.0	0.0	91.5	0.0	0.0	0.0	0.0	0.0	0.0	0.0	0.0	0.0
Tearfulness, sighing	7.4	96.8	0.0	0.0	0.0	0.0	29.6	0.0	48.3	0.0	0.0	0.0
Motor agitation	6.1	0.0	0.0	0.0	0.0	0.0	55.4	0.0	0.0	0.0	0.0	0.0
Withdrawal	5.3	3.2	2.5	0.0	16.9	6.1	0.0	0.0	37.0	0.0	0.0	0.0
Early morning awakening with unpleasant mood	2.0	0.0	3.4	0.0	0.0	0.0	12.0	0.0	0.0	0.0	0.0	0.0
Recent thoughts of death	0.4	0.0	0.0	77.5	0.0	0.0	0.0	3.2	0.0	0.0	100.0	0.0

Suicidal thoughts/ actions	0.0	0.0	0.0	22.5	0.0	0.0	0.0	0.0	0.0	0.0	0.0	0.0
Persistent sad or anxious mood	1.7	0.0	2.6	0.0	2.0	0.0	3.0	0.0	14.7	0.0	0.0	0.0
Wandering												
No occurrence in prior 7 days	87.8	100.0	100.0	0.0	100.0	100.0	0.0	0.0	100.0	100.0	0.0	100.0
Occurred less than daily	3.3	0.0	0.0	0.0	0.0	0.0	31.3	0.0	0.0	0.0	0.0	0.0
Occurred daily	7.3	0.0	0.0	0.0	0.0	0.0	68.7	0.0	0.0	0.0	0.0	0.0
Verbally abusive												
No occurrence in prior 7 days	81.1	100.0	11.7	100.0	100.0	100.0	0.0	0.0	100.0	0.0	100.0	100.0
Occurred less than daily	9.9	0.0	66.9	0.0	0.0	0.0	50.8	0.0	0.0	0.0	0.0	0.0
Occurred daily	7.4	0.0	21.5	0.0	0.0	0.0	49.2	0.0	0.0	0.0	0.0	0.0
Physically abusive												
No occurrence in prior 7 days	87.1	100.0	100.0	100.0	100.0	100.0	0.0	0.0	100.0	0.0	100.0	100.0
Occurred less than daily	7.1	0.0	0.0	0.0	0.0	0.0	60.2	0.0	0.0	0.0	0.0	0.0
Occurred daily	4.7	0.0	0.0	0.0	0.0	0.0	39.9	0.0	0.0	0.0	0.0	0.0
Socially inappropriate behavior												
No occurrence in prior 7 days	82.6	100.0	100.0	100.0	100.0	100.0	0.0	0.0	100.0	0.0	100.0	100.0
Occurred less than daily	7.2	0.0	0.0	0.0	0.0	0.0	45.8	0.0	0.0	0.0	0.0	0.0
Occurred daily	8.5	0.0	0.0	0.0	0.0	0.0	54.2	0.0	0.0	0.0	0.0	0.0
Range of QI												
>160	6.8	0.0	8.7	0.0	10.4	17.1	0.0	12.7	21.4	5.7	0.0	3.4
160–200	23.3	13.8	70.2	0.0	45.0	38.2	29.4	36.6	64.4	0.0	0.0	16.8
201–240	31.1	15.4	0.0	60.0	0.0	27.6	26.9	1.8	0.0	77.7	73.0	0.0
241–270	16.6	31.2	15.5	9.1	30.2	7.3	32.7	15.2	14.2	0.0	0.0	36.2
271–299	9.0	0.0	0.0	23.8	0.0	4.3	0.0	0.0	0.0	12.3	23.4	8.7
300+	10.9	39.7	5.6	7.1	14.4	3.3	6.2	5.1	0.0	0.0	0.0	34.9

Table 6.2. (*Continued*)

	Frequency	1 Major Stroke	2 Depressed	3 CVD Diabetes Dementia	4 Early Dementia	5 Late Stage Dementias	6 Alzheimer's Disease	7 Comatose and Pulmonary	8 Multiple Chronic Disease & Hip Fracture	9 CVD	10 Post-Acute CVD	11 Rehab.
						Activities of Daily Living						
Bed mobility												
Independent	49.7	0.0	100.0	100.0	100.0	0.0	100.0	0.0	0.0	0.0	23.5	100.0
Supervision	3.5	0.0	0.0	0.0	0.0	0.0	0.0	0.0	0.0	0.0	19.5	0.0
Limited assistance	6.7	0.0	0.0	0.0	0.0	0.0	0.0	0.0	0.0	0.0	37.0	0.0
Extensive assistance	8.1	0.0	0.0	0.0	0.0	0.0	0.0	0.0	0.0	22.4	20.1	0.0
Total dependence	31.9	0.0	0.0	0.0	0.0	100.0	0.0	90.1	0.0	77.6	0.0	0.0
Transfer												
Independent	32.4	0.0	0.0	100.0	0.0	0.0	0.0	0.0	0.0	0.0	0.0	0.0
Supervision	4.5	0.0	0.0	0.0	0.0	0.0	0.0	0.0	0.0	0.0	18.8	0.0
Limited assistance	7.9	0.0	0.0	0.0	0.0	0.0	0.0	0.0	0.0	0.0	33.4	0.0
Extensive assistance	11.4	0.0	0.0	0.0	0.0	0.0	0.0	0.0	0.0	0.0	47.8	0.0
Total dependence	43.8	0.0	0.0	0.0	0.0	100.0	0.0	0.0	0.0	100.0	0.0	0.0
Locomotion												
Independent	41.7	0.0	0.0	100.0	0.0	0.0	0.0	0.0	0.0	0.0	0.0	0.0
Supervision	5.2	0.0	0.0	0.0	0.0	0.0	0.0	0.0	0.0	0.0	28.4	0.0
Limited assistance	7.4	0.0	0.0	0.0	0.0	0.0	0.0	0.0	0.0	0.0	40.5	0.0
Extensive assistance	7.0	0.0	0.0	0.0	0.0	0.0	100.0	0.0	0.0	0.0	31.1	0.0
Total dependence	38.6	0.0	0.0	0.0	0.0	100.0	0.0	68.4	0.0	100.0	0.0	0.0
Grooming												
Independent	22.3	0.0	0.0	76.9	0.0	0.0	0.0	0.0	0.0	0.0	0.0	0.0
Supervision	9.1	0.0	0.0	23.1	0.0	0.0	0.0	0.0	0.0	0.0	10.5	0.0
Limited assistance	11.2	0.0	0.0	0.0	0.0	0.0	0.0	0.0	0.0	0.0	48.9	0.0
Extensive assistance	9.3	0.0	0.0	0.0	0.0	0.0	0.0	0.0	0.0	0.0	40.6	0.0
Total dependence	48.1	0.0	0.0	0.0	0.0	100.0	0.0	0.0	0.0	100.0	0.0	0.0
Bathing												
Independent	9.4	0.0	0.0	36.0	0.0	0.0	0.0	0.0	0.0	0.0	0.0	0.0
Supervision	10.4	0.0	0.0	39.7	0.0	0.0	0.0	0.0	0.0	0.0	0.0	0.0
Limited assistance	13.3	0.0	0.0	24.3	0.0	0.0	0.0	0.0	0.0	0.0	39.8	0.0

Extensive assistance	10.5	0.0	0.0	0.0	0.0	0.0	0.0	0.0	0.0	0.0	60.2	0.0
Total dependence	56.5	0.0	0.0	0.0	0.0	100.0	100.0	0.0	0.0	100.0	0.0	0.0
Dressing												
Independent	18.4	0.0	0.0	73.2	0.0	0.0	0.0	0.0	0.0	0.0	0.0	0.0
Supervision	6.7	0.0	0.0	26.8	0.0	0.0	0.0	0.0	0.0	0.0	0.0	0.0
Limited assistance	10.7	0.0	0.0	0.0	0.0	0.0	0.0	0.0	0.0	0.0	49.9	0.0
Extensive assistance	10.8	0.0	0.0	0.0	0.0	0.0	0.0	0.0	0.0	0.0	50.2	0.0
Total dependence	53.4	0.0	0.0	0.0	0.0	100.0	0.0	0.0	0.0	100.0	0.0	0.0
Eating												
Independent	55.2	0.0	100.0	99.9	100.0	0.0	0.0	0.0	0.0	0.0	47.5	100.0
Supervision	8.2	0.0	0.0	0.0	0.0	0.0	36.0	0.0	0.0	13.9	39.9	0.0
Limited assistance	5.2	0.0	0.0	0.0	0.0	0.0	26.1	0.0	0.0	19.4	12.4	0.0
Extensive assistance	5.8	0.0	0.0	0.0	0.0	0.0	37.9	0.0	0.0	30.9	0.1	0.0
Total dependence	25.4	0.0	0.0	0.0	0.0	100.0	0.0	97.3	0.0	35.9	0.0	0.0
Toileting												
Independent	27.7	0.0	0.0	93.3	0.0	0.0	0.0	0.0	0.0	0.0	0.0	0.0
Supervision	4.1	0.0	0.0	6.7	0.0	0.0	0.0	0.0	0.0	0.0	10.8	0.0
Limited assistance	7.5	0.0	0.0	0.0	0.0	0.0	0.0	0.0	0.0	0.0	38.9	0.0
Extensive assistance	9.7	0.0	0.0	0.0	0.0	0.0	0.0	0.0	0.0	0.0	50.3	0.0
Total dependence	50.9	0.0	0.0	0.0	0.0	100.0	0.0	0.0	0.0	100.0	0.0	0.0

[1]Events resident can recall are current season, location of own room, staff names/faces, that he/she is in nursing home.

Source: Nursing Home Case-Mix and Quality Demonstration.

major stroke, there are other classes of residents where limited strokes are important comorbidities. Since the GoM solution allows for partial membership in multiple classes for a person, stroke can be expressed for a person in two ways. First, there is "classic" stroke manifestation in Class 1. However, there are many persons where the late effects of stroke, or limited strokes, are "comorbid" conditions resulting in residual or minimal functional impairment. In this case, stroke may be represented by only partial membership in Class 1, i.e., not all the medical and functional elements of a classic, major stroke are present in the person. Thus, with partial membership in Class 1, some stroke attributes might be combined with persons with partial membership in a "dementia" or "CVD" class.

The model could have distributed stroke in multiple classes rather than concentrate its presence in a single "archtypical" fuzzy class. The reason why it did not distribute stroke was that the joint frequencies of the conditions typical of a major stroke were larger than the joint frequency of stroke with other more diversely constituted symptom or condition patterns. That is, while stroke is certainly likely to occur for other classes, its rate of occurrence with those classes was not large enough to diffuse some of the stroke "prevalence" from Class 1 (in the Poisson or empirical Bayesian form of the model). This suggests that stroke was a comorbid state of less significance clinically in determining a patients characteristics than when it occurred as a major cerebrovascular event.

In the other forms of GoM and the LCM analysis described below we saw this diffusion of stroke over multiple classes. In the "simple" GoM case (i.e., $\Sigma_{l=1}^{L_j} \lambda_{kjl} = 1$ for all variables and classes) stroke is found in four classes. In the latent class model, stroke was found to be present in *all* of the discrete classes. Thus, by allowing the heterogeneity in the complex GoM model one is able to isolate more distinct and coherent patterns of symptoms.

2. *Depressed*. This class experiences a number of problems with social relations, attitudes, and anxiety and is 100% likely to be diagnosed as depressed. It is a class of residents where a number of physical symptoms are reported (e.g., pain and chest pain) and is subject to anxiety disorders (55.7%). The report of physical symptoms suggests somatization-related depressive disorders (Woodbury and Manton, 1989).

These emotional and social relation problems are associated with eating and behavioral problems and weight loss. These residents have either an unstable, precarious, or deteriorating condition. This class uses speech for communication, is anxious, and sometimes verbally abusive. This class of resident is independent in ADLs.

3. *CVD, Diabetes, and Dementia*. In addition to these three conditions this class is socially well integrated, has multiple circulatory diseases, diabetes, hypertension (risk factors for CVD), and hypothyroidism. These persons are stable but require monitoring. They have edema and hearing problems but are able to communicate and participate socially. They have

recently thought of death (with some suicidal ideation) but are generally intact in ADL functioning. They require supervision in bathing. A third of the class need supervision in dressing. This group appears to function well cognitively—except in new situations.

4. *Early and Midstage Dementia.* This class has a high likelihood of moderate dementia (non-Alzheimer's) with moderately impaired cognitive skills. On the short-term memory test, no one remembered all four items; 31% recalled three items; 69% recalled at least one item. Their gait is unsteady; over half need wheelchairs. Few other medical problems were identified. Their condition is stable with no edema or contractures. This person communicates by speech and generally understands (and is understood) by others. This class occasionally manifests need for physical help in bathing, grooming, and toileting though cognitive deficits require supervision of most ADL.

5. *Late Stage Dementia.* This is an unstable, deteriorating, debilitated, totally dependent dementia class with some "Alzheimer's" disease and "other" dementia. It has severe cognitive and memory impairments with *no* recall of any items on the short-term memory test. It is frequently in an unstable state and often has contractures. Communication is often impaired. Body weight is low, and the person is highly impaired in ADL function.

6. *Alzheimer's Disease.* This class has a diagnosis of Alzheimer's disease (100%) (with 52.1% associated with "other" dementia) with severe cognitive impairment and poor memory status. Most residents recalled none of the short-term memory items, 26% recalled one. This class has a 48.6% chance of hallucinations and delusions. This distinguishes it from all other classes. It is also distinguished from other dementia-related classes by having a deteriorating condition. This class does not have edema or contractures but has disturbed cognitive patterns. Such residents are sad and frequently express motor agitation (the only class who did so). This class has a high prevalence of verbally, physically, and socially inappropriate behavior—again unique to the class. Body weight is low. The class is mobile but needs assistance in locomotion, toileting, grooming, bathing, and eating.

7. *Comatose and Pulmonary.* This class of comatose (91%) residents has some dementia but is best characterized by pulmonary problems, both chronic (e.g., emphysema, 90.9%) and acute (pneumonia, 44.5%). It has the highest likelihood of an explicit terminal diagnosis (24.8%), fever (50.8%), respiratory infection (67.9%), shortness of breath (88.1%), and lung aspiration (29.9%). In addition, 14.4% have an Alzheimer's diagnosis and 41% are diagnosed as demented.

The condition of these residents is unstable. This class may have general edema (30.3%) and contractures and pressure sores. It is totally dependent in ADL functioning. The average body weight is low, and, because of the impaired physical state and the comatose designation, a number of communication and cognitive skills were not assessed.

8. *Multiple Chronic Diseases and Osteoporosis*. This class is characterized by multiple chronic medical problems—CVD, peripheral vascular disease, anemia, arthritis, osteoporosis, cataracts with highly impaired vision, cancer, and hip fracture. These conditions appeared to have caused fractures and hospitalization because these persons had short institutional stays (e.g., 44.9% less than 30 days) and 41% were admitted from acute hospitals. They have had falls in the last six months resulting in fractures. This class has problems with nutrition, is anemic, and has joint pain and arthritis. This class is unstable and has edema and contractures with signs of delirium. These residents are sad and withdrawn. Body weight is low, and the class has dependencies in ADL status. The class does not manifest behavior or memory problems.

9. *CVD*. This group has multiple types of CVD (e.g., congestive heart failure, atherosclerotic heart disease, hypertension), along with CVD risk factors such as diabetes and hypothyroidism. Hypothyroidism, known to cause cardiovascular problems, stroke, and possibly dementia (Osterweil et al., 1992), is prevalent for both this and the CVD, Diabetes, and Dementia class (Class 3). Since the chronic effects of hypothyroidism are controllable through medication (Petersen et al., 1990), the medical problems of this class and Class 3 might have been avoided by earlier intervention. The significance of hypothyroidism as a highly prevalent comorbidity was not been identified in less detailed nursing home studies (e.g., the 1985 National Nursing Home Survey). This class is debilitated with a high incidence of urinary tract infection and moderate frequency of pressure sores. This class is stable but needs monitoring. It has localized edema and contractures of lower extremities. Cognitive ability fluctuates and these residents are dependent in ADL function except for eating where 50% need limited assistance or supervision. The class does not have cognitive or behavior problems but is restrained to prevent falls. This may contribute to the development of pressure sores, constipation, and incontinence. More than two-thirds have been in their current facility at least a year and entered from another nursing home.

10. *Post-Acute Convalescence of CVD*. This class has CVD and diabetes, hypertension, and, again, hypothyroidism. It has gait problems and some incidence of urinary tract infection. It has good social interaction. This class is stable but needs monitoring. Edema is a problem. Such residents think of death and are moderately impaired. This class appears, in contrast to Classes 4 and 7, to have less complicated CVD; i.e., it has had a recent acute hospitalization with fewer comorbidities and little associated dementia where CVD was the focal medical problem.

11. *Rehabilitation*. This class is relatively intact with good social interaction, few recorded medical conditions, few nutritional problems, and good cognitive skills—memory and perception. It is stable, has no edema or contractures, no behavior problems or dependency. It has the greatest likelihood of having legal responsibility for self (44.8%) and to be a

self-payor (43.8%) (Manton et al., 1994). These residents receive rehabilitation services five days a week.

The weighted prevalence of these classes in the study population is of interest—even though the study population in the six states are not representative of the U.S. nursing home population in general. The classes *symptom*-weighted prevalence (i.e., $\Sigma_k\, g_{ik}/I = \bar{g}_k$) are Class 1, 5.8%; Class 2, 7.7%; Class 3, 7.1%; Class 4, 16.0%; Class 5, 16.4%; Class 6, 9.0%; Class 7, 2.7%; Class 8, 4.3%; Class 9, 4.5%; Class 10, 3.3%; Class 11, 23.3%. Thus, the most prevalent classes are Class 11, the rehabilitation group, Class 4, the early to midstage dementia group, and Class 5, a late stage cognitively impaired group. The procedure was able to identify classes of residents in terms of health, functioning, and behavior that spanned a wide range of medical conditions.

The 11 classes were generated from an empirical Bayesian/Poisson GoM analysis, In the analysis we examined a series of different K values (i.e., number of classes). $K = 11$ was strongly indicated by the likelihood ratio test. The χ^2 associated with the addition of the 11th class was 9664.2 with 4715 d.f. The Wilson–Hilferty t associated with this test (i.e., $\sqrt{2\chi^2 - 1} - \sqrt{2\,\text{d.f.} - 1}$) was 39.7—a highly significant value. Thus, we needed at least 11 classes to describe the variation of the 111 variables and 407 answers. The addition of a 12th class increased the χ^2 by 3479.8 (with 4715 d.f.). The Wilson–Hilferty t for this addition was *negative*, i.e., -13.7. This is a strong indication that the 12th fuzzy class did *not* significantly contribute to the fit of the data.

In addition to the empirical Bayesian/Poisson GoM model we ran three other types of crisp and fuzzy class models. The crisp set model was the LCM discussed earlier in Chapter 2 which produces discrete set assignment functions (i.e., $g_{ik} = 1$ or 0). A second analysis was the simple GoM model where all $\Sigma_{l=1}^{L_j}\, \lambda_{kjl} = 1$, i.e., there is no differential heterogeneity for any of the J variables relations to the K fuzzy classes. The third analysis used was an extension of the empirical Bayesian GoM model. This was estimated so that, if $\Sigma_{l=1}^{L_j}\, \lambda_{kjl} \rightarrow 0$, a value was derived for the $K-1$ just before 0 was reached.

Testing goodness-of-fit between models, we first used the Wilson–Hilferty t to compare the χ^2 of the LCM (79,074.9 with either 8598 or 4070 degrees of freedom) with that of the simple GoM model (i.e., $\chi^2 = 135{,}361.7$ with $45{,}250 g_{ik} + 4070\lambda_{kjl}$ or 49,320 degrees of freedom). The t for the test where one degree of freedom is subtracted for the classification of each case in LCM (i.e., only one $g_{ik} = 1$; all others are 0 in a crisp set assignment function) was highly significant with $t = 50.1$. A more conservative test, where we did not penalize for the classification of a case in LCM (i.e., we counted only 4070 degrees of freedom for the p_{ijl}) was still highly significant: $t = 21.43$. Thus, even the simple GoM model with no QRFs explained the data better than the LCM with $K = 11$.

We also compared the simple GoM model with empirical Bayesian

Poisson GoM model. The empirical Bayes version has 50,551 degrees of freedom (1221 more due to a $\Sigma_{l=1}^{L_j} \lambda_{kjl}$ being estimated for each class and variable) and produced a χ^2 of 151,066.8. The Wilson–Hilferty t was 127.81 so that the empirical Bayesian model fit the data far better than the simple GoM model.

The GoM model with special adjustments for $\Sigma_{l=1}^{L_j} \lambda_{kjl} \to 0$ had nearly the same χ^2 as the basic empirical Bayesian Poisson model and did not have a substantially different fit.

Thus, the results were clear that the 11 classes were necessary to explain the data and that the empirical Bayes form of the model produced the best description of the data. To examine what this means for the substantive interpretation of parameter estimates we present Table 6.3.

In Table 6.3 we took a selection of 24 medical condition variables from the 111 total variables and compared the four different estimates of the response probabilities (i.e., the p_{kjl} for LCM and λ_{kjl} for the three forms of GoM).

What is interesting, beyond our prior observation that the LCM provided very diffuse estimates of λ_{kjl} for most variables is that, for variables like stroke, the adjustment for heterogeneity was considerable (i.e., 3.76). This is manifest in a greater ability to concentrate partially related conditions in a fuzzy class to describe patient syndromes of a complex nature. That is, in simple GoM, since implicitly all heterogeneity weights (QRFs) = 1.0, stroke is spread out over four fuzzy classes. In contrast, in the empirical Bayesian or Poisson GoM model, stroke is concentrated in Class 1 with a weighted prevalence of 5.93 (i.e., $\bar{g}_k = 0.0593$), which is related to the marginal prevalence of 22.3% by the heterogeneity factor of 3.76. What allows the stroke variable to be represented in such a concentrated fashion in Class 1 in the empirical Bayesian GoM is that a person with less than a major stroke (i.e., not possessing all symptoms associated with a major stroke) could be represented as a convex mixture (using the g_{ik} as weights) of a less symptomatic class with Class 1, or if there is significant comorbidity that a class with, say, CVD could be mixed with Class 1.

Thus, the representation offered by the empirical Bayesian Poisson GoM is very parsimonious in terms of structural (λ_{kjl}) parameters—partly because of the nonlinearity allowed in the simplex. The heterogeneity parameter and the g_{ik} both operate as statistical filters of a nonparametric type to isolate the essential profiles of linked conditions across persons who manifest the conditions to variable degrees.

Another medical condition, cancer, also has interesting behavior in different models. The LCM provides a highly diffuse pattern. The simple GoM produces a λ_{kjl} of 100% for cancer in Class 2 (i.e., $\lambda_{2jl} = 1$). The two forms of the empirical Bayesian GoM show similar patterns with a 100% loading on Class 8, whereas, when the λ_{kjl} were estimated with the question relevance factors (the Empirical Bayes (EB) model) set equal to 0, there was some expression of cancer in several other classes (e.g., Class 10).

The adjustment for the question relevance factors is computationally

Table 6.3. Comparison of λ_{kjl} Estimated from Four Different Models (Latent Class Model, Since GoM, Empirical Bayes GoM Analysis for Zero QFRs) Generating Latent Groups from Distrete Data

Medical Conditions	Frequency	LCM	1 Major Stroke	2 Depressed	3 CVD Diabetes Dementia	4 Early Dementia	5 Late Stage Dementias	6 Alzheimer's Disease	7 Comatose and Pulmonary	8 Multiple Chronic Disease & Hip Fracture	9 CVD	10 Post-Acute CVD	11 Rehab.
Arthritis	30.0	LCM	26.8	36.1	36.4	33.4	21.0	20.2	26.4	30.8	39.2	36.0	23.2
	30.0	S-GoM	4.1	4.2	25.1	45.0	0.0	26.1	22.5	98.4	100.0	13.5	9.2
	30.0	EB-GoM	58.8	0.0	0.0	100.0	0.0	47.8	100.0	100.0	100.0	0.0	0.0
	30.0	EB-GoM+	0.0	0.0	0.0	100.0	0.0	52.5	39.5	100.0	100.0	0.0	26.4
Anemia	11.1	LCM	10.6	12.8	9.7	13.7	7.0	10.6	8.7	14.2	14.9	11.9	8.9
	11.1	S-GoM	0.0	47.4	7.6	3.4	0.9	19.6	5.5	12.9	38.5	7.9	6.0
	11.1	EB-GoM	18.4	0.0	0.0	2.0	0.0	0.4	25.9	100.0	31.9	0.0	0.0
	11.1	EB-GoM+	0.0	0.0	0.0	20.8	2.2	11.9	2.4	17.0	33.6	12.5	24.3
Anxiety disorder	4.6	LCM	4.8	5.8	5.3	15.7	3.1	1.2	4.2	4.5	3.9	6.9	1.1
	4.6	S-GoM	0.0	0.0	0.0	66.9	0.0	0.0	0.0	0.0	0.0	0.0	0.0
	4.6	EB-GoM	0.0	0.0	0.0	65.2	0.0	0.0	0.0	0.0	0.0	0.0	0.0
	4.6	EB-GoM+	0.0	0.0	0.0	60.6	0.0	0.0	0.0	0.0	0.0	0.0	0.0
Alzheimer's disease	13.1	LCM	5.4	0.9	3.5	2.4	24.7	19.2	41.2	12.5	1.3	7.8	30.9
	13.1	S-GoM	0.0	0.0	0.0	0.0	34.0	0.0	62.8	0.0	0.0	0.0	34.8
	13.1	EB-GoM	0.0	0.0	0.0	0.0	56.3	66.4	0.0	0.0	0.0	0.0	53.4
	13.1	EB-GoM+	0.0	0.0	0.0	0.0	0.0	95.0	34.1	0.0	0.0	0.0	17.5
Other dementia	35.7	LCM	35.2	10.9	16.3	31.5	50.3	51.2	49.4	47.4	9.9	40.2	56.7
	35.7	S-GoM	0.0	0.0	10.1	20.1	82.7	51.6	54.0	80.7	0.0	21.4	57.6
	35.7	EB-GoM	0.0	0.0	0.0	0.0	0.0	100.0	100.0	0.0	0.0	0.0	38.8
	35.7	EB-GoM+	0.0	0.0	16.6	0.0	0.0	96.7	100.0	100.0	0.0	0.0	58.5
Aphasia	4.4	LCM	5.5	2.1	1.0	1.7	0.1	20.4	1.9	4.9	3.2	1.9	2.3
	4.4	S-GoM	6.2	0.0	0.0	0.0	0.0	45.9	0.0	0.0	0.0	0.0	0.0
	4.4	EB-GoM	100.0	0.0	0.0	0.0	0.0	0.0	0.0	0.0	0.0	0.0	0.0
	4.4	EB-GoM+	100.0	0.0	0.0	0.0	0.0	0.0	0.0	0.0	0.0	0.0	0.0
Cancer	6.4	LCM	6.2	9.6	6.7	5.6	5.0	4.7	4.5	10.0	8.8	6.7	2.7
	6.4	S-GoM	0.0	100.0	5.7	0.0	0.0	3.7	2.9	0.0	0.0	0.0	0.0
	6.4	EB-GoM	0.0	10.6	11.1	0.0	0.0	0.0	0.0	100.0	0.0	0.0	0.0
	6.4	EB-GoM+	0.0	27.7	11.1	1.3	16.5	0.0	0.0	0.0	4.2	41.6	6.4

Table 6.3. (*Continued*)

Medical Conditions	Frequency	LCM	1 Major Stroke	2 Depressed	3 CVD Diabetes Dementia	4 Early Dementia	5 Late Stage Dementias	6 Alzheimer's Disease	7 Comatose and Pulmonary	8 Multiple Chronic Disease & Hip Fracture	9 CVD	10 Post-Acute CVD	11 Rehab.
Cataracts	6.3	LCM	1.8	9.5	7.0	7.1	5.0	2.6	5.4	9.3	8.5	7.4	6.4
	6.3	S-GoM	0.0	0.0	2.1	5.0	0.0	0.0	0.0	45.8	28.5	0.0	0.0
	6.3	EB-GoM	0.0	0.0	4.4	0.0	0.0	0.0	0.0	100.0	16.0	0.0	0.0
	6.3	EB-GoM+	0.0	0.0	0.0	20.6	0.0	10.7	7.4	16.1	17.1	4.3	3.0
Cerebrovascular accident (stroke)	22.3	LCM	42.8	30.6	10.3	13.4	7.1	44.8	6.3	34.5	23.6	21.8	8.8
	22.3	S-GoM	100.0	0.0	0.0	0.0	0.0	98.3	0.0	0.0	42.3	3.6	0.0
	22.3	EB-GoM	100.0	0.0	0.0	0.0	0.0	0.0	0.0	0.0	0.0	0.0	0.0
	22.3	EB-GoM+	0.0	0.0	0.0	0.0	0.0	0.0	0.0	0.0	0.0	0.0	0.0
Congestive heart failure	24.0	LCM	24.0	26.8	26.3	30.0	13.7	21.8	14.7	25.3	30.3	30.6	18.8
	24.0	S-GoM	0.0	0.0	12.1	41.1	0.0	35.1	0.0	100.0	100.0	6.2	3.5
	24.0	EB-GoM	37.1	0.0	0.0	99.1	0.0	0.0	100.0	100.0	78.8	0.0	0.0
	24.0	EB-GoM+	0.0	0.0	0.0	100.0	0.0	0.0	24.9	98.1	100.0	4.3	36.2
ASHD	21.1	LCM	18.6	21.5	21.4	26.3	16.6	19.9	21.1	23.0	21.4	22.9	20.9
	21.1	S-GoM	0.0	0.0	13.5	35.2	2.5	27.7	11.2	88.7	61.1	9.5	6.7
	21.1	EB-GoM	45.8	0.0	0.0	73.5	0.0	14.1	100.0	100.0	53.1	0.0	0.0
	21.1	EB-GoM+	0.0	0.0	0.0	100.0	0.0	20.4	42.9	100.0	83.1	0.0	28.6
Peripheral vascular disease	8.0	LCM	7.5	13.8	7.3	8.1	4.6	6.9	3.4	12.1	9.9	10.7	4.2
	8.0	S-GoM	20.6	34.1	3.0	14.3	0.0	14.1	0.0	0.0	16.9	11.5	0.0
	8.0	EB-GoM	26.3	13.5	5.7	11.8	0.0	0.0	0.6	82.1	9.1	14.9	0.0
	8.0	EB-GoM+	36.6	7.2	0.0	20.7	0.0	0.0	1.2	11.6	13.9	14.0	0.0
Other cardiovascular disease	20.7	LCM	21.8	19.6	21.6	27.6	17.6	20.0	13.7	22.2	24.2	23.4	15.1
	20.7	S-GoM	5.9	0.0	13.6	33.5	6.4	35.5	7.9	83.5	69.0	0.0	0.0
	20.7	EB-GoM	100.0	0.0	0.0	70.9	0.0	25.3	88.2	53.3	59.1	0.0	0.0
	20.7	EB-GoM+	38.4	0.0	0.0	79.7	0.0	25.8	12.1	100.0	72.8	0.0	24.5
Depression	9.1	LCM	7.2	18.2	6.4	23.0	4.4	4.2	4.7	15.9	10.0	9.7	1.8
	9.1	S-GoM	20.2	0.0	0.0	100.0	0.0	0.0	0.0	0.0	0.0	0.0	0.0

	9.1	EB-GoM	0.0	0.0	0.0	100.0	0.0	0.0	0.0	0.0	0.0	0.0	0.0
	9.1	EB-GoM+	0.0	0.0	0.0	100.0	0.0	0.0	0.0	0.0	0.0	0.0	0.0
Diabetes mellitus	17.6	LCM	19.2	22.1	19.1	20.2	15.5	17.9	8.3	19.8	21.6	18.3	10.1
	17.6	S-GoM	42.8	30.8	15.9	11.0	19.5	20.4	0.0	18.7	36.3	5.7	7.9
	17.6	EB-GoM	39.0	46.1	20.6	17.7	32.9	0.0	0.0	22.6	17.7	25.7	11.1
	17.6	EB-GoM+	41.5	100.0	50.4	4.0	88.7	0.0	0.0	0.0	0.0	76.4	0.0
Emphysema/ asthma/COPD	10.7	LCM	8.5	15.7	11.4	20.4	8.4	4.4	6.1	11.8	15.0	11.9	6.3
	10.7	S-GoM	0.0	0.0	0.0	77.3	0.0	17.8	0.0	0.0	47.2	0.0	0.8
	10.7	EB-GoM	0.0	0.0	10.2	53.2	0.0	0.0	10.6	100.0	12.2	0.0	0.0
	10.7	EB-GoM+	0.0	0.0	0.2	74.6	4.8	4.9	4.9	12.6	23.8	0.0	9.8
Explicit terminal prognosis	0.8	LCM	0.4	1.8	0.2	0.0	0.0	0.6	0.0	3.8	1.1	0.1	0.0
	0.8	S-GoM	0.0	19.7	0.0	0.0	0.0	0.0	0.0	0.0	0.0	0.0	0.0
	0.8	EB-GoM	0.0	0.0	0.0	0.0	0.0	0.0	0.0	31.4	0.0	0.0	0.0
	0.8	EB-GoM+	0.0	2.5	0.0	1.8	0.0	0.0	0.0	0.0	0.0	7.0	1.5
Hypertension	26.9	LCM	27.8	34.6	33.1	30.6	24.6	22.6	16.3	27.0	31.6	30.1	14.6
	26.9	S-GoM	60.6	15.5	30.9	25.9	20.7	33.6	14.6	31.2	53.5	16.0	4.0
	26.9	EB-GoM	100.0	0.0	12.1	57.8	5.3	22.6	45.2	0.0	71.0	0.0	0.0
	26.9	EB-GoM+	100.0	0.0	7.0	54.7	0.0	23.8	21.7	90.3	74.6	0.0	0.0
Hypothyroidism	7.1	LCM	5.2	8.4	6.7	9.7	8.2	6.3	9.9	5.4	7.9	7.2	4.4
	7.1	S-GoM	0.0	76.1	5.5	0.0	20.9	13.1	0.0	0.0	0.0	0.0	0.0
	7.1	EB-GoM	1.2	6.2	2.8	10.1	16.8	2.8	0.0	0.0	14.4	17.4	8.4
	7.1	EB-GoM+	0.0	0.0	5.3	8.6	3.3	2.7	1.0	21.3	11.7	19.2	11.4
Osteoporosis	8.5	LCM	7.8	12.6	8.4	11.2	5.2	6.2	7.7	7.2	10.8	10.8	6.5
	8.5	S-GoM	0.0	15.8	4.8	12.1	0.0	0.0	0.0	56.1	26.5	0.0	0.0
	8.5	EB-GoM	0.0	0.0	0.0	31.3	0.0	0.0	25.7	37.0	24.6	0.0	0.0
	8.5	EB-GoM+	0.0	0.0	0.0	32.4	0.0	0.0	11.2	28.5	23.9	0.2	7.7
Parkinson's disease	6.4	LCM	8.6	7.9	4.6	5.7	4.5	8.1	3.8	7.3	5.0	6.7	8.5
	6.4	S-GoM	0.0	30.6	5.2	0.0	5.3	7.6	0.0	9.5	3.5	7.3	10.3
	6.4	EB-GoM	30.0	0.0	0.0	0.0	0.0	3.3	30.8	0.0	6.7	0.0	0.0
	6.4	EB-GoM+	22.4	0.0	0.0	3.1	0.0	0.0	30.9	0.0	7.8	0.0	0.0

Table 6.3. (*Continued*)

			1	2	3	4	5	6	7	8	9	10	11
Medical Conditions	Frequency	LCM	Major Stroke	Depressed	CVD Diabetes Dementia	Early Dementia	Late Stage Dementias	Alzheimer's Disease	Comatose and Pulmonary	Multiple Chronic Disease & Hip Fracture	CVD	Post-Acute CVD	Rehab.
Pneumonia	1.8	LCM	0.7	2.5	1.2	1.1	1.0	5.0	1.3	2.9	0.8	2.0	0.1
	1.8	S-GoM	0.0	18.8	0.0	0.0	0.0	12.0	0.0	0.0	0.0	0.0	0.0
	1.8	EB-GoM	7.5	0.0	0.0	0.0	0.0	0.0	0.0	52.4	0.0	0.0	0.0
	1.8	EB-GoM+	7.0	0.0	0.0	7.5	0.0	0.0	0.0	0.0	0.0	0.0	7.6
Septicemia	0.3	LCM	0.5	0.6	0.0	0.0	0.0	1.0	0.0	1.1	0.0	0.1	0.0
	0.3	S-GoM	0.4	4.4	0.0	0.0	0.0	1.5	0.0	0.0	0.0	0.0	0.0
	0.3	EB-GoM	1.6	0.0	0.0	0.0	0.0	0.0	0.0	8.5	0.0	0.0	0.0
	0.3	EB-GoM+	0.0	1.6	0.0	0.0	0.0	0.0	0.0	0.0	0.0	0.2	1.4
Urinary tract infection in last 30 days	7.3	LCM	10.8	9.5	2.7	2.3	2.1	12.6	4.6	13.6	8.1	9.7	4.7
	7.3	S-GoM	0.0	100.0	0.0	0.0	0.0	30.6	0.0	1.3	0.0	0.0	0.0
	7.3	EB-GoM	19.2	0.0	0.0	0.0	0.0	0.0	0.0	100.0	0.0	0.0	0.0
	7.3	EB-GoM+	14.2	21.0	0.0	1.7	7.8	0.0	0.0	4.9	4.0	31.6	17.5

difficult because we are trying to produce the λ_{kjl} near (as we approach) a boundary condition. Thus, the current solution may not be optimal. We are examining several alternative search algorithms, such as the so-called "ameoba" algorithm, to get better estimates of the λ_{kjl} in these cases. In addition to algorithmic changes the exact topological significance of zero QRFs is being explored.

6.4. SUMMARY

The analyses in this chapter illustrated operating characteristics of the fuzzy set (GoM) model with different degrees of complexity. Of perhaps greatest interest in the analysis of the nursing home residents is that we could isolate 11 fuzzy classes from an example with $J = 111$ questions and

$$\sum_{j=1}^{J} L_j = 407 \text{ answers} .$$

This poses an interesting question about the properties of the fuzzy set solution. The reason we were able to generate 11 classes may have been because of the large value of J with increased stability of the g_{ik}. Thus, the analysis may, in effect, be operating under the principle of consistency where as J increases the g_{ik} become consistent. What is needed is to examine a wide variety of solutions to identify when J is large enough for a given K to produce g_{ik} (not their moments) that are consistently estimated. Having the g_{ik} themselves be consistently estimated opens a wide range of ancillary types of analyses that can be used to explore the basic data structure (e.g., as in Chapter 4 for aggregate forms of analysis).

These ancillary analyses are potentially of great significance in health policy and actuarial analyses in generating continuous underwriting factors. That is, an expression for an individual's risk can be gotten from a GoM analyses as a weighted (by the g_{ik}) combination of the K basic profile. Such a continuously scored underwriting factor is important for examining the risk in health care organization of having to provide expensive long-term care services. It could also be important in underwriting private insurance for long-term care services. Currently, the Society of Actuaries is investigating the properties of fuzzy sets used as underwriting factors.

REFERENCES

Buhlmann, H. (1970). *Mathematical Methods in Risk Theory*. Springer-Verlag, New York.

Cox, D. R., and Hinkley, D. V. (1974). *Theoretical Statistics*. Chapman & Hall, London.

Efron, B., and Morris, C. (1972). Limiting the risk of Bayes and empirical Bayes estimators—Part II: The empirical Bayes case. *Journal of the American Statistical Association*, **67**: 130–139.

Efron, B., and Morris, C. (1973). Combining possibly related estimation problems. *Journal of the Royal Statistical Society*, No. 3, 379–402.

Efron, B. and Morris, C. (1975). Data analysis using Stein's problems. *Journal of the Royal Statistical Society*, **70**: 311–319.

Johnson, N. L., and Kotz, S. (1972). *Distributions in Statistics: Continuous Multivariate Distributions*. Wiley, New York.

Manton, K. G., Woodbury, M. A., Stallard, E., Riggan, W. B., Creason, J. P., Pellom, A. (1989). Empirical Bayes procedures for stabilizing maps of U.S. cancer mortality rates. *Journal of the American Statistical Association* **84**: 637–650.

Manton, K. G., Cornelius, E. S., and Woodbury, M. A. (1994). Characteristics of nursing home residents: A multivariate analysis of medical, behavioral, psychosocial and service use factors. In review at *Journal of Gerontology*.

Morris, C. N. (1982). Natural exponential families with quadratic variance function. *The Annals of Statistics* **10**: 65–80.

Morris, C. N. (1983). Parametric empirical Bayes inference: Theory and applications (with discussion). *Journal of the American Statistical Association* **78**: 47–65.

Osterweil, D., Syndulko, K., Cohen, S. N., Pettler-Jennings, P. D., Hershman, J. M., Cummings, J. L., Tourtellotte, W. W., and Solomon, D. H. (1992). Cognitive function in nondemented older adults with hypothyroidism. *Journal of the American Geriatric Society* **40**: 325–335.

Petersen, K., Bengtsson, C., Lapidus, L., Lindstedt, G., Nystrom, E. (1990). Morbidity, mortality and quality of life for patients treated with levothyroxine. *Archives of Internal Medicine* **150**: 2077–2081.

Rencher, A. C., and Krutchkoff, R. G. (1975). Some empirical Bayes estimators allowing for varying error variances. *Biometrika* **62**: 643–650.

Robbins, H. (1955). The empirical Bayes approach to statistics. *Proceedings of the Third Berkeley Symposium on Mathematical Statistics and Probability* **1**: 157–164.

Robbins, H. (1964). The empirical Bayes approach to statistical decision problems. *The Annals of Mathematical Statistics* **35**: 49–68.

Robbins, H. (1980). An empirical Bayes estimation problem. *Proceedings of the National Academy of Sciences, USA* **77**: 6988–6989.

Robbins, H. (1983). Some thoughts on empirical Bayes Estimation. *Annals of Statistics* **11**: 713–723.

Robins, L. N. (1981). National Institute of Mental Health Diagnostic Interview Schedule. *Archives of General Psychiatry* **38**: 381–389.

Woodbury, M. A., and Manton, K. G. (1989). Grade of Membership analysis of depression-related psychiatric disorders. *Sociological Methods and Research* **18**(1): 126–163.

CHAPTER 7

Forecasting and Simulation with Fuzzy Set Models

INTRODUCTION

In this chapter we illustrate the use of fuzzy state models to make forecasts of the aggregate behavior of individually complex systems (e.g., human health and senescent changes). Forecasting requires extrapolating parameter estimates beyond the data from which they are derived. Forecasting involves projection or extrapolation of outcomes over time, space, to other populations, or, most generally, to unobserved regions of the parameter space (e.g., simulation studies). We also examine the use of fuzzy set parameters in ancillary computations such as the hypothetical elimination of specific medical conditions or in examining the "embeddability" of discrete-time statistical model approximations in the underlying continuous-time theoretical processes.

These projections help evaluate whether the fuzzy state structure of the model reflects the operation of the stochastic mechanisms generating observations. One can estimate models satisfying maximum likelihood conditions that do not forecast or predict well because the parameters of the model which fit the realized data are not behaviorally isomorphic to the underlying stochastic mechanisms. In this case, subject-matter-based parameter constraints may be necessary to motivate smoothing, or penalty functions, based on Bayesian principles, i.e., mixing information in a priori distributions with the available data. A modeling effort is thus not truly complete until the ability of the fitted model to forecast and predict outcomes is examined. This is ultimately the only way to validate assumptions made about deterministic and stochastic model components.

7.1. THE FORECASTING MODEL

To illustrate the forecasting model, we will use the fuzzy state parameters in forecasts to generalize a multivariate stochastic process model of physiological changes observed in longitudinal studies such as the Framingham Heart Study (Manton et al., 1994). The general form of the model (without yet specifying the characteristics of the state space) has two components. The first is a multivariate process describing the stochastic evolution of "state" variables. In the Framingham study, these are chronic disease "risk factors," which are generally continuous measures of physiological states. A multivariate diffusion process estimated from the Framingham data, for example, might describe changes in the blood pressure, cholesterol, blood glucose, vital capacity, body mass, or hematocrit of individuals over time. The second component is a model of discrete health changes. There may be several different health changes modeled simultaneously, e.g., coronary heart events, stroke, cancer, and death. These discrete changes are described as a stochastic jump process whose parameters are functions of time-varying physiological state variables whose trajectory is governed by the multivariate continuous, or fuzzy state, diffusion processes.

The data examined in this chapter derive mainly from nationally representative longitudinal samples of elderly populations (i.e., the 1982, 1984, and 1989 National Long Term Care Surveys linked to mortality data for 1982 to 1991), much older on average (i.e., 65 to 118) than is usual in the longitudinally followed community populations used in risk factor studies, but which are generally followed for shorter periods of time (e.g., 7 to 10 years compared to, say, 42 years for Framingham). However, because of the generally advanced age of these national survey populations and their rapid rate of health and functional change, the subjects in these surveys may manifest more medical events and functional impairments than in longer-term epidemiological studies (Manton and Suzman, 1992). Indeed, in some studies (e.g., the Duke Old Longitudinal Study; Manton and Woodbury, 1983) the populations, or at least specific subgroups, may be followed to near "extinction," i.e., cohort survival experience and state variable evolution are fully (or nearly so) observed. The data, as in prior chapters, are assumed to be in categorical form—or to be so transformed without serious information loss.

To help explicate the mathematical and statistical properties of the fuzzy state forecasting model, we first consider "crisp set" forecasting models for small numbers of discrete states and their continuous-state limiting form as the number of states increases without bounds (Kushner, 1974). We then illustrate how fuzzy set processes can be embedded in a population forecasting model to represent more complex state spaces and the different behavior of fuzzy and crisp set forecasts.

7.2. PARTIALLY OBSERVED MULTISTATE HEALTH PROCESSES

To estimate transition rates between latent discrete physiological states one can use mathematical tools developed for pharmacokinetic studies known as stochastic compartment models (e.g., Jacquez, 1972). Specifically, in studying drug metabolism, often only the time and dose of drug ingestion and the time, amount, and pathway of metabolite excrection (e.g., respiration or renal clearance), are observed. The rate of drug clearance through the intermediate, unobserved metabolic states must be inferred from the temporal "correlation" of system inputs and outputs. Examples of intermediate states, and their parameters, are rates of conversion of the parent drug to active and inactive metabolites by the liver and other metabolic organs or tissues or the rates of serum and plasma clearance of the drug from physiological compartments, e.g., the central nervous system for lipid soluble metabolites crossing the blood/brain barrier. Such problems are solved with "inverse" estimation techniques, which use ancillary laboratory and animal data to construct a mathematical model of the metabolic processes operating in the intermediate, latent states.

Similarly, chronic morbidity and mortality processes can be examined in populations using "inverse" estimation with a probabilistic structure derived from ancillary epidemiological and clinical studies of disease mechanisms (Manton and Stallard, 1988). In this case the "states" are morbidity or mortality status and the process is represented as transitions between such discrete states. The simplest case involves a single disease and competing mortality risks, where persons enter the population (in a healthy state) at birth and exit the population, because of a disease process causing death, at a specific age. The problem, as in pharmacokinetic studies, is to estimate, from the temporal correlation of age (and cause-specific) schedules of births and deaths, the rate of unobserved morbid transitions and durations in state. This is done using ancillary data to construct a detailed, biologically motivated model of intermediate, latent (in demographic data) health transitions. In mortality analyses we may also have data on sex, marital status, the reporting of multiple medical conditions at death (which can help identify chronic comorbidities such as diabetes and hypertension), and birth cohort. In addition there may be health survey and disease registry data providing data on such risk factor changes as cohort and time trends in cigarette consumption (e.g., from the U.S. National Health Interview Surveys; NCHS, 1985, 1975, 1964) or from the National Cancer Institute's (1981, 1984) estimates of site-specific cancer incidence and patient survival from its Surveillance, Epidemiology and End Results (SEER) program.

The amount and type of data available vary. A first step is to specify the parametric structure of a model of morbid state transitions. Then transition rates must be estimated from vital statistics data on cause specific mortality counts and census data on population counts. Estimation is done "inversely" by using ancillary information to define the functional dependence of each intermediate transition rate on different time parameters and then projecting "backward" (i.e., "deconvoluting") in time from the age (and date) at death distribution to produce waiting times distributions in prior (to death) health states.

For example, the multihit/stage model of carcinogenesis (Armitage and Doll, 1961) suggests that tumor incidence is a Weibull function of age (e.g., $\lambda_1(a) = \alpha a^{m-1}$). The proportionality factor (α_i) can be assumed distributed to represent individual differences in risk (e.g., Manton and Stallard, 1982). The model might represent tumor growth as an exponential function (derived from clinical data; Archambeau et al., 1970; Steel and Lamerton, 1966) with the parameter controlling tumor doubling times assumed to be normally distributed over individuals. The manifest (i.e., postdiagnosis) phase of the disease could be described by data from clinically based studies—such as the National Cancer Institute's (1981) SEER program—which can be used to generate nonparametric estimates of mortality, "incidence," and "cure" rates. Competing risk parameters for causes of death can either be estimated assuming risks are independent of disease state (at best an approximation when only end-point data are available) or assuming that transition rates vary systematically across health and risk factor states when information on intermediate states is available (e.g., Tolley and Manton, 1991).

Estimation of the intermediate state transition functions also depends upon assumptions about the cohort, or period, invariance of biological parameters. Specifically, in a developed country (e.g., the U.S.) there may be a long time series of individual mortality data where the risk for a specific cause of death may be tracked over age (and time) for a large number of "partial cohorts." There is now 40+ years worth of U.S. cancer mortality data for the period 1950–1989+. If we group cohorts into five-year "bands" (e.g., a "cohort" is represented by the average risk of all persons aged 30 to 34 in 1950, "aged" 31 to 35 in 1951, etc.), we may track the experience of, say, nine cohort groups (e.g., cohorts whose mean age is 32.5, 37.5, . . . , 72.5 years in 1950) for 40 years (to age 72.5, . . . , 112.5 years in 1989). There is no obvious biological rationale on which to expect the age dependence of incidence or parameters of the tumor growth function to vary significantly over time or cohort. Thus, parameters like the number of DNA errors (e.g., p53 or cdc mutations) necessary to occur before growth control is lost or the latency time (or its variation) may be hypothesized to be biological constants of the disease (i.e., characteristic of a disease's natural history or "progression" in a "typical" host) across cohorts. Clinical and laboratory data may establish plausible ranges of values for parameters like the variance of disease latency. Such biological "constraints," though incurring a significant (but proportionately small) likelihood penalty in the data used for calibration, may produce more accurate forecasts in independent data (Manton and Stallard, 1992). These substantively derived constraints are applied in the spirit of "Bayesian," or "empirical" Bayesian smoothing of parameters (e.g., Morris, 1983).

If exposure factors vary by cohort, $\bar{\alpha}$ and γ, parameters controlling the mean and variance of individual risks (for, say, a gamma mixing distribution—or more generally for the Dubey (1967) distribution), could vary by cohort. Cross-cohort constraints on these parameters applied in estimation should represent substantive hypotheses about disease behavior (e.g., assumptions about the temporal inertia of pathology in complex biological systems). Once parameters are estimated, projections can be generated by (a) extrapolating the age patterns of disease incidence within each cohort (Manton and Stallard, 1982) and (b) making adjustments to parameters based upon exposure estimates (e.g., assumptions about asbestos exposure over time and cohort; Stallard and Manton, 1993).

The parameters of a compartment model for a cohort can be integrated into a multistate population model. The general Markov form of such a model is (Manton et al., 1992).

$$N_{t+1} = P_t N_t \tag{7.1}$$

where N_t is the distribution of persons in selected health states at time t and the elements of P_t are age and time-specific transition probabilities estimated from compartment or other stochastic models. Elements of P_t, based on biologically rationalized models, illustrate how transitions depend on physiological processes—and meaningful interventions in those processes. Use of such parameters in (7.1) will produce better estimates of the population behavior and outcomes of disease processes than from mortality (endpoint) and population data alone. This approach helps to evaluate forecast uncertainty and to model a wide range of exogenous social, economic and environmental inputs into health processes (Manton et al., 1992).

7.3. EXTENSION TO INFINITELY MANY STATES—MULTIVARIATE GAUSSIAN STOCHASTIC PROCESS MODELS

Above, we discussed how stochastic compartment models can be used to estimate the transitions between *discrete* or categorical health states—even if not observed. In those models the effects of unobserved variables on the individual's risk were represented as a unidimensional time (age) homogeneous factor. A problem in applying those procedures to other types of data, like the Framingham Heart Study, is that often a number of risk factors *are* measured at repeated times. It is impractical to represent the large number of states, and transitions between states, required to represent discrete, high dimensional temporal processes—whether states are measured or inferred. Furthermore, measurements of state variables are subject to error, i.e., systematic, as well as stochastic, uncertainty about the "true" state values. It can be shown that a continuous-time, continuous-state multivariate stochastic process can be generated under certain regularity conditions as the number of states increase to infinity (Gillespie, 1983; Kushner, 1974). Under a Markov assumption, the resulting continuous-state stochastic process is Gaussian; under "regularity" conditions, the process can be represented by stochastic differential equations (see Kushner, 1974). This is a standard approach to modeling high dimensional data. After presenting it we will show how it can be *further* generalized to fuzzy state spaces, i.e., a continuously varying state space with finite number of points constructed from large numbers of discrete measures but with boundary conditions appropriate to the discrete measures.

The multivariate stochastic process we examine (Woodbury and Manton, 1977) describes, for the individual, changes on J *measured* variables in a physiological state space by *two* sets of equations—one representing changes of the continuous measurements (e.g., risk factors) and the other representing discrete health changes. Similar models are used in theoretical models of human aging (Strehler, 1975). We refer to an individual's position in the state space as the value of the measured continuous variables. An individual's position in the health state is determined by the discrete jumps. The equations describing the individual's movement in the state space, with coordinates, z_{ijt} are given by Yashin et al. (1985) (or, more succintly, the vector $\mathbf{z}_{it}$ where i indexes individuals and t indexes continuous time). The system of J equations, with i suppressed, is given by

$$dz_t = (A_0(t) + A_1(t)\mathbf{z}_t)\,dt + D'(t)\,d\omega_t\,, \tag{7.2}$$

where A_0 and A_1 describe the dependence of change ($d\mathbf{z}_t$) on the vector of current values of $\mathbf{z}_t$ and ω_t is a J-dimensional independent Wiener process with unit scale. The scale is given by $D(t)$. The second equation describes the probability intensity $\mu(t, \mathbf{z}_t)$ that an individual with unit scale disappears from the point z_t, again with i suppressed, in the state space,

$$dS(t, z_0^t) = -\mu(t, z_t)S(t, z_0^t)\,, \tag{7.3}$$

where z_0^t is the trajectory of $\mathbf{z}$ from 0 to t and $S(t, z_0^t)$ is the probability of surviving to t given that one has followed the state space trajectory z_0^t.

These two equations describe aging "kinetics" for individuals. Assuming the initial distribution of $\mathbf{z}_0$ is Gaussian, equations (7.2) and (7.3) produce a Fokker–Planck equation to describe the change with time of parameters of the J-dimensional population distribution function $f_t(z)$, or

$$\frac{\partial[f_t(\mathbf{z})]}{\partial t} = -\sum_{j=1}^{J}\frac{\partial[u_{jt}(\mathbf{z})f_t(\mathbf{z})]}{\partial \mathbf{z}_j} + \frac{1}{2}\sum_{i=1}^{J}\sum_{j=1}^{J}\frac{\partial^2}{\partial z_i\,\partial z_j}[\sigma_{ij}(t)f_t(\mathbf{z})]$$

$$[\mu(t, \mathbf{z}) - \mu(t)]f_t(\mathbf{z})\,. \tag{7.4}$$

The first term on the right in (7.4) represents deterministic state changes (u_{jt} refers to elements of A), the second refers to diffusion generated by a multivariate Wiener process (with constant scale), and the third to mortality, i.e., state-dependent loss of persons from the population.

Equation (7.4) can be used to estimate statistical parameters for multidimensional Gaussian processes subject to systematic mortality (Woodbury and Manton, 1977) from longitudinal studies by specifying hazard and dynamic functions. The hazard is assumed quadratic. Thus, for individual i, with physiological variables, $\mathbf{z}_{it}$,

$$\mu(\mathbf{z}_{it}) = [\mu_0 + b_i \mathbf{z}_{it} + \tfrac{1}{2}\mathbf{z}_{it}^{\mathrm{T}} B \mathbf{z}_{it}] . \tag{7.5}$$

The quadratic form is justified in several ways. First, it interacts with the dynamics of physiological variables, the $\mathbf{z}_{it}$ given by equation (7.2), to keep them normally distributed. Second, a quadratic represents the failure process expected if homeostatic forces control physiological changes so that the z_{it} stay in an interior "viable" region unless there are strong exogenous shocks. Third, we seldom have sufficient data to estimate a polynomial response surface of greater than second order. Finally, the quadratic has desirable properties for cause-specific or temporal disaggregation. Other hazard functions with time-fixed covariates may not have these properties; e.g., the logistic or Cox regression estimated for separate causes do not necessarily produce a logistic or exponential when summed. The "closure," under cause aggregation or disaggregation of the quadratic hazard, allows us to write the total hazard, for M specific causes of death, or decrements, as

$$\mu(\mathbf{z}_{it}) = \sum_m \mu_m(\mathbf{z}_{it}) = [\mu_0 + b_i \mathbf{z}_{it} + \tfrac{1}{2}\mathbf{z}_{it}^{\mathrm{T}} B \mathbf{z}_{it}]_m . \tag{7.6}$$

Equation (7.6) shows that each of J variables may have a different relation to each of M causes of death, e.g., cholesterol, may have different relations to specific cancers, types of stroke, and pulmonary and heart disease (Frank et al., 1992; Epstein, 1992; Iso et al., 1989; Neaton et al., 1992). Addition of the m-component quadratic functions yields a quadratic form representing the net effect of each $\mathbf{z}_{it}$ on total mortality.

Equation (7.6) can be generalized by conditioning on discrete, fixed variables to produce a discretely mixed Gaussian process. It can also be made a function of an unobserved age-inhomogeneous process by multiplying it by a Gompertz function ($\mathbf{e}^{\theta t}$), or,

$$\mu(t, \mathbf{z}_{it}) = [\mu_0 + b_1^{\mathrm{T}} \mathbf{z}_{\mathrm{it}} + \tfrac{1}{2}\mathbf{z}_{it}^{\mathrm{T}} B \mathbf{z}_{it}][\mathbf{e}^{\theta t}] . \tag{7.7}$$

In (7.7) the effect of each $\mathbf{z}_{it}$, and its interactions, is made dependent on the age-specific average effect of unobserved state variables through the Gompertz function—controlled by θ, which regulates the age rate of increase in mortality. Actually, θ can represent the effects of several types of unobserved state variables. One might be time or age "fixed" losses of physiological function that we might associate with biological senescence. In this case, one can view the quadratic function of the $\mathbf{z}_{it}$ as reflecting a time-varying, multivariate generalization of the scale and constant terms in the Makeham–Gompertz function (Finch, 1990). A second source of variation that is represented by the θ is the unobserved changes in state variables; i.e., as the time t to $t + \Delta t$ between measurements increases, the fidelity with which z_0^t is represented declines with the lost information being included in θ. This is demonstrated in Section 7.7 along with data that may suggest how the different components of θ might be identified (Manton et al., 1992).

The $\mathbf{z}_{it}$ represent the heterogeneity of individuals—like regression. However, conditional upon $\mathbf{z}_{it}$, individuals are *not* identically distributed as in regression. The average residual age-dependent heterogeneity is represented by θ. As the amount of heterogeneity represented by $\mathbf{z}_{it}$ increases, θ decreases (i.e., $\theta \rightarrow 0$) and the value of the Gompertz multiplier approaches 1. This means that the average effect of age-related unobserved state variables on mortality (one model of biological "senescence") decreases. When θ is 0, all temporal changes of mortality are described by the evolution of the $\mathbf{z}_{it}$ generated by (7.8).

The second component describes change in $\mathbf{z}_{it}$ as a function of past states using the autoregressive process,

$$\mathbf{z}_{it} = A_{0i} + (I + A_1)\mathbf{z}_{i(t-1)} + \mathbf{e}_{it} \,. \tag{7.8}$$

A_{0i} represents initial variability across individuals (e.g., genetic effects), A_1 describes dependence on physiological states at time $t-1$; $\mathbf{e}_{it-1}$ represents diffusion or stochastic variability for each of J state variables for each individual. Equation (7.8) can be generalized to have higher-order temporal effects, or exogenous inputs, e.g.,

$$\begin{aligned} \mathbf{z}_{it} = A_{0i} + (I + A_1)\mathbf{z}_{i(t-1)} + A_2 \cdot \text{age} + A_3(\text{age} \cdot \mathbf{z}_{i(t-1)}) \\ A_4 \cdot \mathbf{y}_{i(t-1)} + A_5(\text{age} \cdot \mathbf{y}_{i(t-1)}) + \mathbf{e}_{it}^{(\text{age}\cdot d)} \,. \end{aligned} \tag{7.9}$$

In (7.9) we not only have individual "tracking" (A_{0i}), and regression on prior state values (A_1), but also the effect of age (A_2), age interactions with $\mathbf{z}_{it}(A_3)$, the effects (A_4) of exogenous variables ($\mathbf{y}_{i(t-1)}$), and age interactions with $\mathbf{y}_{i(t-1)}(A_5)$. Diffusion ($\mathbf{e}_{it}$) is controlled by homeostatic forces whose strength is time (age-) dependent with parameter d.

The quadratic function in (7.7) parameterizes the risk of death in terms of z_{it} whose change over time is governed by (7.8) or (7.9)—or other more general specifications (e.g., the equation is linear in parameters, but age, $\mathbf{z}_{i(t-1)}$, or $\mathbf{y}_{i(t-1)}$, the arguments, may be nonlinear functions). Equation (7.7) performs a similar function in clinical and epidemiological risk analyses except that the $\mathbf{z}_{it}$ are time varying so that the changes of the $\mathbf{z}_{it}$ due to dynamics and mortality selection are to be modeled as well as "risk."

If disease processes generating death are correlated, their representation by "reduced form" equations may hide details of the causal structure. For example, the effects of cholesterol on different diseases is mediated through multiple biochemical and metabolic pathways (e.g., immune stimulation due to CMV-induced vasculitis, oxidation of LDL, activation of local growth factors). To resolve collinearity, and to identify direct, and indirect, influences of $\mathbf{z}_{it}$ we, in effect, need a "structural" equation model. For example, (7.7) can be expanded by explicitly representing the state dynamics (7.9), in the hazard (7.7), for total mortality conditional on age (t), risk factors, ($\mathbf{z}_{it-1}$), cause of death (m), and exogenous factors ($\mathbf{y}_{it-1}$);

$$
\begin{aligned}
&\mu(t, \mathbf{z}_{it-1}, m, y_{it-1}) \\
&\quad = \sum_m [\mu_{0m} + b_m[A_{0i} + (I + A_1)\mathbf{z}_{i(t-1)} + A_2 \cdot \text{age} + A_3(\text{age} \cdot \mathbf{z}_{i(t-1)}) \\
&\quad\quad - A_4 \cdot y_{i(t-1)} + A_5(\text{age} \cdot y_{i(t-1)}) + \mathbf{e}_{i(t)}^{(\text{age}\cdot d)}] \\
&\quad\quad + \tfrac{1}{2}[A_{0i} + (I + A_1)z_{i(t-1)} \\
&\quad\quad - A_2 \cdot \text{age} + A_3(\text{age} \cdot \mathbf{z}_{i(t-1)}) + A_4 \cdot \mathbf{y}_{i(t-1)} + A_5(\text{age} \cdot \mathbf{y}_{i(t-1)} \\
&\quad\quad + \mathbf{e}_{i(t)}^{(\text{age}\cdot d)}]^{\mathrm{T}} B_m[A_{0i} + (I + A_1)\mathbf{z}_{i(t-1)} + A_2 \cdot \text{age} + A_3(\text{age} \cdot \mathbf{z}_{i(t-1)}) \\
&\quad\quad + A_4 \cdot \mathbf{y}_{i(t-1)} + A_5(\text{age} \cdot \mathbf{z}_{i(t-1)} + \mathbf{e}_{i(t)}^{(\text{age}\cdot d)}]] \cdot \mathbf{e}^{\theta_m t}\,.
\end{aligned}
\tag{7.10}
$$

Equation (7.10) makes the dependence of M modes of failure on endogenous and exogenous, time-varying factors, and the interaction of those factors with age and stochastic influences, explicit. With these effects explicit it is possible to identify the structural relation of endogenous and exogenous variables by imposing appropriate parameter constraints in statistical estimation from multivariate longitudinal data.

For example, the coefficients estimated directly for total mortality will differ from those estimated for M causes with the total mortality function being created as the sum of those causes. Certainly, the implications of interventions in risk factors ($\mathbf{z}_{it}$) may be very different; e.g., the relation of total cholesterol to cancer and to CVD may be both significant—but with different signs (e.g., Jacobs et al., 1992). These differences will become evident in (a) using the equations to make forecasts, and (b) using the equations to make estimates of the effects of specific risk factor interventions.

To write the discrete-time likelihood function for a set of observations, let S_{t-1} $(= S_t + \bar{S}_t)$ denote the index set of all individuals alive and present at time t. The parameters of (7.7) and (7.9) are estimated using procedures for a discrete time process with S_t, the set of survivors to t (and $\bar{S}_t$, the set of nonsurvivors), for a longitudinal observational plan whose likelihood may be written

$$L_T = L_1 * L_2 * L_3 , \qquad (7.11)$$

where the three terms are

$$L_1 = \prod_{i \in S_0} f_0(\mathbf{z}_i(0)) \qquad \text{(initial conditions)} ,$$

$$L_2 = \prod_{t=0}^{T-1} \prod_{i \in S_t} f_t(\mathbf{z}_i(t) \mid \mathbf{z}_i(t-1)) \qquad \text{(dynamic equations)} ,$$

and

$$L_3 = \left(\prod_{t=0}^{T-1} \prod_{j \in S_t} e^{-\mu(t, \mathbf{z}_j(t-1))} \right) \left(\prod_{t=0}^{T-1} \prod_{i \in \bar{S}_t} (1 - e^{-\mu(t_1, \mathbf{z}_1(t-1))}) \right)$$

(hazard function conditional on survival to $T-1$) ,

where $\bar{S}_t$ refers to the set of persons who do not survive the interval $t-1$ to t, i.e., the set of persons who die. For m causes of death, L_3 can be factored into,

$$L_3 = L_{31} \cdot L_{32} \cdot \cdots \cdot L_{3m} \qquad (7.12)$$

by exclusively subsetting $\bar{S}_t$ (i.e., deaths over the interval) into m exclusive sets of cause-specific deaths (i.e., $\bar{S}_t = \bar{S}_{t1} + \bar{S}_{t2} + \cdots + \bar{S}_{tm}$);

$$L_{3m} = \prod_t \prod_{j \in S_t} e^{-\mu_{mt}(\mathbf{z}_{it-1})} \prod_{i \in \bar{S}_{t+1} - \bar{S}_{j,t}} e^{-(1/2)\mu_{mt}(\mathbf{z}_{it-1})}$$
$$\times \prod_{j \in \bar{S}_{j,t}} (1 - e^{-\mu_{mt}(\mathbf{z}_{it-1})}) . \qquad (7.13)$$

Maximum likelihood estimates of parameters in (7.7) and (7.9) can be used to construct life tables using the simultaneous partial differential relation of life table and state variable parameters:

$$S_t = S_{t-1} |I + V_{t-1} B_{t-1}|^{-1/2} \exp\left\{ \frac{\mu(t-1, m_t) + \mu(t-1, m_t^*)}{2} - 2\mu\left[t-1, \frac{m_t + m_t^*}{2}\right]\right\} \qquad (7.14)$$

(change in probability of survival $t-1$ to t) ,

$$m^*_{t-1} = m_{t-1} - V^*_{t-1}(b_{t-1} + B_{t-1}m_{t-1}) \tag{7.15}$$

(change in mean of $\mathbf{z}_{it-1}$ due to quadratic mortality selection)

$$V^*_t = (I + V_tB_t)^{-1}V_t \tag{7.16}$$

(change in the variance–covariance matrix of $\mathbf{z}_{it}$ due to mortality).

The equations describing how the estimates m^*_t and V^*_t change among *survivors* to t as a function of risk factor dynamics are (excluding, for the moment, all inputs to the system except the autoregressive terms involving the time-varying risk factors $\mathbf{z}_{it}$) are

$$m_t = A_{0t-1} + (I + A_{1t-1})m^*_{t-1} \tag{7.17}$$

and

$$V_t = \sum_{t-1} + (I + A_{1t-1})V^*_{t-1}(I + A'_{1t-1})\,. \tag{7.18}$$

Equations (7.14)–(7.18) can be used to construct a life table conditional upon a continuous-state multidimensional process that evolves according to (7.9) and interacts with age-dependent mortality (7.7) (e.g., see (7.10)). This process represents dependence between m competing risks through the continuous state physiological processes (i.e., $\mathbf{z}^t_{i0}$). Eliminating a cause of death (say, CVD), assuming dependence, alters the risk factor distribution over time because individuals with elevated risk factors that affect more than one disease will survive longer and elevate other disease risks. For example, smoking affects both cancer and CVD. Eliminating CVD leaves more smokers alive at each age to die of cancer, so the number of smoking-related cancer deaths increases.

This model can be used to simulate risk factor interventions by altering different sets of parameters in equations (7.14)–(7.18) to reflect the impact of specific interventions on a cohort. Simulations can represent time-varying covariate effects for specific cohorts (introduced at specific times) including (a) diffusion, (b) interactions of multiple diseases and (c) time and age changes of state space parameters. The process can be generalized by (a) conditioning it on "fixed" variables and (b) making it "Markovian" by redefining the state space to include higher-order temporal dependencies as in Chapter 4.

7.4. MIXED STATE STOCHASTIC MODELS

The model can be generalized to include discrete states with stochastic transitions because the Gaussian process is the "limiting" case for discrete state processes under specific conditions. For example, one may use age, time since last smoked, and smoking duration, coded in five-year groups, and blood pressure in four intervals, to define a health model that, with 15 age groups, has 1100 discrete states. Including other conditions such as cholesterol and past coronary history increases the number of states to the point where the parametric structure exhausts the data or requires the imposition of unrealistic assumptions, e.g., risk factor independence (Tsevat et al., 1991). A second limitation is computational complexity; i.e., functions of expected numbers in each compartment, or variances of those functions, require n-fold products of transition matrices (see Tolley and Manton, 1991).

Gillespie (1983) shows the problem can be simplified (i.e., the dimensionality of the model parameter space reduced) by representing the compartment model by stochastic differential equations. Kushner (1974; see also Schuss, 1980) illustrates how compartment models, as they increase in dimension, generate systems of stochastic differential equations. Kushner (1974) defines x_N as a Markov chain with transitions at $t_n^N, n = 1, \ldots, N$, for states x_N^k, $k = 1, 2, \ldots, K$. As N increases, assume that $t_{k+1}^N - t_k^N = \Delta^N$ goes to 0 as $N \to \infty$ and $x_N(t) = x_N(t_n^N)$ for $t_n^N < t < t_{n+1}^N$. Also in this case, assume that as $\Delta \to 0$,

$$\frac{1}{\Delta} E[x_N(t + \Delta) - x_N(t) \mid x_N(t) = x] \to m(x, t) , \tag{7.19}$$

$$\frac{1}{\Delta} \mathrm{Cov}[(x_N(t + \Delta) - x_N(t)) \mid x_N(t) = x] \to l(x, t) , \tag{7.20}$$

where $m(x, t)$ and $l(x, t)$ are a vector and matrix of bounded continuous functions. Then, under regularity conditions, if the stochastic differential equation

$$dX(t) = m(X(t), t)\, dt + l(X(t), t)\, d\omega(t) \tag{7.21}$$

has a unique solution, then $X_N(t) \to X(t)$ as $N \to \infty$ in that $E_x f(X_N(t)) \to E_x f(X(t))$ for every continuous function f. In (7.21), $\omega(t)$ is a K-dimensional, independent Wiener process with constant unit scale.

Specifically, consider a model where an individual's state is a function of

1. Age, a_i
2. Length of time smoked, s_i
3. Blood pressure level, h_i
4. Smoking status: $j = 2$ if current smoker; $j = 1$ if not current smoker ($j = 0$ if dead).

The probability of lung cancer death is assumed a function of the total time the person smoked, and not the time since he quit (Peto, 1977). Thus, we assume that the likelihood of restarting smoking or changes in blood pressure is dependent only on current smoking and length of time smoked. Let $X_N(t_N)$ represent age, smoking duration, blood pressure, and smoking at t_N,

$$X_N(t_N) = \begin{pmatrix} a(t_N) \\ s(t_N) \\ h(t_N) \\ j_N \end{pmatrix}. \tag{7.22}$$

If an individual is age a at t_N, then in a short time Δ the individual either ages to $a(t_N) + \Delta$ or dies. If an individual is a nonsmoker, then the individual will either add nothing to smoking duration, become a smoker and add Δ to $s(t_N)$, or die.

Define $\lambda_{j_1 j_2}(x(t))$ to be the instantaneous probability of transitioning from j_1 to j_2 at t_N given that $X_N(t_N) = x(t_N)$, or, symbolically,

$$\Pr[\text{individual goes from } j_1 \text{ to } j_2 \text{ in } \Delta \,|\, x(t), t] = \lambda_{j_1 j_2}(x(t), t)\Delta \,.$$

Then, for $j = 2$ (currently a smoker)

$$E[a(t_n + \Delta) - a(t_N) \,|\, x(t_N)] = \Delta(1 - \lambda_{20}(x(t_N)\Delta) \tag{7.23}$$

$$E[s(t_N + \Delta) - s(t_N) \,|\, x(t_N)] = \Delta[1 - \lambda_{20}(x(t_N))\Delta - \lambda_{21}(x(t)\Delta] \,. \tag{7.24}$$

For $j = 1$ (not a current smoker),

$$E[a(t_N + \Delta) - a(t_N) \,|\, x(t_N)] = \Delta(1 - \lambda_{10}(x(t_N))\Delta) \tag{7.25}$$

$$E[s(t_N + \Delta) - s(t_N) \,|\, x(t_N)] = \Delta\lambda_{12}(x(t_N)) \,. \tag{7.26}$$

For blood pressure, we define two continuous, bounded functions $K_1(x(t_N))$ and $K_2(x(t_N))$, and assume that blood pressure change over Δ is approximately

$$h(t_N + \Delta) - h(t_N) = K_1(x(t_N))\Delta + K_2(x(t_N))\sqrt{\Delta} \tag{7.27}$$

with probability $q(x(t_N))$,

$$h(t_N + \Delta) - h(t_N) = K_1(x(t_N))\Delta - K_2(x(t_N))\sqrt{\Delta} \,, \tag{7.28}$$

with probability $q(x(t_N))$, and

$$h(t_N + \Delta) - h(t_N) = K_1(x(t_N))\Delta \,, \tag{7.29}$$

with probability $1 - 2q(x(t_N))$.

The expected value of the difference is

$$E[h(t_N + \Delta) - h(t_N) \,|\, x(t_N)] = K_1(x(t_N))\Delta \,. \tag{7.30}$$

Thus, as $\Delta \to 0$ we get the system of equations for a given j

$$dX^*(t) = m^*(x(t))\, dt + V\, dw(t) \tag{7.31}$$

where

$$X^*(t) = \begin{pmatrix} a(t) \\ s(t) \\ h(t) \end{pmatrix}$$

$$m^*(x(t)) = \begin{cases} \begin{pmatrix} 1 \\ 1 \\ K_1(x(t)) \end{pmatrix} & \text{if } j = 2 \\ \\ \begin{pmatrix} 1 \\ 0 \\ K_1(x(t)) \end{pmatrix} & \text{if } j = 1 \end{cases}$$

and

$$w(t) = \begin{pmatrix} w_1(t) \\ w_2(t) \\ w_3(t) \end{pmatrix}$$

where the $w_i(t)$ are independent Wiener processes. Following Kushner (1974), **V**, the scale parameters for the diffusion process is a 3×3 matrix, are

$$\mathbf{V} = \begin{pmatrix} 0 & 0 & 0 \\ 0 & 0 & 0 \\ 0 & 0 & K_2(x(t))\sqrt{q(x(t))} \end{pmatrix}.$$

This completes equation (7.31).

Thus for fixed j, we have a system of stochastic differential equations that describe the behavior of $a(t)$, $s(t)$, and $h(t)$. To solve the system, however, we need transitions into smoking, nonsmoking, and death states. Using standard results for Markov chains (e.g., Papoulis, 1984) if $p_j(t)$ is the probability that an individual is in smoking state j $(j = 0, 1, 2)$ and $P(t) = (p_0(t), p_1(t), p_2(t))$, then

$$dP(t) = P(t)\Lambda(x(t))\, dt \,, \tag{7.32}$$

where

$$\Lambda(x(t)) = \begin{pmatrix} -\lambda_{01}(x(t)) - \lambda_{02}(x(t)), \lambda_{01}(x(t)), \lambda_{02}(x(t)) \\ \lambda_{10}(x(t)), -\lambda_{10}(x(t)) - \lambda_{12}(x(t)), \lambda_{12}(x(t)) \\ \lambda_{20}(x(t)), \lambda_{21}(x(t)), -\lambda_{20}(x(t)) - \lambda_{21}(x(t)) \end{pmatrix}.$$

Equations (7.31) and (7.32) define systems of equations similar to those of Woodbury and Manton (1977) except that there is a smoking–no smoking dichotomy in addition to vital status; i.e., there are multiple jumps. In this case, Fokker–Planck equations (7.4) can be generated for each "stratum" (i.e., discrete state) to give the temporal change in the stratified density given initial conditions. Once parameters are estimated, they may be used in P_t to make it a discrete mixture of multiple stochastic processes.

7.5. FUZZY STATE FORECASTS: PARAMETER CALIBRATION

In this section we consider a further generalization of the continuous-state, two-component stochastic process where high (but finite) dimensional discrete state spaces can be projected into lower dimensional fuzzy set state spaces to forecast, for example, health and functional changes at advanced ages. A fuzzy state model differs from a Gaussian process in that (a) the random walk is bounded by the simplex B, (b) there is no constraint to generate a specific state distribution, and (c) though the space is continuous, it is finitely populated and hence characterized only up to the information limits imposed by the density of cases in the space. This implies that the state values for an individual are located at a discrete point, but continuously varying. This also implies that there is a general, continuous state space underlying the observed categorical variables, which, as a set, represent a crude partitioning of the underlying continuous space, and that, by projecting from the J-dimensional measurement space to a K-dimension "fuzzy" state space, the continuous state space can be approximated to the limits of the data.

Thus, for example, multiple measures of specific types of physical and mental functioning in elderly humans are combined to represent the intensity of functioning on K dimensions of functionality, but with a finite degree of resolution of any individual's position in the state space determined by the data. Whether or not this finiteness of the fuzzy state space represents an actual loss of information is dependent on an analogue to the Heisenberg uncertainty principle; i.e., at some finite level, measurement and phenomena interact. For example, the detail of responses available from demented, elderly persons is limited—through the lack of detail in response may itself say something about the individual's cognitive state. Similarly, because of the lower average functional capacity of the elderly frail individuals, there is a limit to the length of interview before the quality of response decays.

As before, for parameter estimation, we assume each individual is described by l_j binary responses to J variables each represented by y_{ijl}, which is predicted by

$$\Pr(y_{ijl}=1)=\sum_k g_{ik}\lambda_{kjl}\,. \tag{7.33}$$

We reorganize the response space in (7.33) to reflect the temporal structure of observations. Suppose that measurements of physiological state denoted x_{ijlte} are made at times t(i.e., y^t_{ijl}) and that between physiological state measurements, t transitions to absorbing states are observed. For each absorbing state transition we define an episode e. With these definitions, the likelihood for observations over t is (see Chapter 5)

$$L=\prod_i\prod_j\prod_l\prod_t\prod_e\left(\sum_k g_{ik(te)}\cdot\lambda_{kjl(te)}\right)^{x_{ijlte}}\,. \tag{7.34}$$

In (7.34), we can describe specific estimable processes by applying constraints on g and λ over t and e (Manton et al., 1987). A generally useful process involves episode constraints on $g_{ik(te)}$ (i.e., $g_{ik(t+)}$, denoted here after as $g_{ik\cdot t}$) and time and episode constraints on $\lambda_{kjl(te)}$ (i.e., $\lambda_{kjl(++)}$). Processes in the G space (i.e., the space containing the $g_{ik(t+)}$) represent temporal changes in the individual's position in the time-invariant fuzzy state space whose boundaries are set by the $\lambda_{kjl(++)}$. For i, we can estimate a K-element vector, $\mathbf{g}_{it}$, one for each time t of measurement. Changes in the $\mathbf{g}_{it}$ reflect changes in the $y^{(t)}_{ijl}$. Because of the dimensional reduction from M (the original measurement space) to G (the fuzzy state space), changes in large numbers of discrete health and functional variables, each possibly measured with error, can be represented by a smaller set K of more reliable (by producing K "averages" from J measures) fuzzy state dimensions.

The structural parameter space Λ can be decomposed into interior and exterior components (these are different from internal and external variables by virtue of their explicit representation of time). Processes in the exterior component, described by Λ^{E}, may represent transitions to A absorbing states, terminating episodes manifest over time. The g^{I}_{ik} and $\lambda^{\mathrm{I}}_{kjl}$ for interior variables J^{I} are simultaneously estimated in the full likelihood. For J^{E}, exterior variables $\lambda^{\mathrm{E}}_{kjl}$ are estimated conditionally upon g^{I}_{ik} estimated only from J^{I}. Exterior variables describe transitions to A absorbing states from which life table functions can be calculated for each of K dimensions with l_t time intervals and $l_A=A$ absorbing states. Semi-Markov processes can be represented in the subspace Λ^{E} by extending life table functions to make current absorbing states dependent upon prior state change; e.g., for a first-order dependency (*net* of the state described by g_{ikt}), this might be $\lambda^{E}_{kl=((l_A \times l_{t_1}) \times (l_A \times l_{t_2}))}$, etc.

The Λ^{E} transitions represent a general competing risk structure with dependence on the K processes over the G space. Since Λ^{E} can describe very general absorbing state processes, this provides a description of dependent competing *processes*, rather than of discrete event risks. The Λ and G spaces are "dual" to one another (Weyl, 1949; Woodbury et al., 1994), so it is possible to embed a temporal structure in either by appropriate parameterization (e.g., Clive et al., 1983; Manton et al., 1986). The selection of G or Λ in which to embed the temporal process depends on (a) substantive considerations (e.g., which transitions are exogenous and which are endogenous—conditionally estimated on G) and (b) computational and statistical issues, (i.e., mapping $J^{I} \times L^{I}_{j}$ responses onto a K-dimensional G process is parsimonious while projections into Λ^{E} are "one-to-one" mappings of transitions to parameters).

Transformation of the Λ^{E} into life table parameters can be used to represent other properties of the absorbing processes such as duration (i.e., life expectancy). Life tables may be formed for (a) K dimensions, (b) individuals by weighting the λ estimated for each of K dimensions by g_{ik}, and (c) for a population with any composition represented by $\bar{g}_K = \Sigma_i \, g_{ik}/N$ (i.e., $\bar{g}_k$ may be estimated from another population or be selected a priori to describe theoretically expected changes in the state distribution).

Fuzzy state life tables are presented in Table 7.1 (Tolley and Manton, 1987) for the California Multipurpose Senior Services Project—a federally supported demonstration (and latter) Medicaid Waiver project evaluating the efficacy of different modes of providing LTC.

In the demonstration study, both control and client groups were enrolled. The waiver project did not retain a control group. Thus, the presence of a control group and the content of the client group change in complex ways over time. To deal with the complex "confounding" of cases and controls over time in this mixed design, we estimated health scores for both populations simultaneously under both the demonstration and waiver condition. To describe the health measures, five dimensions ($K = 5$) were needed (Tolley et al., 1987).

Type 1: "Frail"—Impaired elderly individuals with chronic medical problems (e.g., stroke, epilepsy, and paralysis) and impairment on most ADLs and all IADLs. They are often black, less educated, forced to move, and "confused" on the Portable Mental Status Questionnaire. They do not live alone; otherwise they could not reside in the community.

Type 2: "Widows"—are less physically limited than the "impaired" but they live alone, have heart problems, trouble walking, and trouble in performing IADLs. Since they are less impaired, they can live in the community without personal assistance. The first two types are often

Table 7.1. Summary Life Table Statistics, MSSP, Percent Who Make Transition in 18 Months and Median Time to Transition, Clients and Controls for the Five Pure Types and Blended to Match the Actual MSSP Enrollment in FY 1984. All Observations Begin in the Community

	Type One: Frail		Type Two: Widows		Type Three: Younger Males		Type Four: Alone Females		Type Five: Acute		Blended: MSSP	
Cause	%	Median Time	%	Median Time	%	Time	%	Time	%	Time	%	Time
					Controls							
Nursing Home	3.0	2.3	1.2	2.0	1.1	8.2	0.8	4.6	0.4	2.5	0.8	3.1
Hospital	43.8	2.0	25.5	3.3	22.2	2.4	40.5	2.7	55.8	2.1	42.6	2.3
Death	8.5	1.6	2.1	2.7	2.9	5.0	3.4	1.3	1.6	0.8	3.1	1.6
					Clients							
Nursing Home	5.3	9.8	0.6	10.7	1.2	12.6	0.2	1.4	0.6	1.8	1.1	5.3
Hospital	58.1	3.4	29.3	6.8	41.0	5.4	56.8	4.3	69.8	4.6	58.7	6.1
Death	4.6	2.6	1.0	5.0	0.4	3.9	1.8	1.0	1.4	4.6	1.7	2.8

Note: Estimates are Chiang cause elimination, eliminating reassessment, drop-out, and end-of-study as causes of decrement.

enrolled in the waiver program. However, some controls are "partly" members of these types, which allows comparison of service use between waiver clients and a comparison group.

Type 3: "Younger Males"—IADL impaired tend to be male, younger than the MSSP population overall, and Hispanic. They have difficulty with some IADLs, but not with ADLs. They are often in the comparison group.

Type 4: "Alone Females"—Healthy females live alone and are often in the comparison group.

Type 5: "Acute"—Acutely ill are often bedfast, report feeling "bad," and have four or more "dangerous" medical conditions, e.g., cancer, heart disease, kidney problems, and mental illness. This group is younger, female, and unmarried.

We estimated five sets of life table where l_t is divided into monthly intervals, and where the five modes of episode termination (i.e., $l_2 = A = 5$) are hospital admission, nursing home admission, death, end of study (EOS), and remeasurement. The censoring effects of EOS and reassessment are adjusted using Chiang's (1968) competing risk model. Statistics from three modes of exit (adjusted for EOS and reassessment) are in Table 7.1 for the five groups and, for a weighted average of the five groups, for clients and controls. Each life table is adjusted for health changes described by $g_{ik \cdot t}$. Life table transitions may be used in P_t to make forecasts.

7.6. DIFFERENTIAL EQUATIONS FOR FUZZY STATE PROCESSES: FORECASTING

To generalize (7.4) for "fuzzy states" we modified the diffusion to represent a random walk in a bounded simplex B defined by Λ^{I}. This requires fuzzy set versions of (7.7) and (7.9). First to generalize (7.9), we parameterize $\mathbf{g}_{i(t+1)}$ as,

$$\mathbf{g}_{i(t+1)} = \mathbf{C}_t \mathbf{g}_{it} + \mathbf{e}_{i(t+1)} , \tag{7.35}$$

where $\mathbf{C}_t = \{c_{hkt}\}$ is a $K \times K$ matrix of age-dependent coefficients (see Manton et al., 1991). Estimates of the elements of $\mathbf{C}_t$ are made using the maximum information algorithm discussed in Chapter 4. The diagonals $(1 - c_{(kk)t})$ are age-specific rates of change of the kth disability score over t to $t+1$. Off-diagonals $c_{(hk)t}$ $(h \neq k)$ are the rates at which the kth score at t contributes to the hth score at $t+1$. Since the $g_{ik \cdot t}$ are estimated such that $\Sigma_k \, g_{ik \cdot t} = 1$, a change in c_{hkt} in one equation (i.e., for fixed h) must be compensated for by changes in other equations.

To generalize equation (7.7) to the fuzzy set case, disability incidence (generated for $t+1$ by (7.35)) is related to mortality (since the g_{it} are nonnegative scores summing to 1 over the K dimensions, a constant is not included) as

$$\mu_{t_1}(\mathbf{g}_{it}) = \tfrac{1}{2}\{g_{it}^{\mathrm{T}} B g_{it}\} \cdot \exp \theta t \,. \tag{7.36}$$

In (7.36), mortality is a quadratic function of time-dependent fuzzy state variables, i.e., the $\mathbf{g}_{it}$. Coefficients are estimated using the maximum likelihood procedures in Section 7.3. In forecasts, the g_{it} for a new time are generated from (7.35), while mortality is calculated from (7.36). Note the second term in (7.36), $\exp(\theta t)$, as for the Gaussian case, represents the effect of age-related unobserved variables on the relation of $\mathbf{g}_{it}$ to mortality. Each term in (7.36) is multiplied by $\exp(\theta t)$ so that the effects of $\mathbf{g}_{it}$ are scaled by latent age-related factors. As information in the g_{it} is increased, the amount of information in the latent states decreases, resulting in a decrease in θ. If $\theta = 0$ (and $\exp(\theta t) = 1$), then there is *no* age effect on mortality and all variation in the time to death is determined by g_{it} (i.e., equation (7.35)).

The transition equation (7.35) can be generalized to represent the effects of medical conditions, or risk factors, on the changes in the functional state scores $\mathbf{g}_{it}$. The system of convexly constrained ancillary equations (i.e., the coefficients for each k sum to 1 over the K equations) relating functional status to medical conditions (or risk factors) are (where each set of coefficients may be age or time dependent as indicated by superscript t)

$$\mathbf{g}_{it}^{*} = \boldsymbol{\alpha}^{t} + \boldsymbol{\beta}^{t}\mathbf{z}_{it} + \boldsymbol{\Gamma}^{t}\mathbf{d}_{it} + \boldsymbol{\xi}^{t}(\mathbf{z}_{it} \cdot \mathbf{d}_{it}) + \mathbf{e}_{it}^{(d)} \,, \tag{7.37}$$

where $\mathbf{z}_{it} \cdot \mathbf{d}_{it}$ denotes an outer product. In this equation, $\boldsymbol{\alpha}^{t}$, representing "tracking," and $\mathbf{e}_{it}$, representing stochasticity in a bounded space, are K-element vectors; $\boldsymbol{\beta}^{t}$ is a $K \times J$ constrained (by the convexity of $\mathbf{g}_{it}^{*}$) regression matrix relating J risk factors (which may include social conditions such as education or income) $\mathbf{z}_{it}$ to the $\mathbf{g}_{it}$; $\mathbf{d}_{it} = \{d_{imt}\}$ is an M-element vector, with $d_{imt} = 1$ if the mth medical condition is present at t, $d_{imt} = 0$ otherwise; $\boldsymbol{\Gamma}^{t} = \{\gamma_{km}\}$ is a $K \times M$ matrix of convexly constrained regression coefficients; and $\boldsymbol{\xi}^{t}$ is a $K \times (M \times J)$ convexly constrained regression matrix relating the interaction of J risk factors and M medical conditions. Thus, γ_{km}^{t} (or β_{jk}^{t}) represents how much the kth dimension is affected by the presence of the mth medical condition (or Jth risk factor) at a specific age T. The ξ_{kmj}^{t} represent the effects of interactions of risk factors and medical conditions on $\mathbf{g}_{it}$. The quadratic can be written with (7.37) modeling $\mathbf{g}_{it}^{*}$ as

$$\mu((\text{age}, l)\mathbf{z}_{it}; m) = \sum_{l} (\mathbf{g}_{it}^{*\mathrm{T}} B \mathbf{g}_{it}^{*})_{l}(e^{\theta t}) \,. \tag{7.38}$$

That is, one models changes in fuzzy state l as functions of medical conditions and risk factors which are convexly constrained to operate through the $\mathbf{g}_{it}$, i.e., within the confines of the convex state space. In this case, convexity constraints are imposed by the projection to B.

The trajectory in (7.38) can be altered to simulate health and functional changes by eliminating all (or part) of the medical condition indicated by d_{it} (possibly changing (7.37)). Then the altered $\mathbf{g}_{it}$ can be used in (7.35) to show how prevention of a condition alters the processes of disablement. The effects of $\mathbf{d}_{it}$ may be estimated using a maximum information algorithm (Kaufmann and Gupta, 1985). Below we use sample reweighting to model interventions in the $\mathbf{g}_{it}$ to simulate a cohort using the 1982 and 1984 NLTCS (and mortality from 1982 to 1986). The simulated cohorts will be examined both by altering $\mathbf{g}_{it}$ to reflect changes in income and by comparing the simulations to results using fuzzy set dynamics calculated directly from changes between the 1982, 1984, and 1989 NLTCS and 11,000 deaths occurring between 1982 and 1991.

With estimates of dynamic and mortality coefficients, as observed or as modified by simulating changes as in (7.38), we can produce active life expectancy (ALE) forecasts (i.e., the time expected to be spent without disability) for a cohort using equations that represent the interaction of disability and mortality over age. The equations, developed for the stochastic process in Section 7.3, are altered because the $\mathbf{g}_{it}$ are *bounded* (i.e., they define a stochastic process on the $K-1$ dimensional simplex $B = L_B \cap M$, where L_B is parameterized by the G and Λ spaces). This requires generalization of the diffusion component which, in the prior Gaussian form, assumed variables were *not* bounded.

The age specific survival function, l_t, is

$$l_{t+1} = l_t |I - \mathbf{V}_t B_t|^{-1/2} \exp\left\{\frac{\mu(t_1, \boldsymbol{\nu}_t) + \mu(t_1, \boldsymbol{\nu}_t^*)}{2} - 2\mu\left(t_1, \frac{\boldsymbol{\nu}_t + \boldsymbol{\nu}_t^*}{2}\right)\right\}, \qquad (7.39)$$

where $l_{t_0} = 1$ for initial age t_0 and $\boldsymbol{\nu}_t$ and $\mathbf{V}_t$ are age-dependent means and covariances of $\mathbf{g}_{it}$. In (7.39), age-dependent estimates of coefficients from the integrated quadratic hazard (i.e., $B_t = B\exp[\theta(t + 1/2)]$, using a mid-point approximation) adjust survival for risk associated with the K age-specific average of the $\mathbf{g}_{it}$ (ν_t), and with the variance–covariance matrix of $\mathbf{g}_{it}$ ($\mathbf{V}_t$).

The changes in $\boldsymbol{\nu}_t$ and $\mathbf{V}_t$, due to mortality, are denoted by $\boldsymbol{\nu}_t^*$ and $\mathbf{V}_t^*$ respectively. The mean $\boldsymbol{\nu}^*$ is determined by

$$\boldsymbol{\nu}_t^* = \frac{\boldsymbol{\nu}_t - \mathbf{V}_t^* B_t \boldsymbol{\nu}_t}{\sum_k \langle \boldsymbol{\nu}_t - \mathbf{V}_t^* B_t \boldsymbol{\nu}_t \rangle_k}, \qquad (7.40)$$

where the new mean $\boldsymbol{\nu}_t^*$ of the $\mathbf{g}_{it}$ is calculated for persons expected to survive the interval. In (7.40) the bracketed term in the denominator

"$\langle \mathbf{a} \rangle_k$" extracts the kth component of the vector $\mathbf{a}$. The new variance is

$$\mathbf{V}_t^* = (I + \mathbf{V}_t B_t)^{-1} \mathbf{V}_t \tag{7.41}$$

where $\mathbf{V}_t^*$ is the variance of $\mathbf{g}_{it}$ at t for the set of persons expected to survive the interval. Mortality in all three equations is affected by $\boldsymbol{\nu}_t$ and $\mathbf{V}_t$.

The variance of $\mathbf{g}_{it}$ is altered by a diffusion process (reflecting loss of homeostatic control with age) which depends on time. To describe time-dependent, bounded diffusion we assume that diagonal elements of $\mathbf{V}_t$, reflecting the variance of each g_{it} have, as a maximum, Bernoulli variance $(\nu_{kt}(1 - \nu_{kt}))$ on each of K dimensions. In calculating the diffusion matrix we assume that (a) the correlation matrix $\mathbf{R}$ of the K sets of $\mathbf{g}_{it}$ is constant over age, where $\mathbf{R}$ is estimated from the empirical covariance matrix after conditioning on age, and (b) ratios of the variance of the $\mathbf{g}_{it}$ to the Bernoulli limits. Square roots of these ratios define the diagonals of a matrix $\mathbf{S}$ that rescales the variances. $\mathbf{W}_{t+1}$ is diagonal with diagonal elements equal to $\sqrt{\nu_{k(t+1)}(1 - \nu_{k(t+1)})}$. The covariance matrix for $t + 1$ is

$$\mathbf{V}_{t+1} = \mathbf{W}_{t+1} \mathbf{SRSW}_{t+1} \,. \tag{7.42}$$

With (7.42), diffusion is

$$\boldsymbol{\Sigma}_{t+1} = \mathbf{V}_{t+1} - \mathbf{C}_t \mathbf{V}_t^* \mathbf{C}_t^{\mathrm{T}} \,, \tag{7.43}$$

where $\mathbf{C}_t$ is defined in (7.35). The equation describing how $\mathbf{g}_{it}$ alters the means of the $\mathbf{g}_{i(t+1)}$ is

$$\boldsymbol{\nu}_{t+1} = \mathbf{C}_t \boldsymbol{\nu}_t^* \,. \tag{7.44}$$

These equations produce the parameters of a life table with covariates representing the distributions of K fuzzy state dimensions at t. Constraints in (7.42) map the random walk over B onto a spherical coordinate space (with spherical transformations $\mathbf{W}_{t+1}$), with dimensions scaled by $\mathbf{S}$, to eliminate discontinuities. This is similar to forms of GoM where "curvature" in the parameter space topology (i.e., the parameters generating the probability space L_{B}, from the y_{ijl}) is produced by (a) empirical Bayesian normalization of λ_{kjl} estimates for Dirichelet variability which "curves" faces of B proportional to the degree of unobserved heterogeneity (which may reduce K), (b) Mth-order GoM analysis, where R dimensions of the G space may generate "folds" in the linear probability space to better accommodate cases clustered in regions of "high" density (i.e., portions of the state space may have finer partitioning and be better "resolved" on the continuous metric, due to data being denser in those regions), and (c) quadratic GoM, where probability amplitudes a_{ijl}, rather than probabilities $p_{ijl} = a_{ijl}^2$, are modeled (i.e., the linear probability space L_B is generalized to an L_B^2, or quadratic probability amplitude, metric). Thus, this generalizes the space to either express empirical nonlinearity or theoretical metric constraints. This is different from clustering algorithms, which use a distance function where the Minkowski metric is varied, in that the mapping functions used here are well specified and can be evaluated using maximum likelihood.

This can be seen to have some analogies to modeling of chaotic systems since changes can be nonlinear but are constrained by boundary conditions, and feedback, to remain in well-defined interior regions of a continuous, fuzzy space. The fuzziness of the space helps *dampen* severe changes in the system due to the operation of complex interactions among the J discrete variables, which are synthetic constructs from J^* continuous, unobserved "variables" for which K are the estimable dimensions with resolution related to the density of observations in the space. Thus, instead of the absolutely continuous space assumed in the standard Gaussian diffusion process, we have a constrained continuous, but finitely populated, space whose "gauge" is determined by the amount of data we have to approximate the process.

The ν_{kt} decompose life expectancy into components associated with K types of disability (e.g., "active" life expectancy or impaired or institutional life expectancy) associated with each of K fuzzy states. Projections can be altered by modifying the "risk factor" distribution for a cohort and generating new dynamics for the simulations. Thus, the effects of modifying the distribution of disability by primary disease prevention may be forecast.

Since cohort dynamics are needed for forecasting, we need to estimate, or simulate, a cohort's experience. To simulate a cohort, C_t, estimated from cross-sectional data for a short time series (e.g., 2 years; the 1982 to 1984 NLTCS), must be modified to reflect a reasonable scenario about disability expression during the life of a cohort currently aged x. One procedure alters the sample weights (i.e., inverse of the probability of selection) of persons of specific types to simulate specific disability changes and then to estimate $\mathbf{C}_t$ from the reweighted data. One intervention applied to the 1982 and 1984 NLTCS (Manton et al., 1991) to approximate cohort transitions assumed that, of 80% of the transitions to a disability state from the *nondisabled* population, 50% could be prevented and that, of the remaining 20%, 67% are prevented. These two assumptions reduced disability incidence 53%. This simulates chronic disability prevention in a subpopulation that (a) is younger than the general NLTCS (Medicare eligible elderly disabled) population, (b) has not been disabled long, and (c) has moved to a low level of disability. This is consistent with the observation that, in Geriatric Evaluation Units, young–old persons with lower levels of functional impairment and recent onset of disability are easier to rehabilitate (Rubenstein and Josephson, 1989). However, recent studies (e.g., Fiatarone et al., 1990) suggest rehabilitation of function in even frail, institutionalized persons is possible to ages 86 to 96 and beyond.

The simulated (using the sample reweighting technique) cohort life expectancies are higher than in current (i.e., 1984) period life tables, but match the cohort life expectancy projected for persons age 65 in 1984 (SSA,

1983), i.e., the time of the second NLTCS. They match period life tables projected by Social Security (SSA, 1989) for the approximate midpoint (i.e., 2005) of midrange projections (i.e., 1990 to 2020). The interventions produce age-specific dynamics ($\mathbf{C}_t$) consistent with those estimated from the Medicare component of the S/HMO evaluation on a monthly basis (see Chapter 8).

7.7. IDENTIFIABILITY OF PROCESS PARAMETERS AND "EMBEDDABILITY"

A property of the fuzzy state processes discussed that has both theoretical and substantive consequences, and which is essential to forecasting, is the "embeddability" of discrete-time, estimable approximations of stochastic processes in the underlying, theoretical continuous time processes. Embeddability takes on several additional implications as the "gauge" of the fuzzy state space is determined by the empirical density of cases defining the space itself. Essentially, since it is not possible to continuously monitor human health processes, either over time or the universe of all physiological parameters, in a population, a temporal (and variable) space sampling scheme is necessary; i.e., one can make limited measurements only at fixed times—with times determined to maximize information about the process. Alternatively, one may (a) randomly sample over time or (b) have a process that makes measurements when a critical health change occurs. Additionally, since it is not possible to measure all relevant variables—the fuzzy state space is embedded in a infinite dimensional latent variable space whose behavior may, or may not, be well described.

When measurements are made at fixed times, the issue of "embeddability" is how the intersurvey interval relates to the rate at which events of interest occur. If the interval is too long, changes within the interval are not observed, and transition rates may be "biased" against "short-term" transitions; i.e., the coefficients c_{kk} in (7.35) are misestimated—which, by virtue of the appearance of the $\mathbf{g}_{it}$ in the mortality function, produces coefficients that are misestimated (except for the adjustment by θ). The parameter θ "corrects" some of these biases in the mortality coefficients by modeling the average unexplained time-dependent trend. Hence, $e^{\theta t}$ in the model serves as a "self-correcting" aspect to the life table equations even if intermediate transitions are missed—because mortality selection is, in part, corrected for information loss. Estimates of θ near zero indicate no time-dependent unaccounted-for trend. Thus, no time-dependent bias is observed.

For example, the NLTCS was done in 1982, 1984, and 1989. Thus, there is a 2- and a 5-year intersurvey interval. To deal with different length intervals one can divide intervals into months (i.e., 24 and 60 months) and estimate $\mathbf{g}_{it_m}$ at each measurement time and assume that the $\mathbf{g}_{it_m}$ (i.e., for each intervening month) are the same except when new measurements are made when there is a jump in information. Alternately, for $\mathbf{g}_{it}$ measured at two or more times, the monthly changes producing the $\mathbf{g}_{it_m}$ may be assumed to fulfill assumptions about changes over the interval (e.g., linear or exponential interpolation). With a model created for $\mathbf{g}_{it_m}$ then the health state of persons who die in an interval can be updated to the implied $\mathbf{g}_{it_m}$ in the month before death to improve estimates of the quadratic mortality function. The effects of this temporal "missing" data in estimating health change is demonstrated in (a) the estimated values of θ for the 2- and 5-year follow-ups of the NLTCS being different, representing more changes in disability (even with constant rates) occurring over the 5-year interval and (b) large reductions in θ in the NLTCS (and the S/HMO evaluation where measurements were made when health changed (i.e., t is taken as a random variable)). The mixed observational plan of fixed-time health assessments and continuously monitoring service change and mortality in administrative files, providing indicators of underlying changes, is a third strategy. That is, if the quadratic hazard can be estimated for a short interval (e.g., 3 months), then that hazard can be used to improve estimates of g_{it_m} for persons who die over the full (i.e., 24 or 60 month) interval. This approach was used in imputing the disability status of nonresponders in the 1982 and 1984 NLTCS (Manton et al., 1991).

We can illustrate the consequences of embeddability, which involves questions of identifiability and missing data, and temporal sampling vis-a-vis an important topic, i.e., whether the human survival curve is becoming rectangularized. In this review we are interested in identifying the data conditions necessary to *validate* a specific failure model, not simply to determine if a model fits the data. Specifically, in addition to, $\mathbf{z}_{it}$, define $\mathbf{x}_{it}$ as a vector of values for *m unmeasured* state variables. Known risk factors (e.g., cholesterol) *not* measured and factors relevant to survival not yet identified, whether intrinsically or extrinsically controlled, are both in $\mathbf{x}_{it}$. $\mathbf{z}_{it}^*$ and $\mathbf{x}_{it}^*$ are "fixed" over age. $\mu_{\mathbf{z}}(t, \mathbf{x})$ is the hazard at t which is a function of $\mathbf{z}_{it}$ and $\mathbf{x}_{it}$. $f_{\mathbf{z}}(t)$ is the marginal density of ages at death ($T(i)$) for individuals with $\mathbf{z}_{it}$. Effects of $\mathbf{x}_{it}$ are assumed averaged over all persons with a particular $\mathbf{z}_{it}$. In mortality data with no observed risk factors, $\mathbf{x}_{it}$ is the "frailty" or heterogeneity distribution (e.g., Vaupel et al., 1979). The density of $\mathbf{x}_{it}$ conditional on $\mathbf{z}_{it}$ is $g_z(\mathbf{x})$. $F_{\mathbf{z}}(t, \mathbf{x})$ is the conditional cumulative distribution of $T(i)$ given the realizations $\hat{\mathbf{x}}_{it}$ of $\mathbf{x}_{it}$ and observed $\mathbf{z}_{it}$. $\mathbf{z}_0$ are values of $\mathbf{z}_{it}$ producing the lowest mortality. As $\mathbf{z}_{it} \rightarrow \mathbf{z}_0$, mortality declines. The existence of $\mathbf{z}_0$ is *required* for curve squaring; i.e., if mortality always declined as $\mathbf{z}_{it}$ changed, then mortality could always be reduced (and life span would be, in effect, "infinite"). If all persons achieve $\mathbf{z}_0$, there is no variation in $T(i)$ due to $\mathbf{z}_{it}$. There is still variation due to $\mathbf{x}_{it}$. Whether $\mathbf{x}_{it}$ determines "irreducible" variance in the $T(i)$ depends on $\mathbf{x}_{it}$ being "controllable." The identifiability issue is that, since we do not know $\mathbf{x}_{it}$, its determination of heterogeneity in $T(i)$ is due to mechanisms with unknown plasticity. Hence, without controlling for $\mathbf{x}_{it}$, we cannot determine if "limits" to life expectancy exist. Although the existence of $\mathbf{z}_0$ is required by curve squaring, its existence does *not* imply the hypothesis; i.e., mortality could be independent of age at $\mathbf{z}_0$. Even if mortality is independent of age, it may be so high that there is little probability of observing long-lived individuals.

The density of $T(i)$ given $\mathbf{x}_{it}$ and $\mathbf{z}_{it}$ is

$$f_{\mathbf{Z}}(t, \mathbf{x}) = h_{\mathbf{Z}}(t, \mathbf{x}) \exp\left(-\int_0^t h_{\mathbf{Z}}(\tau, \mathbf{x})\, d\tau\right). \tag{7.45}$$

In (7.45), both $\mathbf{z}_{it}$ and $\mathbf{x}_{it}$ may change with age. From (7.45), the density of $T(i)$, conditional on $\mathbf{z}_{it}$, but with $\mathbf{x}_{it}$ integrated out, is

$$f_{\mathbf{Z}}(t) = \int h_{\mathbf{Z}}(t, \mathbf{x}) \exp\left(-\int_0^t h_{\mathbf{Z}}(\tau, \mathbf{x})\, d\tau\right) g_{\mathbf{Z}}(\mathbf{x})\, d\mathbf{x}\,, \tag{7.46}$$

or, more compactly,

$$f_{\mathbf{Z}}(t) = h_{\mathbf{Z}}(t) \exp\left(-\int_0^t h_{\mathbf{Z}}(\tau)\, d\tau\right), \tag{7.47}$$

where $h_{\mathbf{Z}}(t) = \int h_{\mathbf{Z}}(t, \mathbf{x}) g_{\mathbf{Z}}(\mathbf{x})\, d\mathbf{x}$. $f_{\mathbf{Z}}(t)$ is the density of $T(i)$. Both the density $g_{\mathbf{Z}}(\mathbf{x})$ of $\mathbf{x}_{it}$ and the hazard may take any form. Thus,

***Definition* 1.** T is subject to (*hard*) *curve squaring* if, for every *fixed* path $(\mathbf{y}_{ij})$, $i \in A$ (A is the set of all individuals), there is a $t_0 = t_0(\mathbf{y}_{ij} < \infty$ such that

$$\lim_{\mathbf{z}(t,i) \to \mathbf{z}_0} \lim_{\varepsilon \to 0} \frac{h(t_0, \mathbf{x}(t_0, i)) - h(t_0 - \epsilon, \mathbf{x}(t_0 - \epsilon, i))}{\epsilon} = +\infty\,. \tag{7.48}$$

The path y gives the trajectory or "sequence" of unobserved factors $\mathbf{x}_{it}$. Thus, for individual i there is a sequence $\mathbf{x}_{it}$ over age t of factors affecting the likelihood of death represented by $\mathbf{y}_i$. Under the definition, if hard curve squaring exists, there will be a maximum lifetime possible, denoted, say, by $t_{\max}$. $t_{\max}$ is the maximum age at death for a given path $\mathbf{y}_i$. Controlling all variation due to $\mathbf{x}_{it}$ identifies $t_{\max}$ due to mechanisms common to all members of a species. If $t_{\max}$ does not exist in this sense, then there is no species-specific limit to life expectancy—there are only life expectancy limits based upon the known state of the individual at given times, i.e., $\mathbf{z}_{it}$. In this case, the search is for additional $\mathbf{z}_{it}$ controlling life expectancy rather than for a $t_{\max}$. That is, $t_{\max}$ is a synthetic construct based upon the $\mathbf{z}_{it}$ known at a given time.

Under Definition 1, if hard curve squaring exists and if all individuals were at $\mathbf{z}_0$, those who have the same $\mathbf{x}_{it}$ and are alive prior to $t_{\max}$ will die at $t_{\max}$; i.e., there is a fixed bound to lifespan even in the most favorable circumstances, and an individual's life span *cannot* be greater than that associated with $\mathbf{z}_0$. The bound on life for these individuals is a function of $\mathbf{x}_{it}$; i.e., the survival curve across a distribution of $\mathbf{x}_{it}$ will not instantaneously go to 0 at a fixed point. There are mechanisms for persons with $\mathbf{x}_{it}$ that determine if they die at t_{max}. These are common for all species members and do not vary conditional on $\mathbf{x}_{it}$. If the $\mathbf{x}_{it}$ cause "irreducible" variation in t_{max}, "curve squaring" is distributed over age. If modifiable elements of $\mathbf{x}_{it}$ are identified and measured, the variability of t_{max} conditioned on $\mathbf{z}_{it}$ is reduced; i.e., more variation in $T(i)$ is attributable to $\mathbf{z}_{it}$

***Definition* 2.** T is subject to (*soft*) *curve squaring* if, for every path $\mathbf{y}_i$, $i \in A$, there is a $t_0 = t_0(y_i) < \infty$ such that for $\mathbf{z}$ and $\mathbf{z}_0$ and every $D > 0$ and $\Delta > 0$ there exists a $t_1 > 0$ such that, if $\tau > t_0 + t_1$, then

$$h(\tau + \Delta, \mathbf{x}(\tau + \Delta, i)) - h(\tau, \mathbf{x}(\tau, i)) > 0 \tag{7.49}$$

and $h(\tau, \mathbf{x}(\tau, i)) > D$.

Thus, once an individual reaches $t_0(\mathbf{y}_i)$, he does not die, but his risk of death increases with age without bound. This allows for two sources of variability in $T(i)$ at $\mathbf{z}_0$. The first, represented in hard curve squaring, is individual variability on $\mathbf{x}_{it}$ in t_0. However, the $\mathbf{x}_{it}$ determining this limit, in addition to fixed genetic factors in (7.49) include stochastic, biological processes where damage mechanisms operate faster than repair. Thus, if free radical damage increases with age, because repair mechanisms (e.g., the p53 gene regulating aspects of DNA repair is "overwhelmed" or itself is mutated) cannot fix all cell damage, there are exogenous processes contributing to the age increase in mortality. The rate of this process is stochastic and depends on changes in the balance of damaging and restorative forces. We will denote this age t_0 as t_{max} even though it is possible to exceed the value of t_{max} before death. Once t_{max} is attained, mortality occurs at random T_i with a density determined by an age increasing hazard. If the hazard is *not* a function of age, neither definition holds. Thus, a test of whether either type of curve squaring is present requires tests of the age dependence of the hazard after conditioning on $\mathbf{z}_{it}$.

For an individual with $\mathbf{x}_{it}$ and $\mathbf{z}_{it}$, the distribution of $T(i)$ for hard curve squaring is

$$f_{\mathbf{Z}}(t, \mathbf{x}) = \begin{cases} h_{\mathbf{Z}}(t, \mathbf{x}) \exp\left(-\int_0^t h_{\mathbf{Z}}(\tau, \mathbf{x})\, d\tau\right), & t < t_0(\mathbf{x}), \\ [1 - F_{\mathbf{Z}}(t, \mathbf{x})]\delta(t - t_0(\mathbf{x})), & t \geq t_0(\mathbf{x}) \end{cases}, \tag{7.50}$$

where $F_{\mathbf{Z}}(t, \mathbf{x}) = \int_0^{-t} h_{\mathbf{Z}}(\mathbf{x}, \mathbf{x}) \exp(-\int_0^x h_{\mathbf{Z}}(\tau, \mathbf{x})\, d\tau)\, d\mathbf{x}$ and $\delta(\mathbf{x})$ goes instantaneously from 0 to ∞ at $x = 0$ with unit integral over the real line. Integrating over $\mathbf{x}_{it}$, the distribution of $T(i)$ is composed of persons dying before, and exactly at, t_{max}, where $t_{max}(\mathbf{x}) < t$. Let $B_{\mathbf{Z}}(t)$ denote the set of $\mathbf{x}$ such that $t(\mathbf{x}) < t$. Then $f_{\mathbf{Z}}(t)$ is,

$$f_{\mathbf{Z}}(t) = \int_{\mathbf{x} \notin B_{\mathbf{R}}(t)} h_{\mathbf{Z}}(t, \mathbf{x}) \exp\left(-\int_0^t h_{\mathbf{Z}}(\tau, \mathbf{x})\, d\tau\right) g_{\mathbf{Z}}(\mathbf{x})\, d\mathbf{x}$$

$$+ \int_{\mathbf{x} \in B_{\mathbf{Z}}(t)} (1 - F_{\mathbf{Z}}(t(\mathbf{x}), \mathbf{x}))\, \delta(t - t(\mathbf{x})) g_{\mathbf{Z}}(\mathbf{x})\, d\mathbf{x}\,. \qquad (7.51)$$

For "soft" squaring, the comparable distribution is

$$f_{\mathbf{Z}}(t) = \int_{\mathbf{x} \notin B_{\mathbf{Z}}(t)} h_{\mathbf{Z}}(t, \mathbf{x}) \exp\left(-\int_0^t h_{\mathbf{Z}}(\tau, \mathbf{x})\, d\tau\right) g_{\mathbf{Z}}(\mathbf{x})\, d\mathbf{x}$$

$$+ \int_{\mathbf{x} \in B_{\mathbf{Z}}(t)} h_{\mathbf{Z}}(t, \mathbf{x}) \exp\left(-\int_0^t h_{\mathbf{Z}}(\tau, \mathbf{x})\, d\tau\right) g_{\mathbf{Z}}(\mathbf{x})\, d\mathbf{x}\,. \qquad (7.52)$$

In (7.52), $h_{\mathbf{Z}}(t, \mathbf{x})$ is a rapidly increasing function of t in the second integral. The second term in (7.51) is replaced in (7.52) by an integral representing mortality risk beyond t_{max}, i.e., the additional time to death component t_1.

***Definition* 3.** Mortality satisfies *condition C*, if $\mathbf{z} = \mathbf{z}_0$ implies that the first integral in (7.51) or (7.52) is *zero* for all t.

Condition C requires that all individuals at $\mathbf{z}_0$ die *at* (*hard* curve squaring) or *after* (*soft* curve squaring) $t_{max}(\mathbf{x})$, i.e., there is *no* mortality prior to t_{max}. Since the $\mathbf{x}_{it}$ are unobserved, mortality, for either hard or soft curve squaring, appears to follow a probability law (i.e., t_{max} has a distribution determined by $\mathbf{x}_{it}$ even if C holds). If C does *not* hold, then some deaths are "premature." To test the hypothesis, an alternative is specified by a hazard where curve squaring does *not* occur, i.e., given $\mathbf{x}_{it}$ and $\mathbf{z}_{it}$, the hazard is not monotone, nondecreasing in t beyond $\bar{t}_{max}$. The simplest case is that the hazard is independent of chronological age (time) beyond $\bar{t}_{max}$. Examples where this pattern has been observed at extreme ages (e.g., over 100) exist for both human population data (e.g., Barrett, 1985; Bayo and Faber, 1985) and in well-controlled animal experiments (Carey et al., 1992; Curtsinger et al., 1992).

Neither hypothesis depends upon the particular hazard function selected. Under the alternative, an estimated hazard, independent of age and positive, *conditional* on $\mathbf{z}_{it}$ *and* $\mathbf{x}_{it}$, is inconsistent with either curve squaring hypothesis. Though the conditional (on $\mathbf{z}$ and $\mathbf{x}$) distribution of T_i is independent of age beyond $\bar{t}_{max}$, the hazard may increase as a function of $\mathbf{z}$ (or $\mathbf{x}$); i.e., $\mathbf{z}_{it}$ may diverge from $\mathbf{z}_0$ due to the dynamics of $\mathbf{z}_{it}$ (or $\mathbf{x}_{it}$). Thus, independent of age, an increase in blood pressure could increase stroke mortality. If mortality increases *only* because of $\mathbf{z}_{it}$, we do not have "curve squaring"; i.e., the appearance of "curve squaring" is produced by changes in $\mathbf{z}_{it}$ (and $\mathbf{x}_{it}$) with age and not mortality risks determined by "age." If $\mathbf{z}_{it}$ is "correlated" with age, conditioning on $\mathbf{z}_{it}$ may eliminate the age dependence of death. Without more information, we do not know if $\mathbf{z}_{it}$ is a process determined by age or one that determines the association of mortality with age. To resolve the correlation one must alter $\mathbf{z}_{it}$ and show that the hazard still depends on age. If altering $\mathbf{z}_{it}$ eliminates the association of age and mortality, then there is not a $t_{\max}$ but a limit that can be extended by modifying $\mathbf{z}_{it}$—assuming the technology modifying $\mathbf{z}_{it}$ does not disrupt other functions; i.e., one must consider the dynamics of $\mathbf{z}_{it}$, cross-temporal interactions among $\mathbf{z}_{it}$ and with $\mathbf{x}_{it}$.

"Fixed frailty" models represent the effects of $\mathbf{x}_{it}$ on human mortality (e.g., Vaupel et al., 1979) as unidimensional and fixed over age with a hazard proportional (with proportionality constant, γ) to $\mathbf{x}$. Let $\mathbf{x}^*$ denote a one dimensional, age-independent random variable. The fixed frailty model using $\mathbf{x}^*$ in place of $\mathbf{x}_{it}$ is the simplest model with unobserved heterogeneity. Let T^* be the age at death assumed to be generated by a constant hazard $\gamma\mathbf{x}^*$. Since the hazard is proportional to $\mathbf{x}^*$, which is *independent* of age, T^* does *not* represent curve squaring. T^* may *always* be constructed to match the T generated by a model with curve squaring.

Lemma 1. *For any T representing "hard" curve squaring for $\mathbf{Z}_{it}$ with a density for the unobserved variables, $g_{\mathbf{z}}(\mathbf{x})$ and a "well-behaved" density $f_x(\mathbf{x})$, there is an age* independent *$\mathbf{x}^*$ with density $g_{\mathbf{z}}^*(\mathbf{x}^*)$ and a failure time random variable T^* based on the fixed frailty model such that T^* has the same distribution as T.*

Proof. From (7.46), the distribution of T^* is a function of $\mathbf{x}^*$,

$$f_{T^*}(t) = \gamma \int \mathbf{x}^* \exp(-\gamma t\mathbf{x})\, g_{\mathbf{z}}^*(\mathbf{x}^*)\, d\mathbf{x}^* . \tag{7.53}$$

The distribution of T^* expanded about $t = 0$, is

$$\begin{aligned} f_D(t) &= \gamma \int \sum_{k=1}^{\infty} \frac{(-\gamma t)^{k-1} y^{*k}}{(k-1)!}\, g_{\mathbf{Z}}^*(\mathbf{x}^*)\, d\mathbf{x}^* \\ &= \gamma \cdot \sum_{k=1}^{\infty} E_{\mathbf{Z}} \mathbf{X}^{*k} \frac{(-\gamma t)^{k-1}}{(k-1)!} , \end{aligned} \tag{7.54}$$

where $E_{\mathbf{Z}}\mathbf{x}^{*k}$ is the kth moment of $\mathbf{x}^*$ about 0. Assume that the RHS of (7.51) is expanded about 0 with the sign of the coefficient of t^k being $(-1)^k$. By equating coefficients for powers of t, the moments of $g_{\mathbf{Z}}^*(\mathbf{x}^*)$ of order 1 and higher can be derived from the expansion (7.51) and (7.54). Hence the result.

Thus, if, for the RHS of (7.51), a Taylor series expansion about $t = 0$ can be generated, a function $g_{\mathbf{Z}}^*(\mathbf{x}^*)$ can be constructed from the coefficients of the expansion such that the density of $\mathbf{z}$, generated using $g_{\mathbf{Z}}^*(\mathbf{x}^*)$ as the density of $\mathbf{x}^*$, is the same as $f_{\mathbf{z}}(t)$ in (7.51) with curve squaring. Consider an example:

1. There are no $\mathbf{z}_{it}$ and one $\mathbf{x}_{it}$.
2. $\mathbf{x}_{it}$ is a gamma random variable with parameters α and β.
3. The upper bound on mortality for any individual, given that $\mathbf{x}_{it} = \mathbf{x}_{\max}$, is $t_{\max} = \mathbf{x}_{\max}$.
4. The hazard is constant; $h(t, \mathbf{x}) = \lambda$ up to $t_{\max}$.

The distribution T, given $\mathbf{x}$, is

$$f_t(t \mid \mathbf{x}) = \begin{cases} \lambda e^{-\lambda t}, & t < \mathbf{x}, \\ e^{-\lambda \mathbf{x}} \delta(t - \mathbf{x}), & t \geq \mathbf{x} \end{cases}. \tag{7.55}$$

Therefore, the distribution of observed life times, is

$$f(t) = \int_0^t e^{-\lambda y} \delta(t - y) \frac{y^{\alpha - 1} e^{-y/\beta}}{\Gamma(\alpha)\beta^\alpha} dy + \lambda e^{-\lambda t} \int_t^\infty \frac{y^{\alpha - 1} e^{-y/\beta}}{\Gamma(\alpha)\beta^\alpha} dy. \tag{7.56}$$

This satisfies "hard" curve squaring; i.e., each person has a realization, $\hat{\mathbf{x}}_{it}$, which determines $t_{\max}$ as $\theta\mathbf{x}$. Prior to $t_{\max}$ each individual has a constant hazard. At $t_{\max}$ the individual dies.

To construct a noncurve squaring hazard for $\alpha = 1$ in the gamma distribution of $\mathbf{x}^*$, we need to construct another $\mathbf{x}^*$. The density of T is

$$f(t) = \left(\lambda + \frac{1}{\beta}\right) e^{-t(\lambda + 1/\beta)}. \tag{7.57}$$

Matching coefficients of powers of t as in Lemma 1 we constrain $\mathbf{x}^*$ as

$$\gamma = 1,$$

$$E\mathbf{x}^{*K} = \left(\lambda + \frac{1}{\beta}\right)^k, \qquad k = 1, 2, \ldots. \tag{7.58}$$

Choosing the density of $\mathbf{x}^*$ to satisfy (7.58) gives the same mortality for $\mathbf{x}_{it}$ without curve squaring as for T with curve squaring. Thus, if heterogeneity ($\mathbf{x}_{it}$) exists, the hypothesis of curve squaring does *not* generate an identifiable, unique marginal density; i.e., the hypothesis is not testable. Though one can construct a model that produces curve squaring to fit the data, an alternative "fixed frailty" model *without* curve squaring can be constructed providing an identical fit.

In the proof, we assumed fixed frailty. The information necessary for identifiability is absent even if there is only *one* $\mathbf{x}_{it}$, i.e., one unobserved, fixed variable. More complex forms of heterogeneity generate an even wider variety of models without curve squaring, producing the same $f_{\mathbf{z}}(t)$. Thus, the results apply to models where more, and possibly time-varying, features of the failure process are described. To resolve the lack of identifiability, information must be available to describe $\mathbf{x}_{it}$, e.g., multiple outcomes of the mortality process for persons with *identical* $\mathbf{x}_{it}$. With the distribution of $\mathbf{x}_{it}$, variation in $T(i)$ can be controlled and curve squaring tested.

Monozygotic (MZ) twins have identical realizations $\hat{x}_{it}$ on genetic factors. Let $T_1 = T(i_1)$ and $T_2 = T(i_2)$ represent life times for such MZ twins. With the exception of $\hat{\mathbf{x}}_{it}$, the mortality of each twin is assumed independent, given $\mathbf{z}_{it}$ (which includes membership of twins). The distribution of $T(i)$ for MZ twins is

$$f_{\mathbf{Z}}(t_1, t_2) = \int h_{\mathbf{Z}}(t_1, \mathbf{x}) h_{\mathbf{Z}}(t_2, \mathbf{x}) \times \exp\left(-\int_0^{t_1} h_{\mathbf{Z}}(\tau, \mathbf{x})\, d\tau - \int_0^{t_2} h_{\mathbf{Z}}(\tau, \mathbf{x})\, d\tau\right) g_{\mathbf{Z}}(\mathbf{x})\, d\mathbf{x}\,. \tag{7.59}$$

Under curve squaring, the joint density is (see (7.51)),

$$\begin{aligned} f_{\mathbf{z}}(t_1, t_2) = {} & \int_{\mathbf{x} \notin B(t_1) \cap B(t_2)} h_{\mathbf{z}}(t_1, \mathbf{x}) h_{\mathbf{z}}(t_2, \mathbf{x}) \\ & \times \exp\left(-\int_0^{t_1} h_{\mathbf{z}}(\tau, \mathbf{x})\, dy - \int_0^{t_2} h_{\mathbf{z}}(\tau, \mathbf{x})\, d\tau\right) g_{\mathbf{z}}(\mathbf{x})\, d\mathbf{x} \\ & + \left[\int_{\mathbf{x} \in B(t_1)} (1 - F_{\mathbf{z}}(t(\mathbf{x}), \mathbf{x}))^2\, g_{\mathbf{z}}(\mathbf{x})\, d\mathbf{x}\right] \delta(t_1 - t_2)\,. \end{aligned} \tag{7.60}$$

Equation (7.52) is generalized similarly. Not all values of t_1 and t_2 are possible. There are many values of $\mathbf{x}$ for which t_1 equals t_2. Vaupel (1990a,b) proposed fixed frailty models for the joint distribution of the mortality of twins. For fixed frailty,

$$\begin{aligned} f_{\mathbf{Z}}(t_1, t_2) &= \lambda^2 \int \mathbf{x}^2 \exp(-\lambda \mathbf{x}(t_1 + t_2))\, g_{\mathbf{Z}}(\mathbf{x})\, d\mathbf{x} \\ &= \lambda^2 \sum_{k=0}^{\infty} E_{\mathbf{Z}} \mathbf{x}^{k+2} \frac{[-\lambda(t_1 + t_2)]^k}{k!}\,. \end{aligned} \tag{7.61}$$

Since t_1 and t_2 appear as a sum, their powers are generated using a binomial expansion. Though we can freely choose $E_{\mathbf{Z}}\mathbf{x}^{k+2}$, we *cannot* freely choose coefficients for t_1 or t_2. Thus, we cannot generate a *fixed* frailty model *without* curve squaring with the same density as for MZ twins *with* curve squaring. It *is* possible to generate non-curve squaring hazards with *time varying* $\mathbf{z}_{it}$ unless further assumptions are made. One such assumption is

Lemma 2. *If* $\mathbf{Z}=\mathbf{z}_0$ and *C holds, the* t_i *of twin pairs for "hard curve squaring" has the distribution*

$$f_{\mathbf{Z}}(t_1, t_2) = \left[\int_{\mathbf{x}\notin B(t_1)} (1 - F_{\mathbf{Z}}(t(\mathbf{x}), \mathbf{x}))^2 g_{\mathbf{Z}}(\mathbf{x})\, d\mathbf{x}\right] \delta(t_1 - t_2)\,. \qquad (7.62)$$

The proof of Lemma 2 follows from (7.60) and C. Lemma 2 states that as $\mathbf{z}_{it} \rightarrow \mathbf{z}_0$, the limiting joint distribution of T_1 and T_2 has all mass on a line $t_1 = t_2$. One cannot generate a non-curve squaring hazard where two independent random variables have such a joint distribution. This singular measure distribution implies that hard curve squaring in MZ twin studies is identifiable *under* C as $\mathbf{z} \rightarrow \mathbf{z}_0$.

For identifiability, twins must be "genetic replicates," (i.e., monozygotic) with mortality independent, given $\mathbf{z}_{it}$ *and* $\mathbf{x}_{it}$. Thus, unobserved *environmental* factors affecting mortality must be identical, i.e., have the same *realization* of $\hat{\mathbf{x}}_{it}$, *not* merely the same distribution. If this is violated, the environmental effect may possibly be identified using dyzygotic (DZ) twins and non-twin pairs raised in identical environments. Even here, Lemma 2 does *not* guarantee identifiability without condition C holding. Since C is unlikely (i.e., for persons with $\mathbf{z}_0$, mortality is 0 before $t_{\max}$), data on MZ twins and on DZ twins and unrelated individuals in the same environment are insufficient to test curve squaring.

Concretely, Johnson and Lithgow (1992) studied the genes of a nematode (*C. elegans*) in attempt to identify which ones might control longevity. They identified one (Age-1) that, when mutated, increased life span by 70%. Though the gene could be precisely located, the physiological mechanisms by which it regulates life span are not yet known. It appears that *C.elegans* with Age-1 mutants have high levels of SOD (super oxide dimutase)—an antioxidant enzyme. Thus, Age-1 may increase life span by controlling the expression of SOD, i.e., higher levels of SOD providing greater protection against free radical damage. However, assuming that the issue of bioavailability to the necessary physiological systems could be resolved, one could exogenously supplement antioxidants and produce the same effect on life span as Age-1. For example, a particular isomer of vitamin E (alpha tocopherol succinate), appears to have potent anticancer effects not shared by other vitamin E isomers (Psarad and Psarad-Edwards, 1992). Thus, extension of life span could be due to (a) genetic alteration of the rate of production of an endogenously produced antioxidant enzyme (SOD) or (b) exogenous supplementation of functionally equivalent, bioavailable antioxidants. Thus, the genetic regulation of physiological processes could be mimicked by appropriate environmental change. For example, several studies have shown that age-related loss of immune function could be reversed by antioxidant intake (Penn et al., 1991a,b; Chandra, 1992). Without measuring features of the physiological processes involved it is impossible to discriminate whether changes in life span are exogenously or endogenously derived.

A concrete human example is differences in Japanese and U.S. life expectancy. There are possibly genetic differences affecting life expectancy (e.g., the role of skin pigmentation in controlling calciferol production, which may affect both osteoporosis and atherogenesis; Moon et al., 1992) and known differences in diet (e.g., the role of phytoestrogens and phytates as antioxidants (Adlercreutz et al., 1991)).

One test of curve squaring is to eliminate population heterogeneity due to $\mathbf{x}_{it}$. Studies of insect cohorts each originating from a homogeneous gene pool and raised under identical laboratory conditions may have little irreducible variability (and C would apply approximately; even in this case variation existed in how often food was changed). This was the case in Curtsinger et al. (1992), where four genetically homogeneous groups of dropsophila were studied. In this study mortality tended to flatten (i.e., cease to rise with age) at extreme ages. Thus, a Gompertz function could not be satisfactorily fit to the data. This extended a study where 1.2 million fruit flies were studied (Carey et al., 1992) to determine the shape of the survival curve at advanced ages. In this larger study, mortality increased to some advanced age and then was constant— or even declined. The advantage of the Curtsinger et al. (1992) study was that the constancy of mortality observed at advanced ages (i.e., no curve squaring) could be shown in a relatively (to the usual study size) large ($n = 11{,}000$) population of generically homogeneous individuals. This eliminated the possibility that genetic heterogeneity and mortality selection produced a population average risk that did not change (or decline) with age. The Carey et al. (1992) study had the advantage of an extremely large population ($n = 1.2$ million), which allowed fine distinctions to be made about the shape of the survival curve at advanced ages. It was estimated that a population of 100,000 would be necessary to adequately discriminate between an empirical distribution of times to death and one generated by Gompertz function (i.e., soft curve squaring). This would be a lower-bounded estimate for a human population study since, in the insect studies, times to death for the entire population are known.

In human populations, neither heterogeneity nor premature death can be wholly eliminated so heterogeneity *must* be modeled. Even in the case of MZ twins, there are problems. Recently, considerable evidence has accumulated to show that chronic diseases (heart and lung disease), to relatively late ages (e.g., 65+), may be due to intra-uterine growth deficiencies (Barker et al., 1991a,b, 1992). In a study of the linkage of schizophrenia and diabetes, where a genetic linkage had been sought, it appears that a deficiency in zinc, which is an important nutrient for fetal growth, may have been operant (Andrews, 1992). Thus, since MZ twins share the same intra-uterine environment, if health effects are due to dysfunction in fetal development due to maternal nutritional deficiencies, they would be equally manifest in MZ twins. In this case, the concordance of effects would be the same for MZ and DZ twins—except for the possible effects of genetic environmental interactions (Weiss, 1990). Even in family pedigree studies, there are likely socio-economic and other traits that tend to be preserved socioculturally in a family over generations.

Thus, a model must test whether the hazard is age dependent *after* conditioning on a multivariate stochastic risk factor process, $\mathbf{z}_{it}$. For i, for any realization $\hat{\mathbf{x}}_{it}$, processes described by (7.7) represent changes of $\mathbf{z}_{it}$ as functions (a) of $\mathbf{z}_{it-1}$, (b) uncertainty, and (c) changes in survival ($dF_{\mathbf{z}}$) as a function of $\mathbf{z}_{it}$. As J increases, the information contained in a data set increases so that the sample size necessary to discriminate between several distributions can be smaller as the state space is made richer. This provides an additional rationale for basing the model on fuzzy state constructs, i.e., the ability to identify nonredundant information as J increases. The fuzzy state model also provides a nonparametric test for the effects of $\mathbf{x}_{it}$ on survival. That is, one can test whether the empirical distribution of time to failure can be generated in a K-dimensional space identified from the $\mathbf{z}_{it}$. If it can, then there is no evidence of $\mathbf{x}_{it}$ affecting survival conditional on $\mathbf{z}_{it}$. In addition, if J is large, then individual trajectories can be consistently estimated and compared (Woodbury et al., 1993).

By assumption, (7.7) and (7.9) can be used to estimate parameters of a mortality process for any fixed trajectory of $\mathbf{x}_{it}$. To test curve squaring, we assume (7.63) for $\mathbf{z}_{it}$ and represent the influence of $\mathbf{x}_{it}$ on mortality by assuming it produces a proportional change $s(\mathbf{x}_{it})$ in $\mu(\mathbf{z})$ which varies with age. At fixed t, the hazard is $\mu_1(\mathbf{z}) = \mu(\mathbf{z})s(\mathbf{x}_{it})$, where $\mu(\mathbf{z})$ is the quadratic form given in (7.6). As noted previously, for any t, the $s(\mathbf{x})$ averaged over values $\mathbf{x}_{it}$ are assumed to be $\mathbf{e}^{\theta t}$ for unknown θ. Thus, the marginal hazard across unknown $\mathbf{x}_{it}$ is

$$\int \mu(\mathbf{z})s(\mathbf{x})g_{\mathbf{Z}}(\mathbf{x})\, d\mathbf{x} = \mu(\mathbf{z})\, e^{\theta t}\,. \tag{7.63}$$

This is the well-known Gompertz hazard that has often been used to study mortality at advanced ages. It differs, of course, in that the exponential, $e^{\theta t}$ which represents the age-dependent force of mortality due to unknown age-correlated variables, is estimated conditionally on a function of known factors $\mathbf{z}_{it}$, where the $\mathbf{z}_{it}$ are time varying. The average of $s(\mathbf{x}_{it})$ represents the dispersion of $\mathbf{x}_{it}$ over age. As the $\mathbf{x}_{it}$ are measured (i.e., $\mathbf{x}_{it} \rightarrow \mathbf{z}_{it}$), the average of $s(\mathbf{x})$ decreases and $\mu(\mathbf{z})$ increases. If all $\mathbf{x}_{it}$ are "known," the average hazard is (7.6). Thus, changes in the parameter θ measure the effect of observed variables on the age dependence of mortality. Thus, it is one test of the "embeddability" of the process of known variables, i.e., a test of how well they represent the "state" of the person. They, likewise, adjust the estimates of the coefficients for the observed variables for the bias of unknown effects, e.g., transitions unobserved between measurement times.

In (7.9) a dynamic model for $\mathbf{z}_{it}$ and a time-dependent model for the *expectation* of the effect of unobserved risk factors ($e^{\theta t}$) on mortality are represented. Without curve squaring, μ depends on age only through $\mathbf{z}_{it}$ *or* the age-specific average of $\mathbf{x}_{it}$ ($e^{\theta t}$). With curve squaring, the hazard increases with age—even conditioned on $e^{\theta t}$. Thus, if curve squaring exists, including $\gamma_0 t$ (soft squaring) or $\gamma_0 \delta(t - t_0)$ (hard curve squaring) in the model should improve the fit of data over that using only $\mathbf{z}_{it}$ and $s(\mathbf{x}_{it})$. To understand how $e^{\theta t}$ and $\gamma_0 t$ differ, realize that both represent interactions of age with $\mathbf{z}_{it}$. When age effects are represented by $e^{\theta t}$, the interaction with each element of $\mathbf{z}_{it}$ reflects the same rate of increase θ, and age-related effects of $\mathbf{x}_{it}$ can be *factored* from $\mathbf{z}_{it}$. If age is included in $\mathbf{z}_{it}$, then each risk factor has a different interaction with age, and the effects of $\mathbf{x}_{it}$ do not factor; i.e., $\mathbf{x}_{it}$ effects on mortality cannot be isolated from $\mathbf{z}_{it}$ whose change is correlated with age. By including $\gamma_0 t$ an age effect is postulated not only for mortality but for each $\mathbf{z}_{it}$ with consequences for dynamics as in (7.9).

We examined the effects of introducing (a) age in $\mathbf{z}_{it}$ and (b) adding time-varying covariates to $\mathbf{z}_{it}$. In the first case, the coefficient γ for age, in the Framingham study, was not significant. Its effects were correlated with Gompertz hazard in (7.63), reducing θ from 8.05 to 5.18%; i.e., about 60% of the average effect of the unobserved factors associated with age were not accounted for by $\mathbf{z}_{it}$ (Manton and Tolley, 1991). In addition, $\hat{\mathbf{z}}_0$ is age variable (i.e., $\hat{\mathbf{z}}_0(t)$) and has an altered relation to the homeostatic point (i.e., the values towards which $\mathbf{z}_{it}$ tend). In the Framingham study, introducing age into $\mathbf{z}_{it}$ distorted the dynamic and mortality functions so that $\hat{\mathbf{z}}_0(t)$ entered physiologically infeasible ranges (e.g., $\hat{\mathbf{z}}_0(t)$ for pulse pressure was negative after age 45; Manton, 1988). The function with $e^{\theta t}$ remained in physiologically feasible ranges. Thus, though one could propose a model with $\gamma_0 t$, the use of an additional statistical criterion (i.e., the reproduction of physiologically meaningful trajectories of all state variables) might require rejection of $\gamma_0 t$ *within* commonly observed age ranges. In the data from the 1982 and 1984 NLTCS, as information for $\mathbf{z}_{it}$ was added into the hazard, θ dropped from 8.1 to 4.0% for males and from 9.4 to 3.6% for females, i.e., far more than when entering the 10 CVD risk factors.

Equations (7.7) and (7.63) show the difficulty of demonstrating curve squaring. If unobserved risk factors are measured, the effect of $e^{\theta t}$ is reduced, and the effect of $\mathbf{z}_{it}$ increases. If the new variables contain individual variability, this predicts $T(i)$ better so that the age constant γ_0 is reduced. If all factors determining age increases in mortality varied across individuals *and* time, then a common species-specific aging mechanism determining $t_{\max}$ would not exist. The use of the fuzzy state variables $\mathbf{g}_{it}$ in place of physiological measurements $\mathbf{z}_{it}$ in the mortality model has several advantages. First, high dimensional, discrete measurement spaces can be projected onto lower dimension fuzzy state spaces. Second, because of the identifiability of B by M the process has well-defined boundaries over time. Third, the condition that individuals have identical realizations can be translated into the $\mathbf{g}_{it}$ for two individuals being the same. Even if no individuals have identical realizations, comparable processes can be estimated conditional on a fixed set of moments of $\mathbf{g}_{it}$ to examine the commonality of the dynamic process elements net of any sampling or censoring effect.

Other models can be used to test the age dependency of the hazard if they represent both $\mathbf{x}_{it}$ *and* $\mathbf{z}_{it}$. If the age dependency of $\mathbf{z}_{it}$ is not represented, the hazard will manifest age dependency due to model misspecification (see, e.g., Ogata, 1980); i.e., the hazard appears to be time dependent because of $\mathbf{z}_{it}$.

Curve squaring could also be examined by comparing the mortality of a population assessed at several times (e.g., Myers and Manton, 1984; Rothenberg et al., 1991). In this case, tests for differences in the density of $T(i)$ are estimated from cross-sectional mortality data at two (or more) times. Under the hypothesis, the density for the later time, given by (7.52), is assumed "closer" to curve squaring (i.e., $\mathbf{z}_{it}$ is closer to $\mathbf{z}_0$) with more of the population living long enough to encounter $t_{\max}$. Lemma 1 shows we can produce fixed frailty hazards with densities identical to the t_i densities of the population at the two times if we specify the distribution of $\mathbf{x}_{it}$. If the distribution of $\mathbf{x}_{it}$ has *not* changed, there is sufficient information to test curve squaring. If $\mathbf{x}_{it}$ changes, it is impossible to determine if life expectancy limits are due to intrinsic or exogenous processes (e.g., influenza epidemics). By using GoM we can use nonparametric estimates of the density of t (i.e., fuzzy set specific life tables), rather than using a specific hazard.

7.8. EXAMPLES

Below we present examples of disability forecasts using the 1982, 1984, and 1989 NLTCS linked to mortality data from Medicare records for the period 1982–1991. Twenty-seven disability measures were subjected to a GoM analysis for persons interviewed in the 1982 and 1984 NLTCS ($N = 12{,}020$). We also analyzed the 1982, 1984, and 1989 ($N = 16{,}584$) surveys. The first analysis produced a six-profile solution where each fuzzy profile or extreme type is defined by $\hat{\lambda}_{kjl}$. The content of the profiles for the 1982 and 1984 NLTCS is described elsewhere (e.g., Manton and Stallard, 1991). The λ_{kjl} coefficients for the 1982–1984 NLTCS were presented in Table 5.2. Consequently, here we only provide short descriptions:

1. "Healthy" have few chronic impairments.
2. "IADL Impaired" have no ADL and few physical impairments (e.g., bending, holding, grasping) but do have IADL impairments associated

with cognition (e.g., problems phoning, managing money, and taking medication).

3. "Physically Limited" have no ADL and few IADL impairments, but moderate physical limitations (e.g., climbing stairs, holding, reaching for and grasping objects).
4. "Moderately Impaired" have bathing, several IADLs, and physical limitations,
5. "Impaired" have ADL and IADL impairments (*not* involving cognitive function) and impairment of upper body function.
6. "Highly Impaired" are heavily dependent on help for ADLs and IADLs.

Thus, the profiles describe six dimensions of disability, e.g., cognitive impairment (Class 2), upper body function (Class 4), lower body function and mobility (Class 5), and mixed or combined disability and frailty. In addition, a seventh discrete class is used to represent persons in institutions (about 5080 persons were institutional residents in 1982, 1984 and 1989). Institutional residence implies a high degree of impairment and loss of social autonomy.

The $\hat{\lambda}_{kjl}$ from the 1982, 1984, and 1989 NLTCS are in Table 7.2. Results from 1982, 1984, *and* 1989 are similar to those from the 1982 and 1984 NLTCS in Table 5.2. The primary differences are in Class 3, which has less impairment of outside mobility in 1989 (Table 7.2). At the bottom of the table are presented the average scores for each of the extreme types. These are also quite similar between the two data sets. Thus, the six dimensions, including the least disabled (Class 1), are reasonably stable constructs over time with the six descriptions above being valid for both the 1982–1984 and 1982–1984–1989 NLTCS analyses with the fuzzy state space coordinates on $J = 27$ discrete measures held fixed.

In Table 7.3 we present, for males and females the disability transition matrix ($\mathbf{C}_{KK}(t)$) evaluated at age ($t =$) 75 for the cohort *simulations* based on the 1982 and 1984 NLTCS described above (Manton et al., 1991). There are a number of differences in the disability transition matrices for males and females. The proportion remaining active is modestly higher for males (92.7 versus 92.2), while the proportion of females entering institutions is higher (i.e., 71.1 versus 67.9). In addition, the proportion returning to healthy states is fairly high, while the persistence in intermediate disability states is low. Thus, these matrices may better reflect the effects of short-term disability transitions. This matrix is combined with the quadratic function in Table 7.4 in a series of partial differential equations to generate a fuzzy state life table functions.

In Table 7.4 are quadratic hazards evaluated at age 75 (i.e., where $e^{(75.0)}$) for both males and females. The ratio between the mortality risk in the most (6) and least (1) disabled class is large (nearly 18 to 1) for both sexes. The θ

Table 7.2. Estimates of Response Profile Probabilities ($\lambda_{kjl} \times 100$) for the Combined 1982, 1984, and 1989 NLTCS ($N = 16{,}485$)

		Profiles					
Variable	Observed Frequency	Healthy	Moderate Cognitive Impairment	Mild Instrumental & Physical Impairment	Serious Physical Impairment	Moderate ADL & Serious Physical Impairment	Frail
ADL—needs help							
Eating	7.0	0.0	0.0	0.0	0.0	0.0	55.2
Getting in/out of bed	39.9	0.0	0.0	0.0	0.0	100.0	100.0
Getting around inside	39.9	0.0	0.0	0.0	0.0	100.0	100.0
Dressing	19.4	0.0	0.0	0.0	0.0	0.0	100.0
Bathing	43.1	0.0	0.0	0.0	0.0	100.0	100.0
Using toilet	21.7	0.0	0.0	0.0	0.0	49.9	100.0
Bedfast	0.8	0.0	0.0	0.0	0.0	0.0	5.5
No inside activity	1.5	0.0	0.0	0.0	0.0	0.0	10.2
Wheelchair fast	7.0	0.0	0.0	0.0	0.0	19.9	25.8
IADL—needs help							
With heavy work	71.9	14.5	100.0	100.0	100.0	100.0	100.0
With light work	22.6	0.0	35.5	0.0	0.0	0.0	100.0
With laundry	41.5	0.0	100.0	0.0	100.0	36.4	100.0
With cooking	29.8	0.0	100.0	0.0	0.0	0.0	100.0
With grocery shopping	56.9	0.0	100.0	0.0	100.0	100.0	100.0
Getting about outside	59.1	0.0	61.9	0.0	100.0	100.0	100.0
Traveling	52.9	0.0	100.0	0.0	100.0	100.0	80.3
Managing money	26.8	0.0	100.0	0.0	0.0	0.0	100.0
Taking medicine	23.5	0.0	100.0	0.0	0.0	0.0	100.0
Making telephone calls	16.0	0.0	87.3	0.0	0.0	0.0	85.5
Function Limitations							
How much difficulty do you have							
Climbing (1 flight of stairs)							
None	18.6	53.5	0.0	0.0	0.0	0.0	0.0
Some	29.1	46.6	88.5	33.8	0.0	0.0	0.0
Very difficult	31.4	0.0	11.5	66.2	50.7	73.0	10.9
Cannot at all	21.0	0.0	0.0	0.0	49.3	27.0	89.1

Bending, e.g., putting on socks							
None	43.5	100.0	100.0	0.0	0.0	100.0	0.0
Some	27.9	0.0	0.0	100.0	0.0	0.0	0.0
Very difficult	18.0	0.0	0.0	0.0	100.0	0.0	0.0
Cannot at all	10.6	0.0	0.0	0.0	0.0	0.0	100.0
Holding a 10 lb. package							
None	29.6	84.2	0.0	0.0	0.0	0.0	0.0
Some	18.1	15.9	58.6	38.9	0.0	24.9	0.0
Very difficult	15.9	0.0	41.4	61.1	0.0	30.3	0.0
Cannot at all	36.4	0.0	0.0	0.0	100.0	44.7	100.0
Reaching overhead							
None	56.1	100.0	100.0	0.0	0.0	100.0	0.0
Some	21.2	0.0	0.0	100.0	0.0	0.0	34.3
Very difficult	13.9	0.0	0.0	0.0	76.8	0.0	14.1
Cannot at all	8.8	0.0	0.0	0.0	23.3	0.0	51.6
Combing hair							
None	71.6	100.0	100.0	0.0	0.0	100.0	0.0
Some	16.0	0.0	0.0	100.0	42.77	0.0	33.68
Very difficult	7.0	0.0	0.0	0.0	57.23	0.0	11.54
Cannot at all	5.4	0.0	0.0	0.0	0.0	0.0	54.8
Washing hair							
None	55.8	100.0	100.0	0.0	0.0	100.0	0.0
Some	14.8	0.0	0.0	100.0	0.0	0.0	0.0
Very difficult	9.4	0.0	0.0	0.0	100.0	0.0	0.0
Cannot at all	20.0	0.0	0.0	0.0	0.0	0.0	100.0
Grasping an object							
None	66.0	100.0	100.0	0.0	0.0	100.0	24.6
Some	20.3	0.0	0.0	100.0	0.0	0.0	34.3
Very difficult	10.1	0.0	0.0	0.0	95.5	0.0	14.8
Cannot at all	3.6	0.0	0.0	0.0	4.5	0.0	26.3
Can you see well enough to read a newspaper	74.3	100.0	0.0	100.0	100.0	100.0	45.4
Mean Scores ($\bar{g}_k \times 100$)							
1982–1984 NLTCS	33.7	11.9	13.5	9.0	16.5	15.2	
1982, 1984, 1989 NLTCS	34.4	11.6	13.1	9.1	16.7	15.3	

Table 7.3. Estimated Transition Parameters × 100 at Age 75, C_{75}, under Disability Prevention Assumption

Class (1982)	Class (1984) 1 Healthy	2 Early Cognitive Impairment	3 Moderately Physical Impairment	4 Heavy Physical Impairment	5 Frail	6 Highly Impaired	Institutional
Males							
1. Healthy	92.72	0.95	0.82	0.48	1.81	1.47	1.74
2. Early cognitive	49.87	18.84	8.22	3.33	6.50	10.80	2.44
3. Moderate physical impairment	43.17	7.56	9.39	4.72	18.73	8.76	7.67
4. Heavy physical impairment	33.37	8.27	5.87	11.36	19.81	12.27	9.04
5. Frail	52.01	3.25	4.18	5.24	20.91	10.48	3.93
6. Highly impaired	37.23	4.12	3.31	3.24	6.49	34.22	11.30
7. Institutional	23.82	0.76	1.68	0.85	4.94	0.00	67.94
Females							
1. Healthy	92.18	1.15	1.46	0.87	1.86	1.03	1.46
2. Early cognitive	40.14	11.26	7.33	5.68	11.59	7.89	16.12
3. Moderate physical impairment	46.31	5.42	12.69	9.18	10.25	7.44	8.70
4. Heavy physical impairment	48.34	4.47	7.28	12.51	6.77	12.98	7.65
5. Frail	40.08	4.87	8.72	7.45	26.58	8.18	4.12
6. Highly impaired	33.43	6.17	5.55	11.17	9.79	24.72	9.17
7. Institutional	20.44	0.90	1.59	2.56	2.54	0.85	71.12

Source: 1982 and 1984 National Long Term Care Survey.

Table 7.4. Estimated Quadratic Mortality Hazard Parameters × 100, at Age 75, $1/2 \cdot B \cdot \exp(75\theta)$*

	Class (1984)						
	1	2	3	4	5	6	
Class (1982)	Healthy	Early Cognitive Impairment	Moderately Physical Impairment	Heavy Physical Impairment	Frail	Highly Impaired	Institutional
			Males				
1. Healthy	3.79	6.17	8.45	0.89	6.55	15.91	9.20
2. Early cognitive impairment		10.05	13.76	1.45	10.66	25.91	14.98
3. Moderate physical impairment			18.83	1.99	14.60	35.47	20.51
4. Heavy physical impairment				0.21	1.54	3.75	2.17
5. Frail					11.32	27.50	15.90
6. Highly impaired						66.82	38.63
7. Institutional							22.33
			Females				
1. Healthy	1.95	3.83	2.77	1.91	3.28	8.27	5.49
2. Early cognitive impairment		7.52	5.43	3.75	6.44	16.24	10.78
3. Moderate physical impairment			3.93	2.71	4.66	11.74	7.80
4. Heavy physical impairment				1.87	3.22	8.11	5.38
5. Frail					5.52	13.92	9.24
6. Highly impaired						35.07	23.29
7. Institutional							15.47

* Lower off-diagonal elements suppressed because they are identical to corresponding upper off-diagonal element.

Source: 1982 and 1984 National Long Term Care Survey

for both groups are reduced from values estimated for the population without disability, i.e., dropping from 8.1 to 4.0% for males (−50.6%) and from 9.4 to 3.6% for females (−61.2%). This changes the general shape of the mortality curve from a highly nonlinear one (e.g., $\theta = 9.4\%$ implies strongly nonlinear age dependence of mortality) to one approximately linear. For example, the time to double mortality with $\theta = 9.4\%$ is 7.7 years, while, conditional on functioning, the $\theta = 3.6\%$ implies a doubling time of 19.6 years. Though the doubling time is long, the absolute risk level of certain groups is very high.

The reduction in θ indicates most effects of age on mortality operate through functional impairment changes with age. This is reasonable for mortality at advanced ages. The samples in the 1982, 1984, and 1989 NLTCS represent persons from age 65 to 118. At advanced ages, physical activity, which is strongly regulated by the 27 disability measures (10 of these are scaled physical performance measures), is a major determinant of risk factor values (e.g., Lochen and Rasmussen, 1992). For example, in males better physical fitness decreased blood glucose, blood pressure, cholesterol, heart rate, and raised BMI and oxidative capacity; Drexler et al., 1992), and the presence of selected medical conditions increased overall mortality (Paffenbarger et al., 1986; Blair et al., 1989; Linsted et al., 1991). Physical activity also affects immune (e.g., natural killer cell activity; an important factor in cancer risks; Kusaka, 1992) and pulmonary function. In one series of institutional patients 56% of deaths in an old (mean age 84.5 years) institutionalized population were due to congestive heart failure, pneumonia, or pulmonary embolism—all of which can be related to inanition and hypostasis. In another series of Alzheimer's patients, 70% of deaths were due to pneumonia—again, related to a lack of physical activity, loss of pulmonary function, and increased infectious disease risks. Thus, the 27 discrete measures reflect well a higher dimension, continuous physiological state space determining the catastrophic failure of the organism. The dynamics of the 27 measures can be parsimoniously represented by the K fuzzy state dimensions so that fuzzy state based partial differential equations can be evaluated to examine state dynamics up to the resolution offered by the data.

The decline in θ was less in analyses of 10 chronic disease risk factors for the 34-year Framingham follow-up. For males, without covariates, $\theta = 9.4\%$ versus 8.1% with risk factors (−14.4%). For females, θ dropped from 10.1 to 8.1% (a 19.6% decline). The proportion of the total χ^2 explained by the risk factors (without the Gompertz hazard) was 61% for males and 45.6% for females. The proportion of the total χ^2 for the 1982 and 1984 NLTCS explained by the $g_{ik \cdot t}$ was 92% for males and 95% for females. Thus, functional measures are better predictors of the age state and mortality risks of individuals age 65 to 118 than are the 10 chronic disease risk factors. It may be that, as impairment emerges, it restricts physical activity, nutrition, and other maintenance functions initiating negative feedback (Drexler et al.,

1992; Harris and Feldman, 1991). This is important for short-term disability transitions in the frail, elderly population. It is important to estimate structural equation systems to determine if effects are due to impairment, or operate through impairment, from risk factors to chronic disease.

One issue in combining the 1982, 1984, and 1989 data to estimate the parameters of a process is that one survey interval is two years (i.e., 1982–1984) and the other is five years (1984–1989). If there are many more unobserved transitions in 1984–1989 than in 1982–1984, then the question of embeddability may be raised. In Table 7.5 we present estimates of θ parameters for the 1982–1984 and 1982–1989 NLTCS data—and for the 1982–1989 data stratified by education.

When the quadratic is re-estimated using the three surveys (and mortality from 1982 to 1991) the θ for males increased to 5.3 from 4.0% and for females to 4.8 from 3.6%. This indicates that, over the five-year interval, additional temporal variation (e.g., intermediate disability change) occurs. The θ, representing the average age-specific effect of unobserved factors affecting mortality, adjusts for this to reduce the bias in coefficients relating mortality to the disability scores. Thus, θ helps improve the model's

Table 7.5. χ^2 Values Associated with the State Variables $\mathbf{x}_t$ and with θ_1, the Total Mortality functions

	Males	χ^2/χ_1^2	Females	χ^2/χ_1^2
1982–1984 National Long Term Care Survey				
1. Full process (θ and $\mathbf{x}_t$)	3206.5	100.0%	4728.1	100.0%
2. $\theta = 0$	2945.9	91.8%	4467.5	94.5%
3. Effect of θ	260.6	8.2%	260.6	5.5%
	($\theta = 4.0\%$)		($\theta = 3.6\%$)	
4. θ alone (no $\mathbf{x}_t$)	1163.6	36.3%	2210.6	46.8%
	($\theta = 8.1\%$)		($\theta = 9.4\%$)	
1982–1984–1989 National Long Term Care Survey				
1. Full process (θ and $\mathbf{x}_t$)	4837.9	100.0%	7493.8	100.0%
2. $\theta = 0$	3821.5	78.9%	6550.5	87.4%
3. Effect of θ	1016.4	21.1%	943.3	12.6%
	($\theta = 5.3\%$)		($\theta = 4.8\%$)	
4. θ alone (no $\mathbf{x}_t$)	1163.6	55.0%	4466.0	59.6%
	($\theta = 8.2\%$)		($\theta = 9.1\%$)	
Low education (8 years or less)				
1. Full process (θ and $\mathbf{x}_t$)	1902.76		3075.9	
	($\theta = 5.6\%$)		($\theta = 5.0\%$)	
High education (9 years or more)				
1. Full process (θ and $\mathbf{x}_t$)	3025.29		5073.7	
	($\theta = 5.1\%$)		($\theta = 2.9\%$)	
$\Delta\chi^2$ due to education stratification	90.2		654.8	

embeddability in the continuous time process when there are more changes in the five-year interval.

Overall, we see that the functional scores in the 1982–1989 data explain 78.9–87.4% of the age dependency (as expressed by the change in χ^2) in the extremely elderly population (i.e., age range 65 to 118; mean age about 80). In Table 7.5 we also present results for 1982–1989 stratified on education. The θ for the two strata average to the θ value for the total populations.

We see that the θ is larger for the low education group (i.e., there are more unobserved factors not well "regulated" by "low" education) for both males and females and that the effect of education on females is much larger ($\Delta\chi^2 = 654.8$; +8.7%) than for males ($\Delta\chi^2 = 90.2$; +1.9%). From the tables we also see ways to decompose the θ into (a) the effect of unobserved factors correlated with age and (b) the effect of unobserved transitions within a longer measurement interval. Specifically, we partitioned the effect of θ into changes between the 1982–1984 and 1982–1989 data (i.e., the effects of a longer measurement interval with more unobserved transitions), the effects due to the $\mathbf{x}_{it}$ (or $\mathbf{g}_{it}$) process, and the effects of an observed variable, education. In this latter case we find that there are (a) larger increases in life expectancy and active life expectancy for low education groups and (b) that more persons were moving into the high education group with time—especially above age 85. These changes in education could be modeled as a discretely mixed, fuzzy state process using the equations for the discretely mixed Gaussian process (with education groups being the discrete states) with the generalized diffusion term.

In Table 7.6 we present the cohort simulations for 1982, 1984 (mortality for 1982 to 1986), and the observed values for the three surveys 1982 to 1989 (and mortality 1982 to 1991).

Life expectancies at age 65 are higher for the 1982–1989 NLTCS data than for the cohort simulation, i.e., for males 15.6 versus 15.4 years and for females 20.9 versus 20.5 years. In general, though, the 1982–1984 cohort simulation did reasonably well in describing the functional trajectories in the 1982–1984–1989 data.

A second modification of dynamics was to reduce the variance in each of the seven categories to *zero* by setting $V_{t+1} = 0$, conditional on $V_{t_0} = 0$. The life table produced (based on the 1982–1989 data) represents the survival of a population of individuals whose initial proportional distributions in the R classes are *identical* to initial means, i.e., g_{ikt_0}, and whose trajectory of future values is described by $\mathbf{C}_t$ in (7.44), with $\nu_t^* = \nu_t$ for all t. This is equivalent to conducting projections with a *crisp* set model where heterogeneity only exists between sets. Since mortality is a quadratic function of $\mathbf{g}_{it}$, Jensen's inequality leads us to expect mortality to be lower in the crisp set projections, i.e., since $E(\mu(\mathbf{g}_{it})) > \mu(E(\mathbf{g}_{it}))$. It is important to determine whether $\mathbf{C}_t$ maintains this inequality with the mortality in (7.36) and (7.37). These life tables are in Table 7.7.

The fuzzy and crisp set projections start out with similar life expectancies

Table 7.6. Simulation, Baseline Cohort Life Tables and Age Specific Mean g_{ik}

Age	Baseline Cohort	Lx	Ex	% of Original Cohort Healthy	Healthy	Moderate Cognitive Impairment	Mild Instrumental and Physical Impairment	Serious Physical Impairment	Moderate ADL and Serious Physical Impairment	Frail	Institutionalized
						Males					
65	1982–1984 cohort simulated	100.0	15.4	92.7	92.7	0.7	1.1	0.8	2.2	1.5	1.0
	Observed 1982, 1989	100.0	15.6	92.5	92.5	3.9	0.7	0.6	1.0	1.0	0.4
75	1982–1984 cohort simulated	66.3	10.7	60.7	91.6	1.3	1.1	0.7	1.9	1.7	1.8
	Observed 1982, 1989	69.1	10.3	59.2	85.6	8.4	0.8	1.0	1.8	1.3	1.2
85	1982–1984 cohort simulated	32.6	6.6	25.6	78.5	4.6	2.0	1.7	3.9	4.2	5.1
	Observed 1982, 1989	32.9	6.1	23.9	72.4	12.2	1.3	2.4	3.6	2.7	5.3
95	1982–1984 cohort simulated	7.1	4.4	4.6	65.3	6.0	2.7	1.4	6.5	7.3	10.9
	Observed 1982, 1989	6.3	3.4	3.5	56.1	14.0	1.6	5.3	6.7	5.4	11.0
105	1982–1984 cohort simulated	0.7	3.5	0.5	68.1	6.0	2.7	1.3	6.3	6.6	9.0
	Observed 1982, 1989	0.3	2.4	0.1	54.5	17.5	0.2	3.5	11.2	3.0	10.2
						Females					
65	1982–1984 cohort simulated	100.0	20.5	91.2	91.2	0.9	1.7	2.2	1.4	1.4	1.1
	Observed 1982, 1989	100.0	20.9	92.3	92.3	3.3	1.0	0.7	1.2	0.8	0.8
75	1982–1984 cohort simulated	80.6	14.2	70.6	87.6	1.5	2.3	1.8	2.7	1.7	2.5
	Observed 1982, 1989	82.6	14.1	69.2	83.8	7.9	1.3	1.2	2.3	1.4	2.2
85	1982–1984 cohort simulated	54.0	8.5	37.1	68.8	4.4	3.6	2.3	5.3	4.0	11.5
	Observed 1982, 1989	55.1	8.5	35.4	64.3	12.2	2.2	2.8	5.6	3.5	9.5
95	1982–1984 cohort simulated	18.2	5.5	8.5	46.8	8.2	4.1	3.1	5.9	8.6	23.4
	Observed 1982, 1989	19.5	5.0	9.0	46.1	11.2	1.9	5.0	6.0	6.2	23.7
105	1982–1984 cohort simulated	3.0	4.5	1.5	49.6	8.5	4.3	3.2	6.1	8.6	19.7
	Observed 1982, 1989	2.5	3.7	1.2	47.2	10.8	1.1	5.0	4.9	5.2	25.8

Source: 1982 and 1984 National Long Term Care Survey

Table 7.7. Life Tables and Age Specific-Means (×100) g_{ik} for Males and Females, with and without Within-Cell Variance Eliminated; 1982, 1984 and 1989 NLTCS

				Class						
Age		l_t	e_t	1	2	3	4	5	6	Institutional
					Males					
65	Fuzzy set	100.0	15.6	92.5	3.9	0.7	0.6	1.0	1.0	0.4
	Crisp set	100.0	15.4	92.5	3.9	0.7	0.6	1.0	1.0	0.4
75	Fuzzy set	69.2	10.3	85.6	8.4	0.8	1.0	1.8	1.3	1.2
	Crisp set	69.2	9.8	83.1	9.0	1.0	1.3	2.1	2.3	1.4
85	Fuzzy set	33.0	6.1	72.5	12.2	1.3	2.4	3.6	2.7	5.3
	Crisp set	32.5	5.3	60.9	13.0	2.0	3.7	4.7	6.8	9.0
95	Fuzzy set	6.3	3.4	56.2	13.9	1.6	5.3	6.7	5.4	11.0
	Crisp set	4.0	2.3	24.1	10.4	2.8	8.6	7.7	15.1	31.2
105	Fuzzy set	0.3	2.4	54.6	17.5	0.2	3.5	11.2	3.0	10.2
	Crisp set	0.0	1.1	5.1	6.2	0.9	4.9	10.0	24.7	48.0
					Females					
65	Fuzzy set	100.0	20.9	92.3	3.3	1.0	0.7	1.2	0.8	0.8
	Crisp set	100.0	20.9	92.3	3.3	1.0	0.7	1.2	0.8	0.8
75	Fuzzy set	82.6	14.1	83.8	7.9	1.3	1.2	2.2	1.4	2.2
	Crisp set	83.3	14.0	81.8	8.4	1.6	1.4	2.7	1.8	2.3
85	Fuzzy set	55.2	8.5	64.3	12.2	2.2	2.8	5.6	3.5	9.5
	Crisp set	56.7	8.0	55.0	12.9	3.2	3.5	7.5	5.7	12.2
95	Fuzzy set	19.5	5.0	46.2	11.2	1.9	5.0	5.9	6.2	23.6
	Crisp set	18.7	4.0	20.4	9.0	3.3	6.1	8.7	13.4	39.2
105	Fuzzy set	2.5	3.8	47.3	10.8	1.1	5.0	4.9	5.2	25.7
	Crisp set	1.1	2.3	5.0	3.8	1.6	3.7	5.2	14.1	66.5

and disability distributions. The trajectories diverge with age. By age 85 male life expectancy is 6.1 years in the fuzzy set projections and 5.3 years in the crisp set projections—a decline of 13.1%. A similar decline is observed for females after age 75. In addition to the differences in life expectancy, there are even larger differences in the distributions of disability. For example, while the prevalence of male active life expectancy is 56.2% in the fuzzy set projections at age 95, it is only 24.1% in the crisp set model. The proportion in the two most severely disabled classes at age 95 is 16.4% in the fuzzy set model and 46.3% in the crisp set model. A similar pattern is noted for females. The reason for these differences is that in the crisp set model a person is in a single category so that, to have the high mortality levels observed at age 95, he must move exclusively into the most severely disabled classes. In the fuzzy set model, a person can begin to increase in disability and have his mortality incrementally increased, due to the interactions of the various disability dimensions in the quadratic function.

The age trajectories of disability are presented in Figure 7.1. We have grouped the two least impaired and two most impaired classes. The age trajectories are wholly different between the fuzzy and crisp set projections. Almost all persons move into the most disabled classes in the crisp set projections in order to achieve the high age specific mortality. This does not occur for the fuzzy set projections where disability is continuously graded at the individual level.

To see what this implies for survival curves, specific to income (stratified at the median for the year) and education (stratified at 8 years of education) groups, we present Figure 7.2. Both variables have a large effect on survival with the effect of income and education being greater for females. Since new elderly cohorts are increasing in income and education, we can expect significantly higher life expectancies.

This can be examined further in Figure 7.3, where we present "phase" plots among survivors to each age; i.e., we present the $\bar{g}_k$ for each age so that the effect of mortality is removed. Disability increases much more sharply with age in the 1982–1984 data than in the 1982–1989 data. Not only is disability less in the 1982–1989 data but the age trajectory rises more slowly from age 75 to 95. Above age 90 to 95 disability remains at a high constant level— similar to the high constant level of mortality found in the Carey et al. (1992) and Curtsinger et al. (1992) experimental mortality studies. The fact that disability reached a "plateau" seems reasonable in that, if disability were to get much higher, since disability is such a strong predictor of mortality, mortality at those advanced ages would become implausibly high. This "plateau" is also consistent with the age trajectories found for CVD risk factors found in the 34-year Framingham follow-up, where, at age 90 to 100, mortality due to the exponential term becomes so high that survivors are highly selected to have good or "optimal" values on their risk factors. Indeed, for some risk factors the trajectory of the mean values "rebound" and begin to improve just like the disability trends in

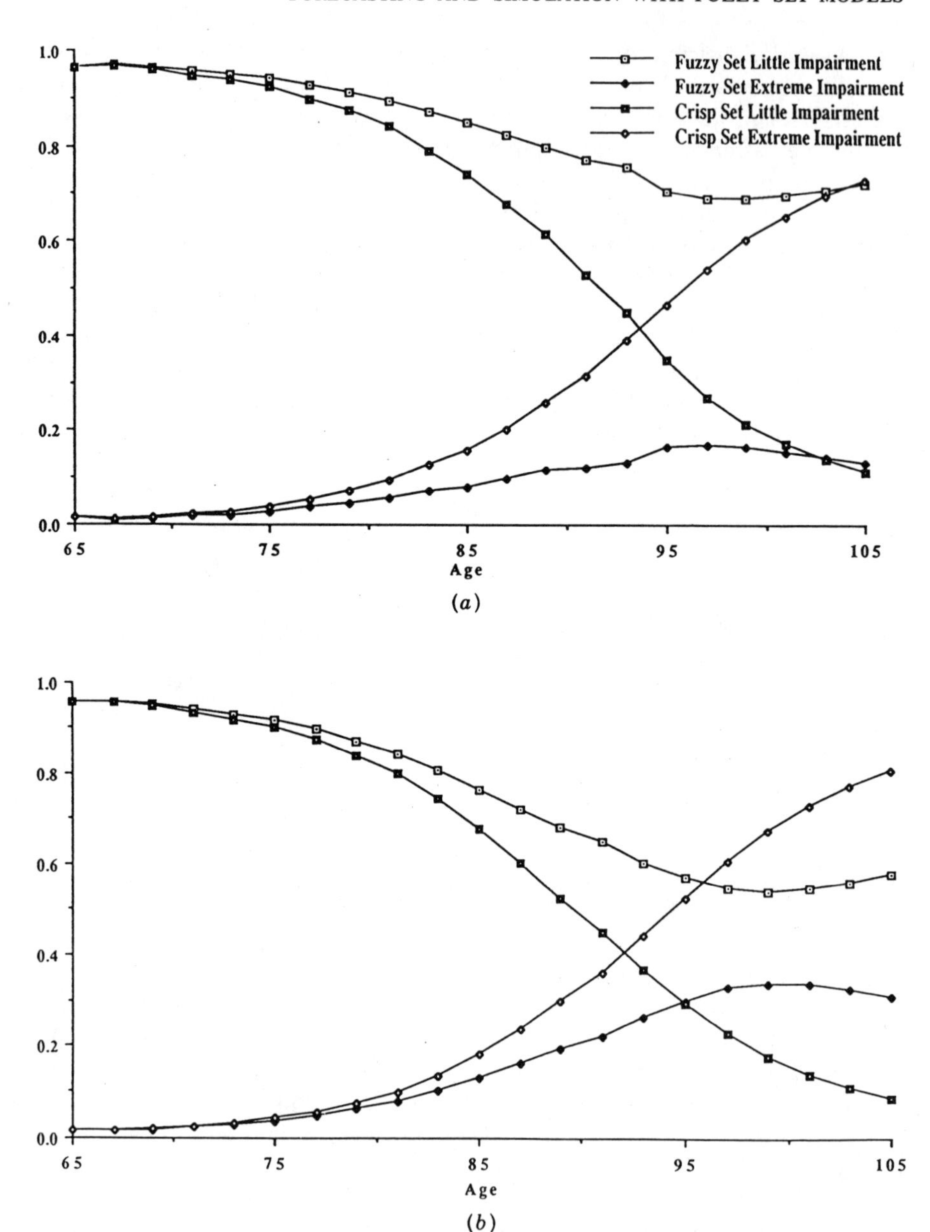

Figure 7.1. Comparing crisp to fuzzy sets. (*a*) Males by health status. (*b*) Females by health status.

Figure 7.3. The fact that a similar pattern is noted for disability and risk factors simply recognizes that the functional state scores are better predictors of mortality risks than risk factors. Thus, for these scores to continue an

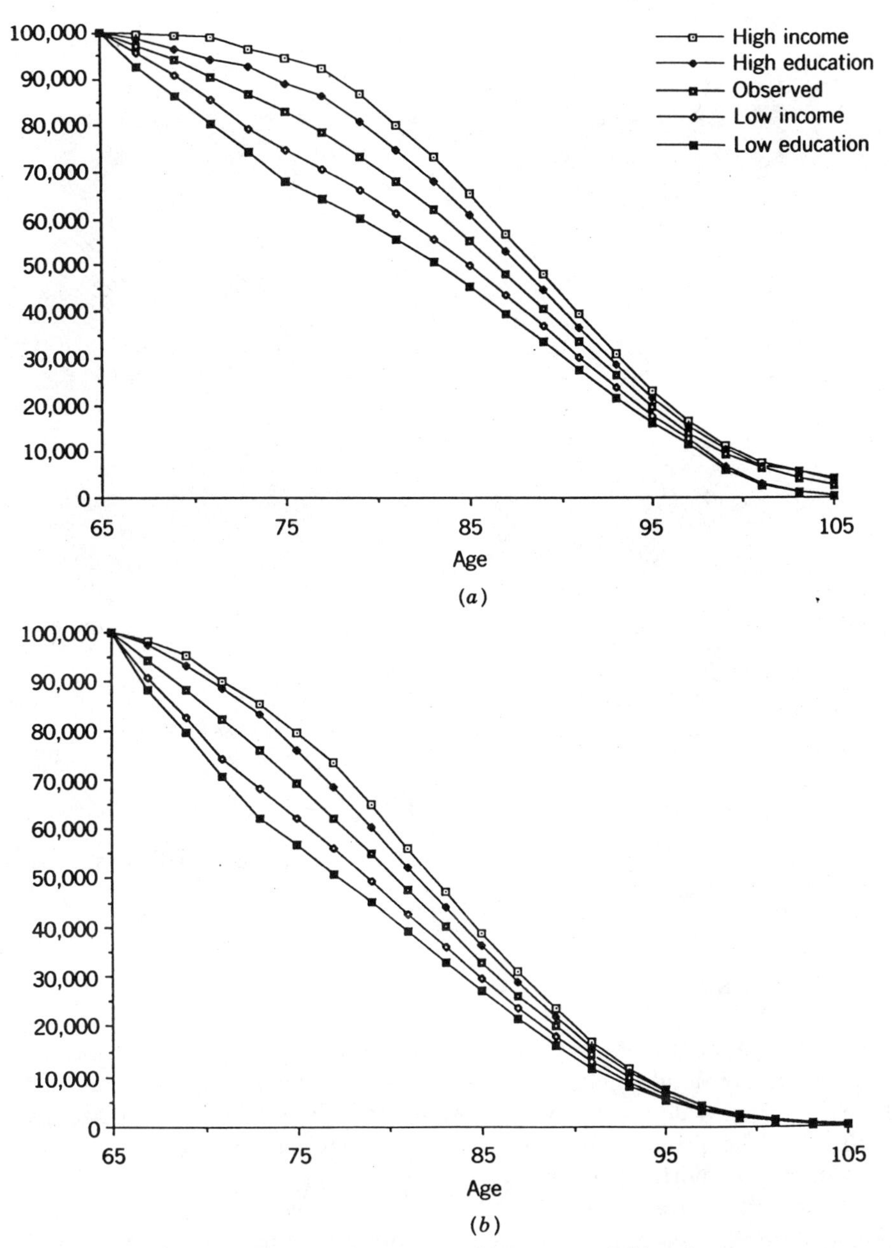

Figure 7.2. Survival curves: (*a*) female, and (*b*) male.

exponential increase would imply a very rapid increase in mortality at latter ages. These nonlinear disability dynamics are reasonable—otherwise we would almost never see survivors to age 100. It also suggests that survival models based on extrapolation of mechanisms observed from age 50 to 90 to

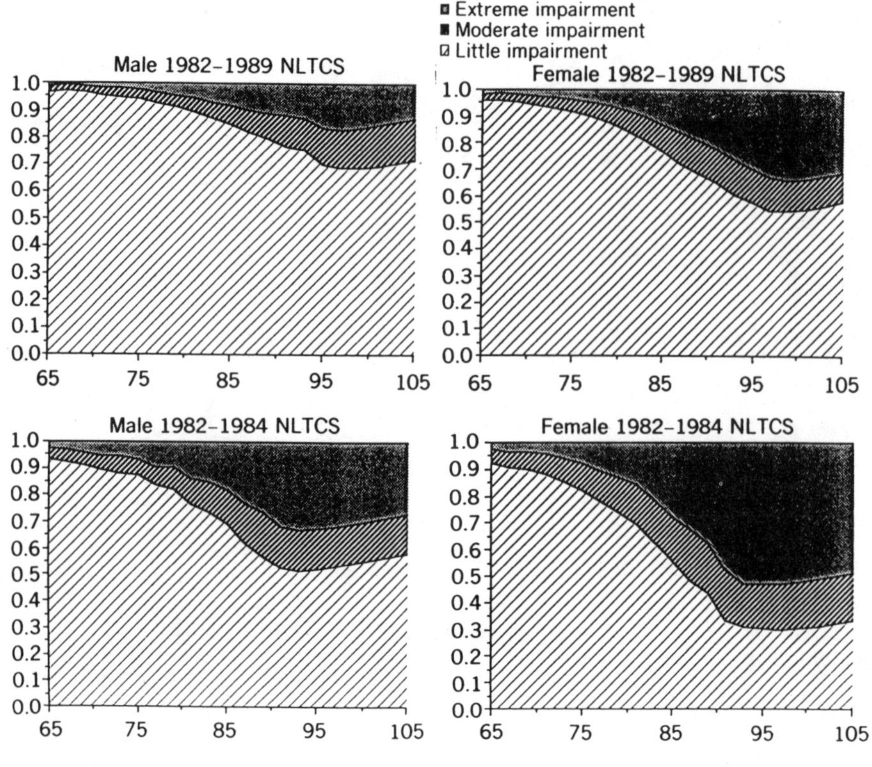

Figure 7.3. Phase plots.

more advanced ages need careful evaluation and possibly adjustment to reflect results in ancillary studies.

7.9. SUMMARY

We examined the use of fuzzy sets to forecast the size of the disabled elderly population. By simulating a cohort by sample reweighting of the 1982–1984 NLTCS we provided a reasonable representation of changes in disability with age. The projections based on these interventions matched cohort projections of mortality made by the SSA actuaries and were similar to results based on the 1982–1989 NLTCS. This suggests that much of the cohort difference in mortality could be accomplished by preserving physical function at latter ages—rather than disease elimination by risk factor control. Changes in the distribution of disability have important policy implications because of the expected rapid growth of the elderly and oldest-old U.S. population. Changes in the rate of growth of the oldest-old population, and in the distribution of disability, have important fiscal consequences because of high per capita LTC needs.

In addition, a comparison of the age trajectory of disability and life expectancy for crisp and fuzzy set models shows they produce different results. Since it is implausible that crisp sets capture all variation in function, this suggests there is a large systematic bias in crisp set forecasts. Since the flow of health costs over time in a health care institution may be viewed as a stochastic process governed by the presentation of different types of cases, we could expect the real time-expenditure patterns to be different in a crisp versus fuzzy set based reimbursement system. Thus, there are other risk-based calculations (e.g., of service reimbursement) that will be similarly biased; i.e., the crisp set model is not a realistic model of population change or of health service use and provider risks.

REFERENCES

Adlercreutz, H., Honjo, J., Higashi, A., Fotsis, T., Hamalainen, E., Hasegawa, T., and Okada, H. (1991). Urinary excretion of lignans and isoflavonoid phytoestrogens in Japanese men and women consuming a traditional Japanese diet. *American Journal of Clinical Nutrition* **54**: 1093–1100.

Andrews, R. C. R. (1992). Diabetes and schizophrenia: genes or zinc deficiency? *The Lancet* **340**: 1160.

Archambeau, J. O., Heller, M. B., Akanuma, A., and Lubell, D. (1970). Biologic and clinical implications obtained from the analysis of cancer growth curves. *Clinical Obstetrics and Gynecology* **13**: 831–856.

Armitage, P., and Doll, R. (1961). Stochastic models for carcinogenesis. In *Proceedings of the Fourth Berkeley Symposium on Mathematical Statistics and Probability*, Vol. IV. *Biology and Problems of Health* (J. Neyman, Ed.). University of California Press, Berkeley, pp. 19–38.

Barker, D. J. P., and Martyn, C. N. (1992). The maternal and fetal origins of cardiovascular disease. *Journal of Epidemiology and Community Health* **46**: 8–11.

Barker, D. J. P., Bull, A. R., Osmond, C., and Simmonds, S. J. (1991a). Fetal and placental size and risk of hypertension in adult life. *British Medical Journal* **301**: 259–262.

Barker, D. J. P., Godfrey, K. M., Fall, C., Osmond, C., Winter, P. D., and Shaheen, S. O. (1991b). Relation of birth weight and childhood respiratory infection to adult lung function and death from chronic obstructive airways disease. *British Medical Journal* **303**: 672–675.

Barker, D. J. P., Meade, T. W., Fall, C. H. D., Lee, A., Osmond, C., Phipps, K., and Stirling, Y. (1992). Relation of fetal and infant growth to plasma fibrinogen and factor VII concentrations in adult life. *British Medical Journal* **304**: 148–152.

Barrett, J. C. (1925). The mortality of centenarians in England and Wales. *Archives for Gerontological Geriatrics* **4**: 211–218.

Bayo, F. R., and Faber, J. F. (1985). Mortality experience around age 100. *Transactions of the Society of Actuaries* **35**: 37–59.

Blair, S. N., Kohl, H. W., Paffenbarger, R. S., Clark, D. G., Cooper, K. H., and Gibbons, L. W. (1989). Physical fitness and all-cause mortality: A prospective study of healthy men and women. *Journal of the American Medical Association* **262:** 2395–2401.

Carey, J. R., Liedo, P., Orozco, D., and Vaupel, J. W. (1992). Slowing of mortality rates at older ages in large medfly cohorts. *Science* **258:** 457–460.

Chandra, R. K. (1992). Effect of vitamin and trace-element supplementation on immune responses and infection in elderly subjects. *The Lancet* **340**: 1124–1127.

Chiang, C. L. (1968). *Introduction to Stochastic Processes in Biostatistics*. Wiley, New York.

Clive, J., Woodbury, M. A., and Siegler, I. C. (1983). Fuzzy and crisp set-theoretic-based classification of health and disease. *Journal of Medical Systems* **7**: 317–322.

Curtsinger, J. W., Fukui, H. H., Townsend, D. R., and Vaupel, J. W. (1992). Demography of genotypes: Failure of the limited lifespan paradigm in *Drosophila melanogaster*. *Science* **258**: 461–464.

Drexler, H., Reide, S. U., Munzel, T., Konig, H., Funke, E., and Just, H. (1992). Alterations of skeletal muscle in chronic heart failure. *Circulation* **85**: 1751–1759.

Dubey, S. D. (1967). Some percentile estimators of Weibull parameters. *Technometrics* **9**: 119–129.

Epstein, F. H. (1992). Low serum cholesterol, cancer and other noncardiovascular disorders. *Atherosclerosis* **94**: 1–12.

Fiatarone, M. A., Marks, E. C., Ryan, N. D., Meredith, C. N., Lipsitz, L. A., Evans, W. J. (1990). High-intensity strength training in nonagenarians. *Journal of the American Medical Association* **263**: 3029–3034.

Finch, C. E. (1990). *Longevity, Senescence, and the Genome*. The University of Chicago Press, Chicago.

Frank, J. W., Reed, D. M., Grove, J. S., and Benfante, R. (1992). Will lowering population levels of serum cholesterol affect total mortality? Expectations from the Honolulu Heart Program. *Journal of Clinical Epidemiology* **45**(4): 333–346.

Gillespie, D. T. (1983). The mathematics of simple random walks. *Naval Research News* **35**: 46–52.

Harris, T. B., and Feldman, J. J. (1991). Implications of health status in analysis of risk in order persons. *Journal of Aging and Health* **3**: 262–284.

Iso, H., Jacobs, D. R., Wentworth, D., Neaton, J. D., and Cohen, J. D. (1989). Serum cholesterol level and six-year mortality from stroke in 350,977 men screened from the Multiple Risk Factor Intervention Trial. *New England Journal of Medicine* **320**: 904–910.

Jacobs, D., Blackburn, H., Higgins, M., et al. (1992). Report of the conference on low blood cholesterol: mortality association. *Circulation* **86**(3): 1046–1060.

Jacques, J. A. (1972). *Compartmental Analysis in Biology and Medicine*. Elsevier, Amsterdam.

Johnson, C. E., and Lithgow, G. J. (1992). The search for the genetic basis of aging: The identification of gerontogenes in the nematode, *Caenorhabditis elegans*. *Journal of American Geriatrics Society* **40**: 936–945.

Kaufmann, A., and Gupta, M. M. (1935). *Introduction to Fuzzy Arithmetic: Theory and Applications*. Van Nostrand Reinhold, New York.

Kusaka, Y., Kondou, H., and Morimoto, K. (1992). Healthy lifestyles are associated with higher natural killer cell activity. *Preventive Medicine* **21**: 602–615.

Kushner, H. J. (1974). On the weak congergence of interpolated Markov chains to a diffusion. *Annals of Probability* **2**: 40–50.

Lindsted, K. D., Tonstad, S., and Kuzma, J. W. (1991). Self-report of physical activity and patterns of mortality in Seventh-day Adventist men. *Journal of Clinical Epidemiology* **44**: 355–364.

Lochen, M. L., and Rasmussen, K. (1992). The Tromso Study: Physical fitness, self-reported physical activity, and their relationship to other coronary risk factors. *Journal of Epidemiology and Community Health* **46**: 103–107.

Manton, K. G. (1988). Measurements of health and disease, a transitional perspective. In *Health in an Aging America: Issues on Data for Policy Analysis*. NCHS Vital and Health Statistics, Series 4, No. 25. USHDDS Pub. No. (PHS) 89-1488. Public Health Service. Washington, DC: USGPO, pp. 3–38.

Manton, K. G., and Stallard, E. (1982). A population-based model of respiratory cancer incidence, progression, diagnosis, treatment and mortality. *Computers and Biomedical Research* **15**: 342–360.

Manton, K. G., and Stallard, E. (1988). *Chronic Disease Modeling: Measurement and Evaluation of the Risks of Chronic Disease Processes*. Charles Griffin, London.

Manton, K. G., and Stallard, E. (1991) Cross-sectional estimates of active life expectancy for the N.S. elderly and oldest-old populations. *Journal of Gerontology* **48**: 5170–5182.

Manton, K. G., and Stallard, E. (1992). Compartment model of the temporal variation of population lung cancer risks. In *Biomedical Modeling and Simulation* (Proceedings of the 13th IMACS World Congress on Computation and Applied Mathematics, July 22–26, 1991, Dublin, Ireland). World Congress on Computation and Applied Mathematics, Dublin, pp. 75–81.

Manton, K. G., and Suzman, R. M. (1992). Conceptual issues in the design and analysis of longitudinal surveys of the health and functioning of the oldest old. Chapter 5 in *The Oldest-Old* (R. M. Suzman, D. P. Willis, and K. G. Manton, Eds.).Oxford University Press, New York, pp. 89–122.

Manton, K. G., and Tolley, H. D. (1991). Rectangularization of the survival curve: Implications of an ill-posed question. *Journal of Aging and Health* **3**: 172–193.

Manton, K. G., and Woodbury, M. A. (1983). A mathematical model of the physiological dynamics of aging and correlated mortality selection: II. Application to the Data Longitudinal Study. *Journal of Gerontology* **38**: 406–413.

Manton, K. G., Siegler, D. C., Woodbury, M. A. (1986). Pattern of intellectual development in later life. *Journal of Gerontology* **41**: 486–499.

Manton, K. G., Stallard, E., Woodbury, M. A., and Yashin, A. I. (1987). Grade of Membership techniques for studying complex event history processes with unobserved covariates. In *Sociological Methodology, 1987* (C. Clogg, Ed.). Jossey-Bass, San Francisco, pp. 309–346.

Manton, K. G., Stallard, E., and Woodbury, M. A. (1991). A multivariate event history model based upon fuzzy states: Estimation from longitudinal surveys with informative nonresponse. *Journal of Official Statistics* (published by Statistics Sweden, Stockholm) **7**: 261–293.

Manton, K. G., Stallard, E., and Singer, B. H. (1992). Projecting the future size and health status of the U.S. elderly population. *International Journal of Forecasting* **8**: 433–458.

Manton, K. G., Stallard, E., Woodbury, M. A., and Dowd, J. E. (1993). Time varying covariates in model of human mortality and aging: A multidimensional generalization of Gompertz. *Journal of Gerontology: Biological Sciences*, in press.

Manton, K. G., Stallard, E., and Singer, B. H. (1994). Projecting the future size and health states of the U.S. elderly population. Chapter in *Studies of the Economics of Aging* (D. Wise, Ed.). National Bureau of Economic Research, Univ. Chicago Press, in press.

Moon, J., Bandy, B., and Davison, A. J. (1992). Hypothesis: Etiology of artherosclerosis and osteoporosis: Are imbalances in the calciferol endocrine system implicated? *Journal of the American College of Nutrition* **11**(5): 567–583.

Morris, C. N. (1983). Parametric empirical Bayes inference: Theory and applications (with discussion). *Journal of the American Statistical Association* **78**: 47–65.

Myers, G. C., and Manton, K. G. (1984). Compression of morbidity: Myth or reality? *The Gerontologist* **24**: 346–353.

National Cancer Institute (1981). *Surveillance, Epidemiology End Results: Incidence and Mortality Data, 1973–77.* NCI Monograph 57. NIH Pub. No. 81-2330, Bethesda, Maryland.

National Cancer Institute (1984). *SEER Program: Cancer Incidence and Mortality in the United States, 1973–81.* NIH Pub. No. 85-1837, Bethesda, Maryland.

National Center for Health Statistics (1964). *Health survey procedure: Concepts, questionnaire development, and definitions in the Health Interview Survey.* Vital and Health Statistics, PHS Pub. No. 1000, Series 1, No. 2. Public Health Service, USGPO, Washington.

National Center for Health Statistics (1975). *Health Interview Survey Procedure, 1957–1974.* Vital and Health Statistics, Series 1, Programs and Collection Procedures, No. 11. DHHS Pub. No. (HRA) 75-1311. USGPO, Rockville, Maryland.

National Center for Health Statistics (1985). *The National Health Interview Survey Design, 1973–84, and Procedures, 1975–83.* Vital and Health Statistics, Series 1, No. 18. DHHS Pub. No. (PHS) 85-1320. Public Health Service, USGPO, Washington.

Neaton, J. D., Blackburn, H., Jacobs, D., Kuller, L., Lee, D. J., Sherwin, R., Shih, J., Stamler, J., and Wentworth, D. (1992). Multiple Risk Factor Intervention Trial Research Group: Serum cholesterol lėvel and mortality findings for men screened in the multiple risk factor intervention trial. *Archives of Internal Medicine* **152**: 1490–1500.

Ogata, Y. (1980). Maximum likelihood estimates of incorrect Markov models for time series and the derivation of AIC. *Journal of Applied Probability* **17**: 59–72.

Paffenbarger, R. S., Hyde, R. T., Wing, A. L., and Hsieh, C. C. (1986). Physical activity, all-cause mortality, and longevity of college alumni. *New England Journal of Medicine* **314**: 605–613.

Papoulis, A. (1984). *Probability, Random Variability, and Stochastic Processes.* McGraw-Hill, New York.

Penn, N. D., Purkins, L., Kelleher, J., Heatley, R. V., and Masie-Taylor, B. H. (1991a). Ageing and duodenal mucosai immunity. *Age and Ageing* **20**: 33–36.

Penn, N. D., Purkins, L., Kelleher, J. Heatley, R. V., Masie-Taylor, B. H., and Belfield, P. W. (1991b). The effect of dietary supplementation with vitamins A, C and E on cell-mediated immune function in elderly long-stay patients: A randomized controlled trial. *Age and Ageing* **20**: 169–174.

Peto, R. (1977). Epidemiology, multistage models, and short-term mutagenicity tests. In *Origins of Human Cancer* (H. H. Hyatt, J. D. Watson, and J. A. Winsten, Eds.). Cold Spring Harbor Laboratory, New York, pp. 1403–1428.

Rothenberg, R., Lentzner, H. R., and Parker, R. A. (1991). Population aging patterns: The expansion of mortality. *Journal of Gerontology: Social Sciences* **46** (2): 566–570.

Rubenstein, L. Z., and Josephson, D. E. (1989). Hospital based geriatric assessment in the United States: The Sepulveda, VA Geriatric Evaluation Unit. *Danish Medical Bulletin: Gerontology*, Special Supplement Series No. 7 pp. 74–79.

Schuss, Z. (1980). *Theory and Applications of Stochastic and Differential Equations.* Wiley, New York.

Social Security Administration, Office of the Actuary (1983). *Life Tables for the United States: 1900–2050.* Actuarial Study No. 89, SSA Pub. 11-115436.

Social Security Administration and Wade, A. (1989). U.S. life functions and actuarial functions. Machine Copy, USDHHS, Baltimore, Maryland.

Stallard, E. and Manton, K. G. (1993). Estimates and projections of asbestos-related diseases among Manville Personal Injury Settlement Trust claimants, 1900–2049. Prepared for the U.S. District Court and Manville Personal Injury Settlement Trust. Duke University, Center for Demographic Studies, Durham, N.C..

Steel, G. G., and Lamerton, L. F. (1966). The growth rate of human tumors. *British Journal of Cancer* **20**: 74–86.

Strehler, B. L. (1975). Implications of aging research for society. *Proceedings 58th Annual Meeting of the Federation of American Societies for Experimental Biology, 1974,* **34**: 58.

Tolley, H. D., and Manton, K. G. (1989). A Grade of Membership approach to event history data. Proceedings of the 1987 Public Health Conference on Records and Statistics. DHHS Pub. No. (PHS) 88-1214, USGPO, Hyattsville, Maryland, pp. 75–78.

Tolley, H. D., and Manton, K. G. (1991). Intervention effects among a collection of risks. *Transactions of the Society of Actuaries* **43**: 443–468.

Tsevat, J., Weinstein, M. C., Williams, L. W., Tosteson, A. N., and Goldman, L. (1991). Expected gains in life expectancy from various coronary heart disease risk factor modifications. *Circulation* **83**: 1194–1201.

Vaupel, J. W., Manton, K. G., and Stallard, E. (1979). The impact of heterogeneity in individual frailty on the dynamics of mortality. *Demography* **16**: 439–454.

Weiss, K. M. (1990). The biodemography of variation in human frailty. *Demography* **27**: 185–206.

Weyl, H. (1949). The elementary theory of convex phoyhedra. *The Institute for Advanced Study, Annals of Mathematics Study* **24**: 3–18.

Woodbury, M. A., and Manton, K. G. (1977). A random walk model of human mortality and aging. *Theoretical Population Biology* **11**: 37–48.

Woodbury, M. A., Manton, K. G., and Tolley, H. D. (1994). A general model for statistical analysis using fuzzy sets: Sufficiency conditions for identifiability and statistical properties *Information Sciences*, in press.

Yashin, A. I., Manton, K. G., and Vaupel, J. W. (1986). Mortality and aging in a heterogeneous population: A stochastic process model with observed and unobserved variables. *Theoretical Population Biology* **27**: 154–175.

CHAPTER 8

Fuzzy Set Analyses of Combined Data Sets: A Model for Evaluation Studies

INTRODUCTION: PROBLEMS IN COMBINING TYPES OF DATA

In this chapter we illustrate how GoM is used to analyze data from multiple data sets in an integrated analysis. Though combining data sets is often necessary, because few individual data sets contain all the information necessary to answer certain complex questions (e.g., changes in health and functioning in the very elderly; Manton and Suzman, 1992), this entails a number of technical and theoretical issues. Some of these issues are more easily managed than others.

First, if information is taken from two or more sample surveys covering *independent* populations (e.g., a sample from the community resident population in the 1984 NLTCS and a sample of the nursing home population in the 1985 NNHS), then the analytic problems—in making statistical inferences from combined parameter sets—may be fairly tractable.

Second, if one links survey data, measured at fixed times, to administrative records for the individual covering service use and mortality (e.g., dealing with "gaps" existing between different samples, differences in the sample structure of surveys) over time, then the analytic problems are again relatively manageable since events in the administrative records are defined exclusively of the survey measurement. The primary question of whether the administrative record of "events' or "outcomes" (e.g., hospital stays) is "representative" of transitions for an individual becomes a question of left or right censoring of the survey (when linkage to survey records is 100%, as in the list sample of the 1982–1984 NLTCS) or the interaction of censoring in the survey with censoring in administrative records (when the administrative record follow-up is incomplete). An example of incomplete follow-up is in area probability surveys such as the Longitudinal Study of Aging, where there is nonresponse to the survey *and* only a partial signing of waivers to allow survey records to be linked to Medicare files.

Third, one can combine data collected from multiple independent, but not population-representative, sites. This is also manageable if the set of measurements on individuals is sufficient to define their "state," i.e., the independent site-specific grouping of individuals contains no information about persons beyond that in the state variables. This is similar to longitudinal event history problems where, with adequate state variable information, the parameters of the sampling plan (e.g., areal clustering of cases) provide no additional information about outcomes of individuals (Fienberg, 1989; Hoem, 1985, 1989).

Using the assumptions given in Chapter 1, conditional on the g_{ik} scores, the responses of individuals are independent, multinomial random variables. In addition the GoM model provides a partitioning of the space of all possible response profiles. Therefore, all differences in multiple samples are contained in the distributions of the g_{ik}. Thus, the GoM model can be used in combining data sets when problems with the sampling strategy (leading to systematic biases in the sample frame) makes combining data problematic. This chapter illustrates this area of application.

8.1. THE MODEL AND CONCEPTS

The model we use for multiple data set analyses is based on the GoM model presented in Chapter 2, or

$$\Pr(y_{ijl} = 1) = \sum_k g_{ik} \cdot \lambda_{kjl} \, . \tag{8.1}$$

To use (8.1) in the situations described above, we need to modify it for the special conditions arising in each situation. The g_{ik}, state space measures which perform the function of "mixing" parameters, can be used to recreate the original population structure to deal with these problems. That is, the degree to which a sample represents a population is determined by the distribution of the g_{ik}.

The first case (i.e., combining independent surveys) utilizes selected properties of the likelihood (2.1). One property is that the most efficient weighting of individual responses is that generated from the maximum likelihood equations (assuming that the distribution is not "contaminated" and it "factors"). In these conditions, sample weights should be equal (i.e., $w_i = 1$) in the maximum likelihood solution. This provides an estimate of the sample g_{ik} distribution.

The second property is that postweighting can be done using, in place of the g_{ik} estimated from the sample, the moments, up to some degree (less than J), of the g_{ik} distribution independently estimated from another population. Thus, we can change the population represented by the sample by reweighting the distribution of g_{ik}. Operationally, while a g_{ik} "prevalence" distribution (e.g., the first moments $\bar{g}_k = \Sigma_i \, g_{ik}/I$) may have been generated in estimating parameters in (8.1), we can "standardize" the g_{ik} distribution to represent the distribution of "another" population (i.e., one that was not used to estimate the parameters) by using that other population's independently calculated $\bar{g}_k$ derived from the λ_{kjl} from the first population and the observed binary responses y_{ijl} from the second. The exogenously produced $\bar{g}_k$ can be used to recombine the λ_{kjl} for specific variables to calculate reweighted response probabilities for J variables in the second population.

For example, if we know the 1985 NNHS covers only persons in nursing homes meeting NCHS criteria (in the 1982, 1984, and 1989 NLTCS a broader definition is used where about 85% of all NLTCS institutionalized elderly persons meet these criteria), then a rescaling of the g_{ik} weights from the NNHS to reflect this shortfall in the institutional elderly population (broadly defined) can be used to close the sample "gap." If the "gap" is not simply a problem in sample coverage, but involves differential response rates for the missing components of the population, then the maximum likelihood imputation procedures presented in Chapter 6 (see Manton et al., 1994a) can be used to adjust the $\bar{g}_k$ to correct the gap if ancillary data (e.g., Medicare administrative records) are available for all sample members.

The second issue, longitudinal data linkage, requires generalizing (8.1) to one of the longitudinal likelihood forms in Chapter 6 (i.e., where individual records have been reorganized along their time axis into fixed time assessments T and episodes e). To resolve systematic censoring of events in the Medicare administrative records, the temporal censoring of events must be explicitly modeled by including each type of outcome recorded in the administrative records as functions of the g_{ik}. Such adjustment can be done in several ways, which we will illustrate in several longitudinal data sets.

A theoretically more complex issue, and one related to the problem of the simulated subsampling of data from the ECA example in Chapter 2 based on systematic differences in symptom expression, is data derived from multiple independent sites that are *not* representative of a common population. This type of design occurs frequently: in the ECA study sponsored by NIMH there were five data collection sites (Baltimore, Maryland; Los Angeles, California; Durham, North Carolina; New Haven, Connecticut; and St. Louis, Missouri: Eaton et al., 1989); in the EPESE study there were four sites (i.e., New Haven, Connecticut; Iowa; East Boston, Massachusetts; and Durham, North Carolina); in the National Channeling Demonstration, 10 sites were chosen (Brown, 1986); in the new cost reimbursement study for nursing homes, six states were selected; in the Social/HMO evaluation, four sites were selected (Portland, Oregon; Brooklyn, New York; Minneapolis, Minnesota; and Long Beach, California; Manton et al., 1994b). The large number of data sets with similar longitudinal structures and the large number of detailed discrete health and functional measurements, but based on nonrandom selection of study sites, exist because it is often impractical to have both intensive survey interviewing and biomedical data collected in a nationally representative survey sample. Exceptions to this are the National Health and Examination Surveys and their various follow-ups. Thus, for budgetary and practical reasons (e.g., being near the facilities of major universities or medical centers, finding sites with specific sets of service available) such data are often collected in a small number of nonrandomly selected sites. These sites may be selected to express as much demographic heterogeneity as is possible, but there are generally too few sites to be representative of the full sociodemographic variation of the U.S. population. For example, to be population representative in the NLTCS required the selection of 173 PSUs, far more areas than in any of the epidemiological or LTC demonstration data sets.

Specifically, we write (8.1) using superscript s ($s = 1, 2, \ldots, S$) to indicate the set of individuals responding in a site, i.e.,

$$\Pr(Y^s_{ijl} = 1) = \left(\sum_k g^s_{ik} \cdot \lambda^s_{kjl}\right). \tag{8.2}$$

In (8.2) we explicitly identify the sets of the g^s_{ik} and λ^s_{kjl} as estimated for *each* site treated as an independent population. We then modify (8.2), by imposing cross-site constraints on the coordinates defining the state simplex boundaries (i.e., the λ_{kjl}), to permit comparisons of g^s_{ik} across sites, or

$$\Pr(Y^s_{ijl} = 1) = \left(\sum_k g^s_{ik} \cdot \lambda^+_{kjl}\right), \tag{8.3}$$

where λ^+_{kjl} is used to represent the extreme profiles for all sites combined. Note that equation (8.3) does not require all of the λ^s_{kjl} values in (8.2) to be equal. Since the λ are chosen as the extreme profiles, they represent a basis within which the distribution of the g_{ik} in each site is represented. When combining multiple studies, each with different distributions of the g_{ik}, a spanning set of extreme profiles covering the data sets from all sites must be selected. This, in effect, produces a set of extreme profiles which represent the *union* of the sets described by the individual extreme profiles. This means that we need to estimate λ^+_{kjl} so that all g^s_{ik} distributions can be represented relative to the same state simplex B^s defined by the combined solution. This is done implicitly by simply applying the maximum likelihood algorithm in (8.3) to the combined data sets. Note that unless $\lambda^1_{kjl} = \lambda^2_{kjl} = \cdots = \lambda^+_{kjl}$, the distribution of the g^s_{ik} estimated using (8.2) will be a transformation of those estimated using (8.3).

Clearly, λ^+_{kjl} may have more classes (i.e., $K^+ > K^s$) than for any single site analyzed separately, simply because more "extreme" λ^+_{kjl} profiles are required to fully span the g^s_{ik} across all sites. Since the g^s_{ik} must sum to 1 for a person, this means that the g^s_{ik} for a site with "less" heterogeneity on the J

measures will be "pulled" towards the center of the simplex (i.e., $\bar{g}_k \rightarrow 1/K^+$). A consequence of this "compression" of the g^s_{ik} distribution for some sites may be the need for a larger K^+ to express the full range of individual heterogeneity.

This strategy of estimating the λ^+_{kjl} to define a simplex B^s that defines a state space that can span the individuals states for any person in the combined set of sites can be extended to following cases in multiple sites over time. For example, if we wish to include several studies, each of which provides data over time, we would combine the data for all studies and times and estimate a general set of extreme profiles for the entire data space. If we denote these profiles as λ^{++}_{kjl}, then all estimated g_{ik} distributions for each study at each time point will be defined with respect to these "general" extreme profiles. The conditional independence assumption of Chapter 1 makes this possible. Thus, the trajectories $g^s_{ik\cdot t}$, independently estimated for each person, can be followed over time for each site; e.g., if trajectories of the $g^s_{ik\cdot t}$ were altered by an experimental condition in a site we need a set of λ^{++}_{kjl} that define a state simplex broad (i.e., with K^{++} fuzzy classes) enough to cover the altered trajectories as well. If the data set contains "comparison" or "control" groups, either explicitly defined or as subgroups represented internally in the sample, then the intervention may be well defined in the state space. For example, in analyses of shipyard or occupational exposures to radiation, the determination of an origin for the biological response function was determined by defining a group of workers exposed to all hazards in the shipyard *except* radiation. Otherwise, one may establish an origin a priori (i.e., define a healthy state with no condition) or from ancillary data.

Note that *no* constraints are imposed on the g^s_{ik} (or $g^s_{ik\cdot t}$) estimates over site or time. Thus, site-specific $\bar{g}^s_k$ are independently generated. All dependencies of the $g^s_{ik\cdot t}$ are represented by the λ^{++}_{kjl}. Thus, the vector values of a set of K^{++} $g^s_{ik\cdot t}$ for any site s at any time t have the identical definition (i.e., the $g^s_{ik\cdot t}$ define the same spot in the state space spanned by λ^{++}_{kjl}) and can be compared directly. For this to work it is necessary that the y_{ijl} given the $g^s_{ik\cdot t}$ are independent; i.e., that data are independent given the $g^s_{ik\cdot t}$ for the common simplex. An analogous condition is considered in Chapter 2, where we deal with the effects of intraclass correlations in the PSUs in a cluster sample. The grouping of individuals (by site or by PSU) will generate a within-class correlation that violates the independence assumption required by the product form of the likelihood (2.2). This can be resolved (a) by using a more general likelihood function with "contamination" built into its parametric structure (e.g., the negative binomial discussed in Chapter 6) or (b) if the measures can explain all of the within-site (or PSU) contamination of individual responses. In practice this may be possible because the geographic filtering processes generating correlations within site (or PSUs) may have well-defined social or economic causes (e.g., income, education, marital status, age, race, sex, family structure—the types of variables frequently used in social area analyses (Goldsmith et al., 1984) and the types of variables that affect choices of location and housing type).

This state space property cannot be so generally exploited in standard multivariate procedures because the individual information is compressed (projected) into a small set of moments (e.g., in principal components analysis the second-order moments in the covariance matrix). In principal components analysis, relations can be distorted by systematic sampling or nonequal covariance matrices within populations that cause problems in defining a covariance matrix with appropriate properties (e.g., Eisenbeis and Avery, 1972; Dempster, 1977). In GoM, since the g_{ik} are independently estimated, confounding with distributional factors can be resolved if the state description is adequate to generate independence of the individual scores over all dimensions.

The GoM model has this property because of constraints on the likelihood estimation represented by the two score functions (3.3) and (3.4). That is, each score function had a tendency to go to a maximum either by extending the range of the simplex spanned by the λ_{kjl}^{++} or by dispersing the $g_{ik\cdot t}^{s}$ as broadly as possible in the state space to explain as much individual heterogeneity as possible. However, since there is "pressure" on the algorithm optimizing the likelihood function to simultaneously increase *both* the $g_{ik\cdot t}^{s}$ and λ_{kjl}^{++} spaces and because both are mathematically linked in the score functions through the $p_{ijl}(= \Sigma_k\, g_{ik\cdot t}^{s}\, \lambda_{kjl}^{++})$ terms, the maximum likelihood value is achieved when the "extensions" of the two spaces comes into equilibrium (i.e., $\partial L/\partial\lambda_{kjl}^{++}$ will increase less than the $\partial L/\partial g_{ik\cdot t}^{s}$ will decrease and vice versa). This results because the λ_{kjl}^{++} are constrained over cases and the $g_{ik\cdot t}^{s}$ over variables. Given that each is conditionally maximized given the other parameter set, one achieves $g_{ik\cdot t}^{s}$ and λ_{kjl}^{++} solution spaces that are jointly maximized.

Though the state-conditional estimation may resolve certain problems in an analysis, it may be that we wish to exploit some of that information in generating model estimates. For example, often epidemiologists and biostatisticians estimate hazard functions with covariates. The crude hazard function is like a reduced-form equation in econometrics; i.e., it may express the relation between exogenous factors and outcomes (or the time to specific outcomes) but not the causal pathways (e.g., the role of cholesterol, free radicals, hematocrit, serum Lp(a), sialic acid, immunologically active cells, fibrinogen, growth factors, platelets, blood pressure, and smoking in generating atheromas). To do this we need to estimate the structural equation where parameter constraints are imposed to represent the system structure (e.g., Chapter 5).

This can be done in the current problem, where site populations are defined by generalizing the GoM model one step further. Specifically, in Chapter 2 (Section 2.3.2) we discussed how LCM generalized contingency table models, and GoM generalized LCM in such a way that there is an independent heterogeneity component added at each stage which can be tested for significance (i.e., the model's parameter spaces, and the composite hypotheses about them are "nested"). We also showed that the empirical Bayesian form of GoM generalized the basic form of the model. Here we generalize GoM by allowing the dimensionality of the case space (say R) to be different than the dimensionality of the variable space (Woodbury et al., 1992, 1994). This can be done by recalling that the second-order factoring of the GoM model can be written as

$$\Pr(Y_{ijl} = 1) = \sum_k \Big(\sum_r \gamma_{ir}\phi_{rk}\Big)\lambda_{kjl} ,$$

where

$$\Pr(\xi_{ik} = 1) = \sum_r \gamma_{ir}\phi_{rr}$$

and where g_{ik} is the realization of ξ_{ik}.

If $R < K$, then there are natural constraints on the solution to insure identifiability. However, $R \geq K$ may also be identified if there are cross-temporal replications of measurements. Thus, for measures made at multiple times y^t_{ijl}, we can write

$$p^t_{ijl} = \Pr(Y^t_{ijl} = 1) ,$$

and

$$p^t_{ijl} = \sum_k \sum_m g_{im} C^{mt}_k \lambda_{kjl} ,$$

(see Chapter 3). For g_{im}, if we substitute

$$\sum_r \gamma_{ir}\phi_{rk} ,$$

constraining variation over t, we get

$$\begin{aligned} p^t_{ijl} &= \sum_k \sum_m \Big(\sum_r \gamma_{ir}\phi_{rm}\Big) C^{mt}_k \lambda_{kjl} , \\ &= \sum_k \sum_r \gamma_{ir}\Big(\sum_m \phi_{rm} C^{mt}_k\Big)\lambda_{kjl} , \\ &= \sum_k \sum_r \gamma_{ir} V_{kr\cdot t}\lambda_{kjl} . \end{aligned}$$

In this last expression, $V_{kr\cdot t}$ represents the relation of r groups of individuals, each group having similar trajectories, to k time-independent extreme variable profiles. The $V_{kr\cdot t}$ may be time varying or time constant, i.e., $V_{kr\cdot +}$. This allows us to pick out subgroups that may be changing in different ways without specifying a specific distribution for the r groups (Woodbury et al., 1992).

This property of the likelihood (and its exploitation in the model extension) has implications for missing data and for the identification of effects in a systematically selected sample when the parts of the population not represented are systematically missing. For example, in an application of the GoM generalization to the 1982, 1984, and 1989 NLTCS, a particular group r was characterized by being missing over time. This approach also has applications not only when sets are systematically selected, but also where the individual has to make a choice to enroll in a particular plan or program. Thus, it has implications for the analysis of data when the comparison and experimental groups are not randomized but systematically filter into one group or the other due to a decision process which may be "static" or "dynamic." In principle, this is no different from the problem of dealing with the contamination of a distribution due to a geographic sorting of cases by a housing decision process.

Specifically, since the likelihood requires the state space to expand over the maximum range possible (constrained ultimately by the time-generalized "possible" measurement space $M \otimes T$), consistent with the distribution of parameters in the complementary space, the λ_{kjl}^{++} will span the largest possible space to bring as many responses as possible within the boundaries of the polytope B_T^s. The expansion of the simplex is constrained by (a) the limits of the measurement space ($M \otimes T$) and (b) the distribution of the $g_{ik \cdot t}^s$. Once the state space is defined, a new case may emerge at any point within the space. Thus, in a study plan with nonrandomized case and control populations, there is nothing to restrict a case from emerging at any point in the simplex B_T^s, whether it is a case or control. Indeed, if there is a strong selection bias in, say, the "case" group such that no "healthy" people are selected, but there are healthy people existing in the "control" group, there is no reason why a new healthy "case" group member could not end up in the region of the state simplex B_T^s populated by control group members. That is, if the $g_{ik \cdot t}^s$ values for a case and a control are equal, we would expect them to have identical outcomes conditional on their state vectors except for the experimental condition. If outcomes are different when $g_{ik \cdot t}^s = g_{ik \cdot t}^s$ for a case and control and conditional independence holds for the likelihood, then the difference in outcomes must be due to the experimental intervention.

To make this explicit, we can write

1. *State Estimation*

$$\Pr(Y_{ijl} = 1) = \sum_{k^{++}} g_{ik^{++} \cdot t}^s \cdot \lambda_{kjl}^{++} \tag{8.4}$$

where J = health, functional and all other variables necessary to describe the state of each person.

2. *Outcome Estimation Conditional on State Characteristics*

$$\Pr(Y_{ij^*l} = 1) = \sum_{k^{++}} {}^*g^s_{ik^{++}\cdot t} \cdot \lambda^{(c)}_{kjl} \tag{8.5}$$

In (8.5), a set of outcome variables j^* are estimated with the state variables held fixed (i.e., ${}^*g_{ik^{++}\cdot t}$). The $\lambda^{(c)}_{kjl}$, where $c = 1$ represents cases and $c = 2$ represents controls, are the responses or outcomes for a person in a given state. The test of the effect of the intervention is a test of the null hypotheses $\lambda^{(1)}_{kj^*l} = \lambda^{(2)}_{kj^*l}$, i.e., that, conditional upon state, the outcomes (which could be measures of survival, service use, or costs) do not differ by experimental condition. The alternative hypothesis is that the response probabilities differ. This is tested by examining the likelihood ratio test for each of the j^* measures where a test of the pooled response (i.e., $\lambda^{(*)}_{kj^*l}$) is made against the sum of likelihood values for the $\lambda^{(c)}_{kj^*l}$.

This may be compared with two other approaches for evaluating the effect of a nonrandomized experiment. One is the use of matched cases and controls; i.e., randomization of cases and controls may break down but for cases and controls one can determine if responses differ. Thus, $g^{(1)}_{ik\cdot t} = g^{(2)}_{ik\cdot t}$ implies that the states of cases and controls are *exactly* matched (except for stochastic influences) on the J variables used to calculate the $g^{(c)}_{ik\cdot t}$. The model is more general, however, in that we do not need to have individuals exactly matched; i.e., if the λ^{++}_{kjl} are commonly defined, then there is an explicit mathematical relation between the $g^{(1)}_{ik\cdot t}$ and $g^{(2)}_{ik\cdot t}$ that can be used to make "functional" comparisons. That is, one can take the values of $g^{(1)}_{ik\cdot t}$, and their associated outcomes, and ask what the response would be for $c = 2$ if that person had the same value on the state vector. Thus, *exact* matches may be generated for the model by equating individual mixing parameters through an estimated function. This approach is important if the state space is of high dimension (i.e., J is large). In this case no "exact" matches may exist.

An alternative approach, due to Heckman et al. (1985), used regression to control for factors affecting the chance of entering a program but which are not associated with the outcomes. This is done by setting up an equation system where the error in the "choice" (i.e., the equation describing the probability of being a case or control) and "outcome" (i.e., change in health or service use) equations is identifiable, i.e., that not all variables in the outcome equation are used in the choice equation. In this approach, the ability to isolate choice from outcome effects requires (a) good proxy variables and (b) a correctly specified model. Using the regression, however, one has a restricted parametric form in that sampling is assumed done from the same distribution to estimate the outcome equation with choice effects assumed "conditioned out." The conditioning reflects, in standard estimation, the assumption of a multivariate normal distribution. If the conditional distributions are not equivalent, then coefficient estimates cannot be compared. This is not a problem in the GoM model since neither normality nor any direct equivalence of the distribution of the cases and controls is assumed. The only assumption is that the state dimensions are equivalent in terms of the J coordinates λ_{kjl}, an assumption that can be fulfilled even if distributions of individuals differ over sites.

A complication can emerge if the site distribution is affected by differential nonresponse (censoring) rates over sites or that nonresponse on specific variables (item nonresponse) occurs in specific sites. This would be a problem with the Heckman and Hotz (1987) procedure if, say, choice factors were not measured for a case. An example of this condition was the comparison of the health service use of S/HMOs with FFS populations where, for the initial health screening interview (the "HSF") and the Medicare administrative data, there was a partial confounding with the experimental condition (Manton et al., 1994a). To make this explicit consider the schematic of the data matrix in Figure 8.1.

In Figure 8.1, nearly all (98.3%) of the S/HMO enrollees filled out a "mail back" health screening form (HSF) because that form is part of the required application process. However, because of enrollment in an HMO before the S/HMO, about 20% of S/HMO enrollees did not have prior Medicare service use in the FFS ("fee-for-service") sector. In contrast, those in the FFS (standard Medicare) population had 100% prior Medicare use data (they were sampled from Medicare lists) but only about a 70% completion rate on the HSF (applied by telephone) due to survey nonresponse (from analysis of the NLTCS, we can expect nonresponse to be concentrated among very old frail persons and females; Manton et al., 1991a).

There are several solutions to this problem with the GoM model, each with its own strengths and limitations. One is to use the standard GoM model where $S = 2$ (i.e., S/HMOs versus FFS; $S = 8$ if the four S/HMO sites are analyzed as separate populations) and all data from the HSF are

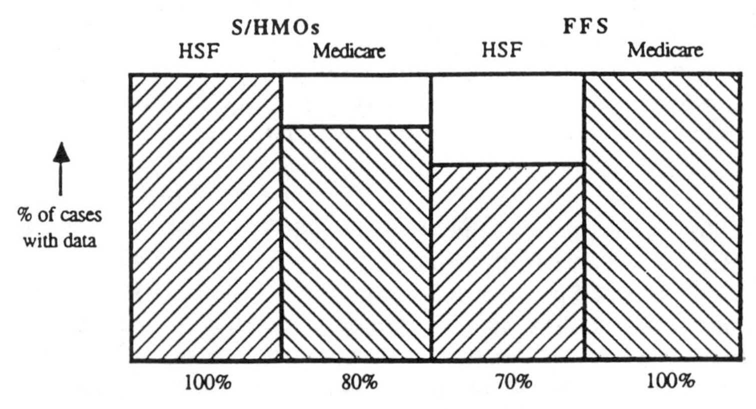

Figure 8.1 Schematic of data matrix.

treated as "internal" (i.e., the g_{ik}, or state variables, are defined from only health and functioning data from the HSF) and all Medicare data are external (i.e., defining the event or outcome space). In this case where $\Sigma_l^{L_j} y_{ijl} = 0$, the variable for that person is "missing" and one gets an appropriate marginal likelihood form from equation (8.1) (under the assumption that data is missing "randomly").

The second approach is to include an additional $L_j + 1$ category in each of the J variables so that, when $\Sigma_{l=1}^{L_j} y_{ijl} = 0$, we have $\Sigma_{l=1}^{L_j+1} y_{ijl} = 1$. This expands the dimensionality of the measurement space so that if information is missing "systematically," that information is used to define the K classes (i.e., the structural parameters are λ_{kjl}^{++} where a common set of variables is used over site, time, case/control status, and data presence/absence). In Berkman et al. (1989), missing data in the New Haven EPESE study were generated by (and thus was correlated due to) a process associated with age decrements in cognitive function. The pattern of missing data defined an additional extreme profile associated with dementia.

A third strategy is to include Medicare variables as internal variables along with the HSF variables so that information *contained* in the Medicare data, but *not* contained in the HSF data could help define profiles. For example, if health and service use were strongly correlated and, for clients in the FFS, persons with missing HSF data were more impaired than average, then they would also presumably have a higher than average use of Medicare services. Thus, the joint information on HSF and Medicare data would produce the g_{ik} for persons with missing HSF data from the information on the Medicare data. If persons with Medicare data, but without HSF data, are different from persons with HSF data, then the g_{ik} estimated from the combined HSF/Medicare data set would be different from that estimated from data restricted to the HSF.

A final approach is to use ancillary data (e.g., Medicare data) to make allocations of the g_{ik} for nonrespondents from the estimated relation of the g_{ik} to the Medicare outcomes for respondents. If the Medicare measures are the same for nonrespondents as for respondents, then we can *initially* assume that the g_{ik} distribution is the same for respondents and nonrespondents. If not, then we have information, using the maximum likelihood imputation procedures discussed in Chapter 4, to estimate the moments of the missing g_{ik} for nonrespondents using the relations estimated for individuals with both sets of data (Manton et al., 1991a). All of these procedures are variations of the MIP (Orchard and Woodbury, 1971). The Heckman and Hotz procedure is a special case where the individual state information is not used—only a few selected moments of their distribution adjusted for selection bias by first-order conditioning (i.e., only the means are adjusted).

One tendency that investigators are subject to, in statistical analyses, is to statistically overcontrol." Thus, while using the joint information on HSF and Medicare data eliminates the bias of nonresponse to the extent that prior Medicare service use is informative about health dimensions, it may not serve other analytic purposes. For example, by generating the g_{ik} from the HSFs, one can study the differences in the distribution between persons with and without Medicare data used as external variables to see what the magnitude of the differences is between different data presence/absence defined subgroups. Suppose we found that g_{ik} calculated only from the HSF data strongly predicted mortality and adverse health changes over time. It is not clear that including the Medicare data (unless there is an "optimal" or very efficient allocation of money to health needs) would necessarily improve the analysis of health characteristics. Indeed, if money and resources are poorly allocated to objective health needs, the additional data could adversely effect the estimation of health dimensions and individual health scores (i.e., they could become less predictive of an independent health outcome like mortality). Thus, if the HSF data do represent health status adequately, then the g_{ik} should be estimated from them alone; i.e., they directly capture the health state of the persons, which may only be imperfectly correlated with the Medicare services actually delivered.

If there is differential nonresponse on health variables between S/HMO and FFS clients, this will affect the estimated aggregate response of the *combined* population but, unless an entire health dimension is missing (i.e., the state space spanned by the λ_{kjl}^{++} is deficient in rank), the relations of service use with each of the g_{ik} values jointly estimated for S/HMOs and FFS will be appropriate since they are estimated conditionally on g_{ik} appropriate for explicit points in the simplex solution (i.e., the vertices defining the simplex where $g_{ik} = 1$). The, the question of "nonresponse" is a moot one applying only in the context of estimating the population distribution parameters (which are not informative for the arbitrary sites) since the relations for individuals can be well estimated and thus may well describe individual health processes and change.

This procedure, for survey data, can be considered in the context of two general approaches to sample theory. In the first, finite population theory, the population is assumed to be fixed. Sampling variation results from the sampling process. In such sampling, individual variation does *not* exist, i.e., sample variation $\rightarrow 0$ as the proportion of the population sampled increases (Potthoff et al., 1992, 1993). In this context, systematically missing data are a problem since the population parameters are not unbiasedly estimated—and they are the only random variables. In superpopulation theory the sample plan is "fixed" and variation results from the stochastic processes operating at the individual level (Hoem, 1985, 1989). If relations are estimated conditionally on the parameters of these individual level processes, then population composition does not matter in estimation of the likelihood of specific outcomes unless whole classes of phenomena are not represented (i.e., all persons with $g_{ik^*} > 0$ are missing in the sample where K^* is an important, but missing, health dimension).

Thus, the problem of generating a model of response for an experimental condition when randomization fails must be considered in three ways.

1. The problem of systematic sorting into experimental versus control conditions (this is handled in the GoM model by using "state variables" that are estimated for the individual).
2. The problem of systematically missing data—which can be confounded with the decision to enter the condition.
3. The problem of how to describe the sampling model from which observations are selected.

All three issues must be resolved simultaneously for an analysis. The GoM model, by using moments from the distribution of individuals, will have a better chance of resolving these problems than standard econometric procedures where estimation is restricted to using information only in lower-order moments.

8.2. EXAMPLE A: COMBINATIONS OF MULTIPLE SURVEYS AND ADMINISTRATIVE RECORDS IN ESTIMATING HEALTH COSTS FOR DISABLED PERSONS

This example deals with estimation of costs from multiple, nonoverlapping surveys (e.g., the 1984 NLTCS and the 1985 NNHS), Medicare administrative records, and economic relations generated from Bureau of Labor Statistics (BLS) wage estimates and HCFA office of the actuary estimates of health care inflation. After estimating the g_{ikt} and λ_{kjl} from the combined NLTCS and NNHS survey data, we can determine both how those health and functional dimensions are affected by medical conditions *and* how altering those medical conditions affects costs over time by altering the distribution of disability by type and intensity (Manton and Stallard, 1990, 1991).

First, we need to define ancillary equations derived from the MLE of the λ_{kjl} for "external" cost variables. It is assumed that the cost of a disease, whether for acute or chronic care, is equal to the difference in costs for a person with that disease versus that for a person without the disease. That is, persons with a disease will also have functional impairments and other comorbid conditions that will require services even if the index disease is eliminated. From the model we can determine how much each dimension of disability is reduced (or increased) if the index disease is not present (Manton, 1987). We can also determine the change in costs for specific medical conditions with and without dementia. Thus, a person who, say, is not demented and has diabetes may be better able to follow nutritional restrictions, take medication, and seek preventative care (e.g., seek early treatment for infection) to manage his diabetes (and its sequelae) less expensively (and more effectively) than a person with advanced dementia who requires caregivers to provide LTC services and to provide continuous monitoring of behavior and care. This will probably have a more complicated, costly course of acute care for diabetes and its physical consequences.

To estimate costs, we use K ancillary equations (Manton and Stallard, 1990) where the parameters β are estimated by the GoM maximum likelihood procedure with the g_{ik} held fixed (i.e., only one set of score equations (3.4) in the algorithm is free to vary in the maximization), or

$$E(g_{ik}) = g_k^0 + \text{Age}_i\left[\beta_{k0} + \sum_{m=1}^{M} \beta_{km} \cdot D_{im}\right]. \tag{8.6}$$

In (8.6), D_{im} indicates the presence (1) or absence (0) of one of M medical conditions. The influence of the mth condition on each of the K disability scores (i.e., β_{km}) is estimated under the convexity restriction that (because $\Sigma_k\, g_{ik} = 1$) a change in the β_{km} in one equation is numerically compensated for by changes in the β_{km} in the $K-1$ other equations. By multiplying D_{im} by Age_i the effect of the medical condition on each dimension of disability is made age dependent. Thus, (8.6) simulates the effect of partly (or wholly) eliminating one (or more) medical condition at a specific age. Since alteration of the D_{im} for the mth condition (e.g., elimination by changing D_{im} from a 1 to a 0) does not effect the remaining conditions, the model reflects the dependency of functional disability on multiple conditions; i.e., the "elimination" of a condition adjusts health costs to reflect the remaining medical conditions. Thus, the model identifies (a) direct LTC costs to manage disability, (b) higher acute care costs generated by the complications caused by diseases in managing comorbidity, and (c) increases in costs due to age-related *general* frailty (Manton et al., 1993a).

To illustrate, for a person with dementia, diabetes, and osteo-arthritis, costs are estimated for different services (Manton et al., 1993a). If we "eliminate" (i.e., prevent or cure) dementia, we reduce each cost component to that for a person with *only* diabetes and osteo-arthritis; i.e., the size of the cost reduction expected to be achieved for dementia reduction depends on the other conditions present for the individual. If eliminating a specific medical condition did *not* cause a person to be discharged from a nursing home because another condition was sufficient to require institutional care, his total costs might decline little with the disease elimination. Even if all conditions are "eliminated," health expenditures are not zero, but remain at the (lower) level of the non–chronically disabled population affected by medical conditions not producing long-term disability and with expenses for health maintenance and preventative services (Tolley and Manton, 1985).

Next, we translate the g_{ik} adjusted for morbidity changes into cost estimates for services by estimating, for cost variable h, an ancillary function where the g_{ik} are fixed (Manton et al., 1990), or

$$E(x_h(t)) = \left[\sum_{i=1}^{I} \delta_{it} \sum_{k=1}^{K} g_{ik} \sum_{l=1}^{L_h} \lambda_{khl}(t)\bar{x}_{hl}(t)\right] \Big/ \sum_{i=1}^{I} \delta_{it} . \tag{8.7}$$

In (8.7), $\bar{x}_{hl}(t)$ is the midpoint of the lth level of the hth cost variable for persons aged t. Multiplying the midpoint by the probability $\lambda_{khl}(t)$ that a person of the kth class has the lth expenditure level and multiplying by the individual weight g_{ik} produces the cost estimate $E(x_h(t))$ where t is age. Both $\lambda_{khl}(t)$ and $\bar{x}_{hl}(t)$ are indexed by t; $\delta_{it} = 1$ if $\text{Age}_i = t$ and $\delta_{it} = 0$ otherwise; i.e., estimates not only reflect the health and functioning of the individual but are dependent on chronological age (a proxy for unmeasured health factors).

Disability data are derived from survey interviews or from patient assessments in LTC demonstration studies, and are linked to Medicare (and possibly Medicaid) administrative records. Thus, cost estimates must also be able to reflect w_i, i.e., the sample weight—the inverse of the sample selection probabilities. Because of the additivity of estimates over the K classes, due to the convexity constraints on the g_{ik}, the adjustment for sample weights requires

$$E(x_h(t)) = \left[\sum_{i=1}^{I} \delta_{it} w_i \sum_{k=1}^{K} g_{ik} \sum_{l=1}^{L_h} \lambda_{khl}(t)\bar{x}_{hl}(t)\right] \Big/ \sum_{i=1}^{I} \delta_{it} w_i . \tag{8.8}$$

The w_i may be adjusted further if there is systematic nonresponse (Manton et al., 1991a). This is important if nonrespondents are sicker, use more health services, *and* are closer to death (i.e., have higher mortality risks) than respondents (NCHS, 1966; Corder and Manton, 1991; Manton et al., 1991a). Even a 5% nonresponse rate may produce significant bias because nonresponders may have several times the expense of responders. The NLTCS sample is a list sample of the U.S. elderly Medicare eligible population with 100% follow-up. Consequently, if correctly weighted and adjusted for nonresponse, national Medicare expenditures are reproduced.

The overall relation of medical conditions to disability and to costs may be written (by substituting (8.7) into (8.8)) as

$$E(x_h(t)) = \left[\sum_{i} \delta_{it} w_i \sum_{k} E(g_{ik}) \sum_{l} \lambda_{khl}(t)\bar{x}_{hl}(t)\right] \Big/ \sum_{i} \delta_{it} w_i , \tag{8.9}$$

$$= \left[\sum_{i} \delta_{it} w_i \sum_{k} \left[g_k^0 + t\left[\beta_{k0} + \sum_{m} \beta_{km} \cdot D_{im}\right]\right]\right.$$

$$\left. \times \sum_{l} \lambda_{khl}(t)\bar{x}_{hl}(t)\right] \Big/ \sum_{i} \delta_{it} w_i . \tag{8.10}$$

In (8.10) the influence of medical conditions on disability, and, thereby the relation to medical expenditures, is explicit. To estimate costs for other factors (e.g., informal care hours), additional data are required. A unit (e.g., an hour) of informal care may be valued as (a) the cost to the caregiver of forgoing job wages (opportunity cost) and/or (b) explicit costs charged to the individual (or to a government agency) if services had to be purchased (substitution costs).

The relation in (8.10) can be extended even further to represent changes in the prevalence of medical conditions due to the alteration of chronic disease risk factors (i.e., the risk factors for D_{im}). This more general model requires the further substitution in (8.10) of risk factor dynamic and hazard functions of the type discussed in Chapter 6. Thus (8.10) is a set of functional relations that can be "driven" over time by a multivariate stochastic process (Manton et al., 1991c).

Costs for informal caregivers depend on their demand wage rate (e.g., Headen, 1992). Demand wage rates for caregivers vary significantly, being different for (a) highly educated caregivers who are labor force eligible and command high wage rates, (b) minorities, women, and less trained persons with low wage rates, and (c) elderly caregivers with low implied wage rates. We used the 1989 median wage of $10.00 per hour for persons over 65 estimated by the BLS in our calculations of informal caregiver "opportunity" costs. Provision of services by formal caregivers is assumed to cost 50% more (a business has to make a profit, pay taxes, buy equipment and insurance, incur travel expenses, etc.). These are crude estimates of wage rates and could be further improved.

In addition to the 1982–1984 NLTCS, the 1985 NNHS, the Medicare Part A and B expenditure records, and inflation factors estimated by HCFA (USDHHS, 1988), data from a specialized informal caregiver survey conducted in 1982 in conjunction with the 1982 NLTCS was used. The six classes derived from a GoM analysis of the pooled samples of the 1982–1984 NLTCS (Manton et al., 1990) are briefly described (see Table 4.2) as

1. No or light impairment (elderly persons possibly with acute illness but little or no chronic disability),
2. Light physical impairment and early dementia,
3. Light impairment and early cardiopulmonary problems,
4. Moderate physical impairment and cardiopulmonary problems,
5. Moderate physical impairment and hip fracture and self care needs,
6. Severe self-care and physical impairments with a high level of cognitive dementia.

These six classes had excellent predictive validity for morbidity, mortality, and both acute and LTC health service use. For comparison, an ADL summary score (1–6 ADLs impaired) is associated with a two-year relative mortality risk of 5:1 (i.e., the ratio of the mortality rate for those with 5 or 6

ADLs impaired and those with no chronic disability; Manton, 1988). The g_{ik} scores show a relative risk (between most and least impaired classes) nearly four times as large (Manton and Stallard, 1990).

The relation of 29 medical conditions to institutional residence (i.e., for $K = 7$) was determined from the 1985 NNHS. For all respondents to the NNHS current resident sample (weighted to their population values) a synthetic institutional "discrete" class was created where $g_{i7} = 1$ if i was in the 1985 NNHS sample and $g_{i(1-6)} = 0$ otherwise. The frequency of institutionalization was based on the estimates from the NLTCS Medicare list sample. That is, the total weighted count of institutional persons was renormed to the number of persons in institutions reported in the NLTCS rather than the NNHS. This was done to close a small sample gap between the NLTCS community and the NNHS institutional samples.

The coefficients from the functions in Table 8.1 (i.e., β_{km}) relate each of 28 (i.e., $M = 28$) medical conditions to each of the seven ($K = 7$) classes. The estimates are made age dependent where (a) nonlinear age effects are represented by estimating separate functions for three age categories (i.e., 65–74, 75–84, 85+) and (b) a linear age term is including in the equation for age variation *within* each discrete age category to represent continuous age variations. Functions for persons aged 65–74 and for persons aged 85+ are presented in Table 8.1.

A term is included in the functions to represent the effect on the K disability classes of being male, along with an age-specific intercept. Here the functions are evaluated at the "midpoint" of the first interval (age 70) and at age 85. For persons aged 70 with no medical conditions, the average g_{ik} on Class 1 is .979 for females and .988 for males. Thus, without any chronic condition, a person in Class 1 at this age has little chronic impairment. Strong membership (i.e., a large g_{ik}) in Classes 2–6, or in an institution (i.e., $g_{i7} = 1$), is rare without medical conditions present at this age. Above age 85, even if none of the 28 conditions is present, the $\bar{g}_k$ for Class 1 is .749, representing the emergence of a significant amount of frailty linked only to advanced age (i.e., unobserved factors associated with age become significant at advanced ages).

The g_{ik} may be interpreted as the prevalence of a disability profile weighted by the number and type of symptoms or disabilities affecting an individual. A g_{ik} of .1 on a given class might reflect 30% of persons having one-third of the symptoms or conditions associated with that class or 15% of persons having two-thirds of the conditions related to that class. Thus, instead of reflecting the number of persons in a discrete category or the probability of a person being in a category, these scores represent the weighted disability burden of the class on the population. By adjusting the λ_{kjl} estimates for differences in individual levels of disability (i.e., the g_{ik}), estimation is made less sensitive to systematic sampling. The frail sixth class has a small negative constant reflecting that it will not occur in the population unless *multiple* comorbidities are present.

It should be clear that, by allowing partial membership in multiple

Table 8.1. Regression Estimates (×100) by Condition and Class

Condition	Class 1	Class 2	Class 3	Class 4	Class 5	Class 6	Institutionalized
			Age 65–74 (evaluated at age 70)				
Constant	97.90	0.16	0.32	0.19	0.38	−0.03	1.09
Males	0.85	0.17	−0.39	−0.39	−0.13	0.13	−0.24
Rheumatism/arthritis	−1381.80	147.00	485.10	176.40	470.40	78.40	19.60
Paralysis	−1734.60	−29.40	−53.90	147.00	455.70	1533.70	−318.50
MS	−1886.50	−127.40	29.40	−127.40	646.80	666.40	803.60
Cerebral palsy	−970.20	4.90	−240.10	142.10	171.50	−58.80	945.70
Epilepsy	−1572.90	88.20	−4.90	−24.50	210.70	798.70	504.70
Parkinson's	−1781.50	78.40	151.90	68.60	147.00	738.90	597.80
Glaucoma	−798.70	558.60	49.00	−107.80	132.30	161.70	0.00
Diabetes	−1068.20	142.10	53.90	93.10	122.50	200.90	450.80
Cancer	−975.10	102.90	−58.80	−34.30	78.40	485.10	401.80
Constipation	−450.80	24.50	88.20	166.60	19.60	362.60	−196.00
Insomnia	−485.10	24.50	220.50	171.50	112.70	132.30	−181.30
Headache	−220.50	122.50	215.60	205.80	−200.90	−19.60	−102.90
Obesity	−122.50	−24.50	93.10	14.70	225.40	−137.20	−49.00
Arteriosclerosis	44.10	77.00	19.60	34.30	−83.30	102.90	−196.00
Mental retardation	−2479.40	823.20	−34.30	−29.40	−274.40	607.60	1381.80
Dementia	3425.10	362.60	−63.70	−107.80	−303.80	563.50	2974.30
Heart attack	−739.90	147.00	−9.80	156.80	151.90	68.60	225.40
Other heart problems	−774.20	−39.20	147.00	137.20	−24.50	4.90	548.80
Hypertension	−749.70	127.40	181.30	98.00	245.00	53.90	34.30
Stroke	−1259.30	−53.90	107.80	−24.50	98.00	588.00	754.60
Circulatory problems	−764.40	122.50	313.60	186.20	240.10	254.80	−352.80
Pneumonia	−357.70	−68.60	−19.60	98.00	68.60	406.70	−127.40
Bronchitis	−176.40	19.60	147.00	112.70	−63.70	−9.80	−24.50
Influenza	−53.90	127.40	49.00	49.00	58.80	−102.90	−122.50
Emphysema	−901.60	171.50	176.40	156.80	107.80	269.50	14.70
Asthma	−137.20	−44.10	83.30	117.60	9.80	−63.70	39.20
Hip fracture	−2425.50	132.30	−49.00	29.40	1499.40	543.90	274.40
Other fractures	−759.50	−73.50	53.90	53.90	485.10	156.80	83.80
			Age 85+ (evaluated at age 85)				
Constant	74.89	2.28	1.58	0.53	3.10	−0.38	18.00
Males	4.83	2.19	−0.67	−0.80	−0.23	−0.16	−5.16
Rheumatism/arthritis	−24.74	4.17	5.44	2.47	8.67	3.23	0.85
Paralysis	−9.18	−6.38	−2.13	−1.45	−1.19	30.01	−9.61
MS	−23.04	3.57	11.14	−1.53	−5.19	−6.63	21.68
Cerebral palsy	−1.70	1.87	−7.48	1.19	−4.25	26.44	−16.07
Epilepsy	−10.63	−0.94	2.72	4.51	−2.30	5.10	1.53
Parkinson's	−21.59	−0.85	−0.60	1.87	−0.77	4.59	17.43
Glaucoma	−10.80	4.00	0.26	0.68	−1.45	1.11	5.95
Diabetes	−13.35	0.51	0.00	−0.34	−0.34	2.72	10.63
Cancer	−17.00	−0.94	−1.28	0.77	−1.79	6.38	13.86
Constipation	−6.21	3.40	0.60	2.64	3.57	7.14	−11.14
Insomnia	−1.87	3.66	3.32	2.47	2.72	3.15	−13.52
Headache	5.27	−0.68	−0.26	4.42	−6.46	3.66	−6.04
Obesity	−7.31	0.17	2.38	1.02	7.91	−0.60	−3.57
Arteriosclerosis	−5.53	4.85	1.02	0.60	−0.51	7.31	−7.74
Mental retardation	7.48	0.43	−2.04	−1.36	−6.04	19.72	−18.28

Table 8.1. (*Continued*)

Condition	Class 1	Class 2	Class 3	Class 4	Class 5	Class 6	Institutionalized
Dementia	−38.42	−0.51	−2.21	−1.11	−5.61	4.51	43.35
Heart attack	−21.34	−1.79	−0.43	0.85	−1.02	−0.94	24.65
Other heart problems	−20.83	−0.51	0.17	0.17	−0.26	0.00	21.17
Hypertension	−9.44	2.04	2.38	0.43	4.00	−0.60	1.28
Stroke	−10.71	−2.04	−3.23	−0.17	−0.94	6.21	10.88
Circulatory problems	−5.44	3.23	4.76	3.74	4.00	7.74	−18.02
Pneumonia	−3.49	1.02	1.02	0.51	3.15	4.51	−6.72
Bronchitis	0.51	0.17	−0.26	2.04	0.51	1.19	−4.17
Influenza	2.13	0.68	1.45	1.70	2.04	0.85	−8.76
Emphysema	−2.21	2.98	−0.17	−0.43	2.89	1.45	−4.59
Asthma	−3.57	−0.94	1.36	1.28	−2.64	6.12	−1.62
Hip fracture	−23.72	−2.64	−0.60	−0.09	8.67	6.38	12.07
Other fractures	−18.70	−1.28	1.70	3.40	2.21	2.89	9.69

groups, we increase both the flexibility of the model to represent disability at advanced age *and* the statistical stability of the score estimates. The ability to represent partial disability (or the inverse, partial *ability*) is important for the very elderly (i.e., the oldest-old; those 85 and over) since, at advanced ages, though the prevalence of some degree of disability is high, this does *not* mean that persons with partial disability are wholly dependent. By using the scores one also minimizes the effects of misclassification in discrete classes (e.g., the error rate in classification is high for persons whose true state is "near" the threshold between two classes). The use of the fuzzy class scores for cost estimation has a third consequence. That is, a $\bar{g}_k = .30$ can arise several ways. One is by having most people who have $g_{ik} > 0$ be near a particular value. The other is to have the g_{ik} for individuals express considerable heterogeneity. The levels lead potentially to analyses based on a principle of second-order stochastic dominance, i.e., the balancing of the value of conditions with higher average payoffs (or costs) with high risk (i.e., a wide distribution of the g_{ik} about their $\bar{g}_k$) versus conditions with a lower expected payoff (i.e., smaller $\bar{g}_k$) but with lower risk (i.e., less heterogeneity of the g_{ik} around the $\bar{g}_k$). The g_{ik} lead naturally to this type of risk versus expected payoff analysis.

8.2.1. Costs of Acute and LTC Services

Medicare Part A and B expenditures, Medicaid and privately paid nursing home costs, "out-of-pocket" payments for "LTC" care, and shadow prices and costs for informal care were estimated for 1991 to determine the cost of dementia and associated disability and morbidity. The 1984 costs were inflated to the projected program costs of Medicare and Medicaid in 1991 (USDHHS, 1988). Cost components (and the total) are in Table 8.2.

Medicare A and B costs were estimated to be $108.3 billion in 1991—$65.5 billion for females, $42.8 billion for males. The increase from 1984 is mostly due to health care cost inflation, which is double the base inflation

Table 8.2. Cost Components by Functional Class, Baseline for 1991

Cost	Total	1	2	3	4	5	6	Total Dependent Population	Institutionalized
				Males ($000,000s)					
Medicare Part A	26,787	14,749	465	988	731	1,992	3,830	8,007	4,031
Medicare Part B	15,988	10,533	369	516	387	870	1,329	3,471	1,984
Medicare Parts A & B	42,775	25,292	834	1,504	1,118	2,862	5,159	11,478	6,014
Medicaid nursing home	7,038	—	—	—	—	—	—	—	7,035
Private nursing home	6,338	—	—	—	—	—	—	—	6,338
Out-of-pocket LTC	3,003	279	504	115	153	160	1,792	2,723	—
Number of hours (in millions)	6,887	4,705	547	350	214	376	695	2,182	—
Informal care	68,868	47,052	5,467	3,502	2,135	3,559	6,954	21,816	—
Substitution	103,302	70,577	8,200	5,253	3,203	5,639	10,430	32,725	—
				Females (%000,000s)					
Medicare Part A	37,512	14,249	1,206	826	1,424	3,583	7,413	14,452	8,802
Medicare Part B	28,020	14,554	645	1,045	1,212	2,318	2,692	7,922	5,544
Medicare Parts A & B	65,532	28,803	1,851	1,871	2,646	5,901	10,105	22,374	14,355
Medicaid nursing home	19,811	—	—	—	—	—	—	—	19,811
Private nursing home	17,840	—	—	—	—	—	—	—	17,840
Out-of-pocket LTC	12,508	3,386	659	318	153	1,613	6,370	9,122	—
Number of hours (in millions)	5,518	2,126	727	534	400	621	1,190	3,392	—
Informal care	55,180	21,264	7,265	5,340	4,002	6,213	11,096	33,916	—
Substitution	82,770	31,896	10,898	8,010	6,003	9,320	16,644	50,874	—

Source: 1982 and 1984 National Long Term Care Survey.

rate. Some expenditure growth is due to a 14% increase in the size of the elderly population from 1984 to 1991.

Medicaid nursing home costs (estimated on a monthly basis from the 1985 NNHS) inflated to 1991 dollars were $19.8 billion for females and $7.0 billion for males, a total of $26.8 billion. Private nursing home costs (from the 1985 NNHS) are estimated to be $6.3 billion for males and $17.8 billion for females, a total of $24.1 billion. Overall, non-Medicare nursing home costs were $50.9 billion in 1991.

From the 1984 NLTCS, estimates of the hours of informal care delivered to persons with specific disability levels were made (Manton, 1989). For females 5.5 billion hours of informal care are estimated to be provided in 1991, equaling $55.2 billion in opportunity costs or $82.8 billion in substitution formal care costs. Estimates combined data from the 1984 NLTCS on (a) the number of caregivers a person with a given disability type and level had and (b) the number of days of care per week delivered to each person by all caregivers with (c) estimates from the 1982 NLTCS "caregiver" survey of the number of *additional* hours of care given a person due to his disabilities. The dollar value of an informal caregiver hour from BLS estimates for persons aged 65+ was $10 in 1989. Since it is more expensive for a business to provide an hour of formal care to substitute for informal care, we assumed an "indirect" cost of 50% to apply to formal care services substituted for informal care. Estimates for males were larger: 6.9 billion hours of informal care per year, i.e., $68.9 billion in opportunity costs, or $103.5 billion in substitution costs.

The estimated 12.4 billion hours of care provided to 32 million elderly persons in 1991 represent an average of 387.5 hours of informal care for each person over 65 per year. This is concentrated among the elderly chronically disabled population. Though the large population of males in Class 1 required a large absolute number of caregiver hours (i.e., informal care provided to "acutely" ill elderly males without *chronic* disability), severely impaired males (Class 6) consumed 4.1 times more care per capita (33.7 hours versus 8.2 hours per week). For females, care was more concentrated in the impaired population. Severely impaired females (Class 6) consumed 1800 hours per annum compared to 147 hours per annum for the lightly impaired, i.e., 12.2 to 1.

Out-of-pocket expenditures were projected (with Medicare inflation factors) to be $3.0 billion for males in 1991 with 91% expended by males with a chronic disability. For females the costs were higher (due to the higher life expectancy of females who have to get along after a spouse dies): $12.5 billion. A large proportion of these expenses (51%) were for frail females in Class 6. Thus, out-of-pocket payments in 1991 for LTC services *excluding* private pay nursing homes and out-of-pocket expenses for acute care total $15.5 billion in 1991. Summing out-of-pocket and informal care "costs" produced similar expenses for males and females (about $71.9 versus $67.6 billion) because of a substitution of (out-of-pocket payments for) formal care for informal care by females. The out-of-pocket estimates are

Table 8.3. Per Capita Annual Expenditures for Medicare Parts A and B (in 1991 dollars)

Age Group	Total	1	2	3	4	5	6	Total Dependent Population	Institutionalized	Ratio of Class 6/1
						Males				
65+	3,278	2,292	2,497	5,371	6,620	8,136	13,011	7,495	12,311	5.7
75+	4,234	2,680	1,958	3,568	10,067	7,430	11,434	6,775	13,301	4.3
85+	5,356	2,381	872	3,758	5,739	7,432	10,255	5,749	13,097	4.3
						Females				
65+	3,451	1,986	3,576	2,759	5,884	7,326	15,283	7,190	10,439	7.7
75+	4,133	2,142	2,711	2,393	5,013	5,483	13,961	6,248	9,378	6.5
85+	5,207	2,062	2,165	2,350	7,083	4,704	11,415	5,916	8,325	5.5

close to the $14.7 billion in copayments for Medicare Part B (assuming a 1/3 copay for Supplementary Medical Insurance; USDHHS, 1988).

Per capita expenditures vary by type for Medicare Part A and B and are shown in Table 8.3. Class 6 has 5.7 times the Medicare cost of non–chronically disabled elderly males at ages 65+ 4.3 times the costs at ages 85+. Differences are larger for Part A; the severely impaired have 7.2 times the expenses of non–chronically disabled persons; the differential decreases with age and is 6.1 for males age 85+.

These expenditures were recalculated assuming the elimination of dementia and are shown in Table 8.4. For males, Part A and B expenses were reduced $2.2 billion. Savings for nursing homes were $2.9 billion for Medicaid and $2.6 billion for privately paid ones. Out-of-pocket savings are small ($28 million) because informal care hours *increased* (probably provided by a female spouse) because of the discharge to the community of elderly males with multiple comorbidities. Implicit costs of informal care increased $1.9 billion. The overall net savings projected for males due to the elimination of dementia is $5.8 billion.

Females cost savings were greater: for Medicare Part A and B, $4.8 billion; for Medicaid nursing home use, $7.2 billion; for private nursing home use, $6.5 billion. Out-of-pocket LTC costs increased $0.7 billion while the increase in informal care "implicit" costs was $2.2 billion. Elimination of dementia reduced costs of $15.6 billion for females in 1991. Combined with $5.8 billion for males, eliminating dementia would reduce costs $21.4 billion in 1991.

8.3. EXAMPLE B: COMPARISONS OF POPULATION FOR DIFFERENT EXPERIMENTAL CONDITIONS WHEN THE POPULATIONS ARE NOT RANDOMIZED

There has been interest in the comparison of cases and controls in LTC demonstration projects. Often, these have been designed as randomized

Table 8.4. Cost Components by Class, Assuming Elimination of Senile Dementia

Cost	Total	1	2	3	4	5	6	Total Dependent Population	Institutionalized
				Males ($000,000s)					
Medicare Part A	25,246	15,052	444	1,011	764	2,113	3,542	7,875	2,319
Medicare Part B	15,339	10,751	352	526	401	918	1,229	3,425	1,163
Medicare Parts A & B	40,585	25,803	796	1,537	1,165	3,031	4,771	11,300	93,482
Medicaid nursing home	4,122	—	—	—	—	—	—	—	4,122
Private nursing home	3,712	—	—	—	—	—	—	—	3,712
Out-of-pocket LTC	2,975	299	498	118	160	175	1,725	2,676	—
Number of hours (in millions)	7,073	4,855	549	370	227	409	664	2,218	—
Informal care	70,730	48,547	5,487	3,699	2,273	4,088	6,636	22,183	—
Substitution	106,094	72,821	8,230	5,549	3,409	6,132	9,953	33,274	—
				Females ($000,000s)					
Medicare Part A	34,392	14,841	1,149	858	1,476	3,791	6,628	13,902	5,648
Medicare Part B	26,391	14,996	616	1,067	1,252	2,430	2,439	7,804	3,591
Medicare Parts A & B	60,783	29,838	1,765	1,925	2,728	6,221	9,067	21,706	9,239
Medicaid nursing home	12,628	—	—	—	—	—	—	—	12,628
Private nursing home	11,372	—	—	—	—	—	—	—	11,372
Out-of-pocket LTC	13,258	4,103	689	347	165	1,861	6,094	9,155	—
Number of hours (in millions)	5,738	2,259	734	571	428	686	1,059	3,478	—
Informal care	57,376	22,592	7,343	5,713	4,275	6,858	10,594	34,785	—
Substitution	86,065	33,887	11,015	8,570	6,413	10,288	15,891	52,177	—

Source: 1982 and 1984 National Long Term Care Survey.

controlled trials (e.g., "Channeling"; Manton et al., 1993b). In other study designs, randomization was not rigorously enforced (e.g., California MSSP). In LTC demonstrations, it is difficult to maintain randomization over time, due to selective mortality and a time-varying history of service interventions and responses to service interventions.

We present a case-mix adjustment procedure that, if the case-mix scores are informative about selection, will control for either enrollment or disenrollment bias or for differential rates of functional and health change. The procedure has two advantages over other bias-controlling techniques. One is that we can control for differences as they emerge through time. This is an advantage over procedures (e.g., Heckman et al., 1985) that deal with a fixed-time intervention (i.e., a pre-postdesign). Second, the GoM procedure is a semimetric procedure (i.e., it is based on direct estimation of moments rather than on estimating parameters of specific distribution functions with fixed moment relations) so that it is not as sensitive to assumptions about distributional forms.

8.3.1. National Channeling Demonstration

The National Channeling Demonstration Project ("Channeling") evaluated whether two interventions affected six outcomes: hospital, nursing home, and formal community care use, receipt of informal care, mortality, and the "well being" of clients and caregivers. Eligible persons had to be 65 or over and meet disability and unmet need criteria.[1]

One intervention ("case management") improved access to case management. The second "a financial control" enhanced case management by purchasing services with supplemental funds to meet needs determined by case managers with limits to expenditures for a person (Brown, 1986). Subjects were randomly allocated to treatment and control groups for each intervention. Each intervention was conducted in 5 of 10 sites.

This randomized control trial (RCT) examined the effect on outcomes of a program designed to *increase* the level of case management over that typical in the community (Brown, 1986, p. 7). The "intervention" is thus *not* case management and community services but a program to stimulate use of those services. The potential to increase services is greater for the financial control model (hereafter "financial management") where supplemental services could be purchased (Applebaum et al., 1986). A prior evaluation

[1] Applicants had to be least moderately disabled in two or more ADLs; severely disabled in three or more IADLs *or* severely disabled in two IADLs and one ADL. Cognitive or behavioral difficulty could count as a severe IADL disability. To guard against Channeling being used as a substitute for existing community services, clients had to have at least two unmet ADL or IADL needs that were projected to continue for at least six months. Applicants had to reside in the community or, if institutionalized, be certified as likely to be discharged within three months. Finally, clients had to live in the demonstration's catchment area, and, for the Financial Control Model, to have Medicare coverage.

concluded that no *global* effect on outcomes was produced by the *additional* services generated by either intervention (Grannemann et al., 1986).

We re-analyzed Channeling data using a longitudinal event history form of GoM (see Chapter 7) applied to compare cross-temporal changes under the two experimental conditions modified to reflect characteristics of the data and the interventions, e.g., 72.5% of people in case management received services during the first six months, 58.3% in the second six months (Applebaum et al., 1986); 20.5% of controls received case management in the first six months, 14.5% in the second six months. The ability of community services to change outcomes is mediated by the case manager's skill in identifying needs and to select efficacious services. In the first six months, under case management only one site (Middlesex County) provided significantly more medical/personal care for clients than controls. Significantly more services were provided by financial management. Service *differences* between treatment and control groups in the five sites ranged from 36.8% in Cleveland to 14.7% in Greater Lynn County (2.5 to 1) *after* controlling for site and individual characteristics. Because some controls received case management, and not all persons in the treatment group, the program effect may be contaminated at the site level if there is a higher probability of receiving case management among those *controls* with higher needs. The *levels* of service received by controls and cases also varied between sites (Applebaum et al., 1986, p. A.3). In financial management sites 76.0–89.4% of clients received case management compared to 7.6–36.6% for controls. Differences in the levels of case management may interact with site differences in the availability of community services to effect outcomes.

To organize our analysis we posited three mechanisms affecting outcomes: (1) the "intervention" designed to increase the use of case management, (2) the ability of case management, once applied, to enrich the set of services beyond what a client would get on his own, and (3) the services (e.g., hospital, home health, informal care) that directly affect "outcomes" defined as changes in client health, functioning, and survival. The first mechanism, the "intervention," *was* exclusively delivered to cases. In the second mechanism, however, both cases *and* controls *received* case management (neither controls nor cases under the basic model received financial enhancement), through cases received it at a higher rate. It is the effect on health outcomes of differences in service levels, produced by increased levels of case management induced by the intervention (or case management with additional resources), that was evaluated (Brown, 1986). Thus, the interventions necessarily operated through intermediate mechanisms to alter outcomes. Intermediate stages were not randomized in the original design. Consequently, the program might not increase the amount of case management received and yet case management still might have a beneficial impact on outcomes for individuals who did receive that service (whether or not in the program).

In the analysis, we examined the randomization process and measurement characteristics of the original design (Brown, 1986). Clients were confirmed to be randomized at screening (i.e., there were few significant case-control differences in 53 variables). The baseline interview showed that, after two weeks, there was evidence of randomization failure. Specifically, two regressions were estimated, one for cases and one for controls, to determine if the probability of loss to follow-up was related to the original 53 screen variables. The two regressions were significantly different at the 0.05 level, not taking into account a large difference in the constants due to the nonresponse rate being 10% higher in controls. Because randomization did not hold for the baseline interview which was to be used to develop care plans, the evaluation used variables from the screen and "comparable" variables from the in-person baseline interview. Medical conditions and other important variables (e.g., Short Portable Mental Status Questionnaire—SPMSQ) only available on the baseline were *not* used (Brown, 1986). In our analyses we used data from the in-person baseline interview instead of the screen data. We controlled for differences in the distribution of attributes between case and control populations by using a multivariate event history model of individual characteristics and their change (Manton et al., 1986b, 1987).

The GoM procedures allowed hypotheses about stages two and three to be evaluated, i.e., to examine multiple related measures of the efficacy of the services delivered for *specific* health and functional subgroups as identified in the analysis of individual health characteristics. This changes the goal of the analysis from determining if the process used to stimulate the delivery of case management (and indirectly other services) changed the average impact on outcomes across intervention groups to determining the most effective (in terms of modifying outcomes) sets of services to provide to persons in specific health and functional classes. Changes in the interrelated outcomes of the service delivery system are modeled as temporal processes rather than fixed effects of treatment on independent outcomes within a fixed interval. This exploits the heterogeneity of client characteristics, changes in their characteristics, and site and service differences to identify individually efficacious caregiving options—rather than the ability of the intervention program to change levels of case management or other services in a community. The sensitivity and power of tests is increased because multiple measures on individuals, over time, are used in the multivariate model.

When the 10 sites were selected, there was no specific reference population from which sites were assumed drawn. We assumed that observations from all sites are drawn from a "superpopulation" where stochasticity emerges, not from sampling, but from individual-level processes. Superpopulation models of sampling are discussed in Cassel et al. (1977) and sampling for event history models in Hoem (1985, 1989). This approach to dealing with a nonrepresentative set of sites relies on the model, and the ability to describe an individual's health state, and *not randomization*, to deal with individual heterogeneity.

The purpose of covariates in the event history model is not simply to guard against randomization failure. We are interested in the effects of covariates on outcomes for individuals in their own right. To apply superpopulation concepts, a description of the individual's state is needed on which to conditionally estimate transition rates. If a person's state and its changes are "well" described, then site selection can be ignored because it is conditionally noninformative. In this case, "site" is a *proxy* variable representing an arbitrarily (conditional on case-mix scores) defined *set* of clients. If individual health characteristics *and* their change with time are well modeled, parameters appropriate to *individual* processes can be used in designing and evaluating hypotheses. Thus, the reference population for the 10 sites is a common superpopulation distribution of individual responses, where site-specific sampling is "noninformative" conditional on individual characteristics as reflected in case-mix scores (scores may be thought of human capital input to a nonlinear, dynamic health service system (Manton et al., 1991b; Hoem, 1985, 1989).

To evaluate "efficacy" we need a criterion which is *not* a measure of service use. Hospital and nursing home use are part of the third stage since "community" and "institutional" services interact, over time, to affect client health. This is explicit if service "outcomes" are modeled as "endogenous," i.e., as components of a *system* where causality is "reciprocal" with "feedback." Such effects cannot be represented if "outcomes" are assumed to be independent. If the system structure is nonlinear, the behaviors of endogenous variables are functions of the values of exogenous characteristics, e.g., health and functional status. Thus, we evaluated endogenous service characteristics for each of K profiles of health characteristics representing human capital "inputs" to the system (Manton et al., 1991b).

The dynamic behavior of a multivariate endogenous system of variables cannot generally be examined in an RCT because "time-varying" covariates are usually *not* analyzed (Kalbfleisch and Prentice, 1980, p. 126). Usually, in an RCT, a single outcome (e.g., time to death) is modeled as a function of treatment (a case) or nontreatment (a control) with biases due to individual differences controlled by randomization. If changes in individual characteristics are modeled, the treatment effect may be "hidden"; i.e., the treatment may affect mortality by altering time-varying covariates so that conditioning on those covariates eliminates aggregate survival differences. Modeling time-varying covariates is, however, necessary to describe the processes *producing* changes. To test whether mechanisms by which interventions change "outcome" are adequately "described" one determines if, after *conditioning* on time-varying covariates, all outcome differences are eliminated. If not, there are unmeasured factors affecting outcome.

The cost of individual services is often not as important as the system operating within overall budgetary constraints. Mortality is a useful global criterion because it is (a) easily measured, (b) clearly undesirable, and (c) frequently follows service use. We determined if, controlling for health, mortality varied (a) over treatments or (b) over differences in service use. Thus, institutionalization of a physically robust but cognitively impaired person might be appropriate, while cognitively intact but frail persons with adequate informal care resources might be managed at home—if mortality outcomes are unaffected (or improved). This avoids the assumption that nursing home care is necessarily of high quality—an assumption implicit when measuring the cost effectiveness of community services against nursing home care costs as a "standard." If nursing home care is *not* adequate for a person with a given set of health attributes under existing reimbursement (e.g., higher mortality is manifest), then nursing home costs are not appropriate to use in evaluating the cost effectiveness of community care. Different conclusions would result if an "appropriate" level of nursing home costs is assumed. It was shown that, even in a "high" benefit state like New York, certain subgroups (e.g., patients with dementia and behavior problems) received too few services (Manton et al., 1990).

Of 6326 persons screened in the 10 sites (with referral from multiple sources), 5626 (88.9%) received in-person baseline assessments, including persons in both community and short-stay hospitals. The population had an average age of 80 at baseline and was more frail and cognitively impaired than the U.S. elderly population but was *not* exceptionally frail relative to the U.S. institutional elderly population. Nursing home use by controls was lower than assumed by the study design. Follow-up interviews were conducted at 6 ($N = 4189$; 74.5%), 12 ($N = 3634$; 64.6%), and 18 months ($N = 1409$; a 50% sample). The "baseline" interview provided measures of the health and functioning of clients and was the initial "assessment" for case managers. At baseline, Channeling staff interviewed clients; evaluation staff interviewed controls. A re-interview of 400 clients by evaluation staff suggested little bias in this procedure. After the "baseline," assessment and follow-up were separate processes. Linkage was made to Medicare and Medicaid records. Private expenses were tracked. After the last interview only Medicare and Medicaid data were collected. Vital status was not followed after 18 months.

Two problems clearly presented by the analysis was (a) left censoring (i.e., the decline in from 6326 to 5626 persons) and (b) right censoring (i.e., the rapid loss of persons over follow-up). Left and right censoring can be dealt with if, conditional on the g_{ik} (a) the probability of selection is independent of site and (b) the probability of death, or loss to follow-up, is independent over individuals. Condition (a) requires that the λ_{kjl}^{++} are defined generally enough to represent all frail elderly subgroups, and their changes over time, in the U.S. population. This requires that individuals partly representing those subgroups (at some stages in the process of health change) exist in at least one site. If all subgroups are at least partially represented (the g_{ik} may be less than 1 and still identify a class), then the

λ_{kjl}^{++} define a $K-1$ dimensional space within which all prevalent health and functioning problems can be described as they change over time. Condition (b) requires that the g_{ik} predict an individual's risk of death and other censoring events well enough that these events are independent after conditioning on g_{ik}.

In (8.1), after generalization to represent events and reassessments over the 18 months of follow-up (see Chapter 6), constraints were imposed to define specific processes. The $g_{ik \cdot te}$ were constrained over e (i.e., $g_{ik \cdot t}$ is a fixed value for each episode e in an interval t to $t+1$) and $\lambda_{kjl \cdot te}$ over t and e (the same $K-1$ dimensions apply at all times). The λ_{kjl}^{++} are assumed constant over site time *and* intervention so that the $g_{ik \cdot t}$ can be compared across all conditions. The $g_{ik \cdot t}$ are updated with new information at t to reduce attrition bias, which differed for cases and controls.

Transition parameters are estimated *conditionally* on the $g_{ik \cdot t}$, for each episode type (i.e., hospital, community, nursing home) and treatment group. As described in Chapter 6, each variable (say j_2) was constrained to describe a transition matrix for each of the K classes (and for each treatment group) where the coefficients $\lambda_{kj_2(l_1 \times l_2)}$ are stratified on "time" (l_1) and "mode" of exit (l_2). The $\lambda_{kj_2(l_1 \times l_2)}$ probabilities are similar to probability densities in multidecrement tables where l_2 are decrements in each of K classes. From the $\lambda_{kj_2(l_1 \times l_2)}$, life table parameters were calculated. The life table survival probability for the Kth class, being a function of the $\lambda_{kj_2(l_1 \times l_2)}$ for all time intervals after x, will be smooth. Standard competing risk adjustment of life table parameters (Tolley and Manton, 1987) can eliminate right censoring (e.g., lost to follow-up) to control attrition within the K classes. To this end, one of the l_2 modes of exit defined is "end of study" (i.e., "EOS"); either follow-up ends or a new assessment is made and health variables updated. Another exit is death. Since the risk of death, or EOS, occurring before a change in service use is determined *within* each of the K classes, these two competing risks are correlated (i.e., dependent) through the health status measures (i.e., $g_{ik \cdot t}$). Since the $g_{ik \cdot t}$ are time variable, the competing risk dependence can change with time/age.

In the longitudinal form of the GoM likelihood, individual health and functional characteristics are used to identify K fuzzy classes and the g_{ik} scores relate individuals to classes. Conditional on the $g_{ik \cdot t}$, K sets of transition variables were estimated representing the interaction of hospital use, nursing home use, community retention, EOS, and death over time within 6-month follow-up periods (the $g_{ik \cdot t}$ are updated at each 6-month assessment). These "outcomes" compete over time, with persons capable of moving into, and out of, outcome states (except death and EOS). Health and functional status are exogenous for each six-month period.

The $\lambda_{kj^*l}^{++}$ were estimated for J^* "external" (or "event" or "cost") variables representing social and economic resources consumed over time. The $\lambda_{kj^*l}^{++}$ for external variables are estimated conditionally on $g_{ik \cdot t}$; e.g., the distribution of informal care (in hours/week) is estimated for each of K

classes. For a class we examine the interaction of hospital, nursing home use, and mortality under each treatment and interactions with the level of informal care (not specific to treatment). If all differences in service use between sites are explained by health inputs and treatment group gradients in case management, then the distribution of site within a class should be the same as observed overall. If the site distributions differ between classes, site characteristics modify "within" treatment effects.

In Table 8.5 we present λ_{kjl}^{++} estimates for 27 health and functioning measures from cases in 10 sites ($S = 10$), in both intervention and nonintervention groups, and over time ($T = 3$). An examination of the marginal frequency showed that the population was frail with an average of 4.3 (of 13) illnesses and 7.9 (of 13) impairments. The six classes can be described as (Manton and Stallard, 1991):

1. The "acutely ill" with no cognitive impairments but multiple illnesses (e.g., anemia, heart disease, cancer, and broken bones; an average of 8.8 conditions) and limitations on 8 of 13 activities with moderate informal care (9 hours/week). This class, in separate analyses, was found likely to be female, have low levels of education and income, rent housing, live in rural areas, be unmarried, and Medicaid eligible. It was concentrated in eastern Kentucky (45.8%; an area with few hospital services), Houston, and Miami, was relatively old (80.2 years; near the mean age for Channeling), more likely white or Hispanic and less likely to be in financial management.

2. Those with "stroke and paralysis" had poor functioning (11 impairments, including bed or chairfast) and multiple illnesses (6.0, including heart disease, hypertension, diabetes, stroke, and paralysis). In a separate analysis, this class is found likely to be black (51.3%), young (mean age 75.2), female, married, poorly educated, urban (e.g., Baltimore, Cleveland, and Philadelphia), and in a treatment group. They use an average of 13.7 hours of informal care per week.

3. The "chronically ill" are relatively functionally intact (4.7 limitations), with multiple illnesses (e.g., arthritis, heart disease, high blood pressure, and diabetes; *no* stroke or paralysis) and, in a separate analyses, were likely to be female, have low income and education, be widowed, and live in eastern Kentucky and Miami. Moderate (8.2 hours) amounts of informal care were used per week. The group was less likely in financial management and more likely to be a control than a case.

4. The "relatively unimpaired" had no cognitive impairment, few (0.9) illnesses (arthritis and cancer), and limited only in bathing and selected instrumental activities (e.g., housework, meal preparation, and travel). Independent analyses showed this class to be young (mean age 76.7), more likely white, female, educated, urban, and use only 4.6 hours/week of informal care.

5. The "frail" had mild to moderate cognitive impairment consistent with

Table 8.5. Demographic and Service Use Characteristics of the Six Case-Mix Groups

	Population Frequency	Profiles					
		1	2	3	4	5	6
		Age					
64–69	11.6	9.7	22.9	13.9	18.0	1.7	6.4
70–74	16.2	3.1	23.7	16.0	28.2	5.8	13.1
75–79	21.0	37.8	29.7	21.3	20.5	17.2	14.3
80–84	23.4	22.1	21.6	32.9	17.8	26.7	22.9
85–89	17.5	23.5	2.2	12.5	11.3	30.2	22.9
90+	10.3	3.8	0.0	3.6	4.2	18.4	20.4
Mean age		80.2	75.2	78.6	76.7	84.1	82.3
		Race					
Black	21.9	0.0	51.3	32.1	8.3	19.5	22.6
White	73.9	89.9	45.2	62.1	91.0	76.1	73.0
Hispanic	4.2	10.2	3.5	5.7	0.7	4.4	4.4
		Sex					
Male	28.2	19.1	28.4	20.5	20.6	26.7	41.5
Female	71.8	80.9	71.6	79.5	79.4	73.3	58.5
		Marital status					
Married	31.1	26.9	49.1	3.7	18.1	24.0	51.1
Widowed	56.4	54.5	43.7	79.0	62.2	66.3	41.3
Divorced	4.7	13.9	5.4	8.4	5.7	1.4	1.3
Separated	2.6	3.4	1.8	5.4	2.7	1.6	2.0
Never married	5.2	1.4	0.0	3.4	11.3	6.6	4.3
		Education					
None	12.7	9.5	1.6	15.3	0.1	16.2	26.4
Elementary	49.5	71.7	65.5	66.9	28.1	54.0	41.8
High school	27.7	13.4	27.0	12.3	52.7	23.1	21.2
College	10.2	5.4	6.0	5.6	19.1	6.7	10.7
		Place of Residence					
Owns home	40.1	25.2	44.8	33.8	35.2	36.8	51.9
Rents home	45.6	64.8	38.5	61.6	60.4	47.8	21.9
Rent free	13.3	10.0	15.0	3.8	3.8	14.5	25.0
Institutionalized only	0.3	0.0	0.4	0.0	0.4	0.7	0.0
Other	0.7	0.0	1.4	0.8	0.2	0.2	1.2

Table 8.5. (*Continued*)

	Population Frequency	Profiles 1	2	3	4	5	6
		Metropolitan Area					
Large city	49.4	36.1	65.9	57.6	45.2	43.1	49.7
Large suburb	17.9	14.0	13.1	9.3	22.2	29.8	13.0
Medium city	8.7	0.0	6.0	10.9	13.1	8.7	8.3
Medium suburb	5.7	0.0	1.6	0.0	8.1	8.2	8.3
Small city	6.5	2.8	3.7	1.9	8.0	7.7	8.8
Small town	3.5	6.9	1.9	4.0	3.2	2.5	4.0
Rural	8.2	40.0	7.8	16.4	0.0	0.0	7.7
Other	0.1	0.2	0.0	0.0	0.2	0.0	0.1
		Income					
<$500/mo	56.6	73.3	46.5	79.2	53.6	62.9	43.8
$500–1000/mo.	33.9	26.7	43.7	20.8	37.1	29.8	37.3
>$1000/mo.	9.6	0.0	9.8	0.0	9.4	7.3	18.9
		Area					
Baltimore	10.5	0.0	20.0	11.4	12.1	10.1	7.4
Eastern Kentucky	8.3	45.8	6.0	19.4	0.0	0.0	7.8
Houston	11.5	17.8	13.5	11.6	15.7	9.1	6.7
Middlesex	10.0	0.6	7.7	5.2	12.0	10.4	14.6
Southern Maine	7.6	3.4	2.4	0.0	12.7	11.0	8.5
Cleveland	8.9	9.1	12.7	9.2	4.5	8.9	10.0
Greater Lynn County	10.1	0.2	9.3	3.3	14.5	18.7	7.0
Miami	12.5	29.1	0.0	16.2	14.5	16.4	8.0
Philadelphia	14.6	0.0	23.8	10.7	9.8	9.6	23.8
Rensselear County	6.0	0.0	4.3	2.9	10.7	7.9	5.4
		Model-Treatment Cross-Tabulation					
Case control	19.9	40.4	18.5	25.2	17.0	13.2	18.8
Case treatment	28.1	35.9	33.4	30.5	27.6	22.8	25.8
Financial control	18.7	0.9	12.9	22.5	20.7	24.7	19.7
Financial treatment	33.4	22.8	35.2	21.8	34.7	39.3	35.7
		Hours/Week for Informal Help					
1–9	29.1	10.9	22.5	59.3	35.2	24.7	23.1
10–39	15.2	1.9	19.2	16.1	10.5	15.4	20.6
40–79	3.8	4.6	7.1	2.5	0.4	3.1	5.8
80–159	1.9	1.9	2.6	0.0	0.1	1.9	3.8
160+	0.8	1.9	0.8	0.3	0.2	0.3	1.7
No Informal Help	49.1	78.9	47.9	21.8	53.6	54.7	45.1
Average Hours		9.2	13.7	8.2	4.6	9.0	16.2

Source: National Channeling Demonstration data.

problems in meal preparation, taking medication, money management, phone use, and travel (Manton and Soldo, 1985). Despite having few illnesses (1.3), mortality is high because of advanced age (mean age 84.1). This class is concentrated in Lynn County, Miami, and southern Maine and is more likely to be in financial management or a control group.

6. The "severely cognitively impaired" were "relatively" (to the overall sample) more male (41.5%), had a high monthly income, were married, and homeowners. It is the second-oldest group (mean age 82.3) consuming the most informal care (16.2 hours/week). It is likely to be in financial management and to live in Middlesex County or Philadelphia.

We calculated transition variables for each class to determine the proportion of cases and length of stay (LOS) in hospital, nursing home, and community episodes. Right censoring due to EOS was eliminated with competing risk adjustments (Chiang, 1968; Tolley and Manton, 1987). Mortality was *not* eliminated but was an outcome whose interaction with other events we wish to describe.

The parameters for hospital episodes are presented in Table 8.6. Discharges from hospitals to nursing homes, home, or death are described by their probability and the hospital LOS (in days, in parentheses). The "acutely ill" have an 8.6% chance of entering a nursing home after an average hospital LOS of 21.6 days if the person is a case management *control*; an "acutely ill" person in the intervention is *more* likely to go to a nursing home (10.6%) after a stay of 21.0 days.

The rows labeled "All" are weighted so that the case mix for each transition (e.g., hospital to nursing home) is *standardized* to a common g_{ik} distribution (i.e., the last row in Table 8.5) so clients and controls can be compared without being confounded by differences in the case mix for different populations. Life tables are calculated for each of the K classes to describe health status dependence between the different types of hospital discharge.

A major goal of the analysis was to determine if the failure to detect a treatment effect was due to case-mix differences (and the interaction of case-mix differences with site-specific service availability) between case and control populations. Case managers could have made large changes in service use for persons with different health status but yet produce little overall difference because they tried to stay within total budget constraints.

Overall, discharges to nursing homes differed little in duration for either case (20.9 and 20.9 days) or financial (25.3 and 24.5 days) management. The longer LOS for financial management appeared due to service differences between sites (i.e., in financial management sites per capita hospital bed supply is 16% higher; nursing home bed supply 14% lower; annual charges for financial management controls are $1800 higher than case management controls). Across discharge type, LOS declined (i.e., 0.8 days for nursing homes; 0.6 days for community; 1.1 days for death) with financial man-

Table 8.6. Probability and Time to Exit in Days (in Parentheses) from Hospital by Destination and Class

Class	Case Management Control ($N = 1583$)	Case Management Treatment ($N = 2249$)	Financial Control Control ($N = 1588$)	Financial Control Treatment ($N = 2814$)
	Discharge to Nursing Home			
1	8.6	10.6	1.9	13.2
	(21.6)	(21.0)	(29.2)	(29.4)
2	3.6	4.0	13.3	17.4
	(13.4)	(13.9)	(24.4)	(22.1)
3	11.1	7.3	6.4	11.2
	(15.8)	(19.7)	(25.5)	(22.9)
4	10.5	14.5	17.3	7.5
	(29.3)	(23.8)	(33.6)	(27.8)
5	28.4	18.1	31.2	25.8
	(17.2)	(18.2)	(24.6)	(23.6)
6	8.1	8.8	15.8	11.3
	(32.5)	(22.3)	(14.2)	(21.7)
All	12.4	11.1	14.8	14.5
	(20.9)	(20.9)	(25.3)	(24.5)
	Discharge to Community			
1	74.6	71.1	80.8	67.3
	(17.3)	(17.2)	(17.9)	(20.0)
2	85.4	87.6	63.5	61.8
	(12.9)	(17.0)	(25.8)	(13.1)
3	83.4	87.2	90.7	82.2
	(12.0)	(15.0)	(12.2)	(13.6)
4	77.5	73.0	79.0	90.0
	(14.5)	(12.4)	(14.0)	(15.9)
5	47.4	58.2	43.9	55.6
	(10.4)	(14.9)	(23.0)	(17.0)
6	88.0	84.9	78.4	81.9
	(14.6)	(11.9)	(13.4)	(13.4)
All	75.0	76.3	73.6	73.5
	(13.6)	(14.5)	(16.0)	(15.4)
	Death			
1	16.9	18.3	17.3	19.3
	(16.0)	(26.2)	(16.0)	(17.0)
2	11.0	8.4	23.2	20.8
	(21.4)	(16.8)	(24.3)	(18.9)
3	5.5	5.5	2.9	6.6
	(16.5)	(25.4)	(15.1)	(15.0)

Table 8.6. (*Continued*)

	Case Management		Financial Control	
Class	Control ($N = 1583$)	Treatment ($N = 2249$)	Control ($N = 1588$)	Treatment ($N = 2814$)
4	12.0	12.5	3.7	2.6
	(9.9)	(19.6)	(12.0)	(4.2)
5	24.2	23.7	24.9	18.6
	(21.7)	(12.1)	(22.7)	(22.9)
6	3.9	6.3	5.9	6.8
	(29.2)	(12.3)	(12.3)	(16.2)
All	12.5	12.6	11.6	12.0
	(18.3)	(18.1)	(19.8)	(18.7)

Note: Values in parentheses indicate average time to exit in days.
Source: National Channeling Demonstration data.

agement—a trend not evident with case management. The probability of the three hospital discharge types changed with a 10.5% reduction (12.4% to 11.1%) in discharges to nursing homes under case management—deferred admissions returned home. Both interventions produced expected, *but* small, overall effects.

Large differences, however, were observed *between* the K case-mix classes. The "acutely ill," with few economic resources and poor functioning, were more likely to be discharged to nursing homes in both interventions—as was the "stroke and paralysis" class with multiple illnesses and severe impairment. Effects for the classes were larger in the financial management intervention. However, neither intervention could meet the needs of these two classes in the community despite the availability of significant (9 and 13.7 hours/week) informal care. The "acutely ill" class lives in rural areas and is less likely to be married while the "stroke and paralysis" class lives in cities and is more likely married. Both have low levels of education and moderate to low incomes. The lack of intervention effects for these two classes is due both to their poor medical status and to limited social and economic resources which are not adequate to support those groups at home (in contrast to the "cognitively impaired"). Thus, increased nursing home use by these classes appears appropriate given the limitations of the interventions—especially since the community services provided were generally not medically intensive.

The "chronically ill" (widows; living aline and poor, though relatively functionally intact) were less likely to be institutionalized under case management. Under financial management they were more likely to be institutionalized and less likely to return home. This is because the class is prevalent in eastern Kentucky, a case management site (19.4%; versus 8.3% of all episodes) with the second-lowest nursing home, and lowest hospital, bed supply.

The "relatively unimpaired" were less likely to use nursing homes under financial management because their limited care needs could be met at home (11% increase). Though they were less likely to be married, with little informal care available, a high level of education (and lack of cognitive impairment) apparently allowed them to function in the community. The opposite was noted for case management where the nursing home use of this class *increased*. The additional resources in the financial management intervention therefore *did* make a difference for persons with limited social resources *if* health and functional impairment was moderate *and* the client was educated and cognitively intact. This effect is important because the group is prevalent (20.8%) and, given its relative young (average age 76.7), will survive for a relatively long time and, potentially, in a relatively functional state.

The "frail" are the oldest group with moderate cognitive impairment. For both interventions, they are *less* likely to enter a nursing home (i.e., from 28.4 to 18.1% for case management and 31.2 to 25.8% under financial management; declines of 36% and 17%) though they have the highest level of institutional use. Community services were effective for this class because the number of medical conditions is small and most limitations are instrumental activities (e.g., shopping, cooking, managing money) where the community services most often delivered (e.g., homemaker) help. This group does not have inside mobility, toileting, or eating problems which make home care difficult.

The "severely cognitively impaired" are most prevalent (25.7%) and showed a decline (from 15.8 to 11.3%; 28%) in *institutionalization* under financial management, but a small increase under case management. They have adequate economic and social resources (i.e., high income, likely to be married, have assets (home ownership) and the highest level of informal care (16.2 hours per week)) to benefit from the supplemented services available under financial management because they did *not* have extensive, acute care needs—in contrast to classes 1 and 2. Because of their greater economic resources, they did not qualify for additional services in the basic intervention.

These findings suggest that, for persons leaving hospitals, large changes in class-specific nursing home use (and discharge home) were hidden either by site differentials in service availability (e.g., the "chronically ill" or "relatively unimpaired" classes) or by the level of medical acuity, and the availability of informal care and other social resources (e.g., the "acutely ill" versus "frail" classes). Overall costs did not change because case managers redistributed services across functional classes according to perceived needs under budgetary constraints.

The services delivered under case management were often social rather than medically intensive. It is possible (e.g., Greene et al., 1989; Gibbins et al., 1984) that more medically intensive home care could have changed the discharge estimations of the "acutely ill" and "stroke and paralysis" classes.

The fact that the "frail" and "cognitively impaired" classes, with a high degree of cognitive problems and institutional risk, could stay at home suggests the efficacy of community services targeted to specific groups (e.g., cognitively impaired persons with adequate social and economic resources). Overall mortality changes, and within class, are small. The largest decline is for the "frail" under financial management who are more likely to return home after a shorter stay.

In the analysis of nursing home residents (using the same life table methodology), case management significantly increased overall hospital use (from 41.5 to 55.8%). The "acutely ill," "stroke and paralysis," "chronically ill," and "cognitively impaired" increased hospital use while, in contrast, the "unimpaired" and "frail" classes showed increased discharge rates to home. Financial management was able, overall, to return more nursing home residents home (a decline from 63.2 to 53.1%). Higher hospital admission rates for the "unimpaired" and "frail" classes under financial management are associated with longer nursing home stays. Financial management reduced hospital use for the "stroke and paralysis" and cognitively impaired," classes with informal care resources. A factor affecting hospital use for nursing home residents is the 20% longer hospital LOS in financial management sites. Case management sites had lower hospital use initially, so that their increases, and management site decreases, converged; i.e., discharge rates went from a 21.7% difference (63.2 versus 41.5%) to a 2.7% difference (i.e., 55.8 versus 53.1%). Thus, the intervention produced consistent behavior for the two sets of sites. Mortality showed a small decline under case management and a small increase for financial management.

In analyzing transitions from the community, there was little change in hospitalization rates. Differences emerged for the six health classes. Under case management, the "acutely ill," "chronically ill," "relatively unimpaired," and "cognitively impaired" showed increases. Under financial management, increases were limited to the "frail". For nursing homes the overall admission rate (for six months) dropped for both interventions—as intended. Case management produced a decline from 6.4 to 5.2%; the decline for financial management was smaller—6.3 to 5.6%. This implies a reduction of 54 stays for case management and 37 stays for financial management—each of about 100 days. The classes showed different changes. The "acutely ill" were more likely to go to nursing homes, the "unimpaired" less likely. The "frail" and "cognitively impaired" showed increased nursing home use under case management but not under financial management.

Overall, both interventions showed declines in mortality. An increase for the "frail" in case management reflected mortality reductions in nursing homes (23.2 to 1.6%); i.e., case managers did not send terminal individuals to institutions, but kept them at home.

The analyses suggest that confounding of case mix with site, treatment type, and attrition may account for a lack of significant overall effects for the intervention. Channeling reduced nursing home use—though more effectively in some classes. While reducing nursing home use, case management did *not* increase mortality. Hospital use by community residents increased in both interventions. The responses of the six classes suggest that greater amounts of medically intensive home care might produce larger changes in outcome. The reduction of differences in service use between case management and financial control sites suggested that similar case management effects occurred in both sets of sites and in areas with different levels of services.

8.3.2. An Examination of Differences in Active Life Expectancy between an Experimental and a Control Group

Next we analyze differences in duration-weighted outcomes between persons in the standard elderly Medicare fee-for-service (FFS) population and those who entered Social/HMOs (capitated organizations where benefits to include LTC services). In this analysis we extend the evaluation to temporal changes in the average duration of active (i.e., disability free) life in the experimental (S/HMO) and control (FFS) groups. The analysis (Manton et al., 1994c) is enhanced over that of two earlier studies of enrollment bias and disenrollment and mortality, because changes in health were assessed over three years using a supplementary form (the Client Assessment Form, CAF)—baseline health status was assessed using the health screening form (the HSF).

Three features of the analysis are of interest. One was that the CAF was delivered when health status changed. Thus, getting a CAF indicates health change. Hence, measurement and health changes were correlated in complex ways that required the modification of the event history model to a continuous time approximation (e.g., Chapter 6). Second, we used the active life expectancy (ALE) measures, calculated from a two-component dynamic model described in Chapter 7, as the standard for comparison. That is, the basic active life expectancy model presented in Chapter 7, is used here to make a comparison of the health dynamics in an experimental intervention, with the health dynamics under baseline conditions. Finally, we illustrate how the g_{ik} can be used to eliminate the effects of left censoring in making comparisons. That is, one does not need to have ALE estimated for the total population with the correct $\bar{g}_k$ weights. The effects of composition, and hence of left censoring, can be controlled using either a common weighting for the intervention and control groups (a type of standardization) or by comparing ALE for specific classes (i.e., ALE measures for the K fuzzy classes for $g_{ik} = 1$).

The analysis employs data drawn from HSF and CAF interviews delivered to persons in all four S/HMO sites. The data are complex because the CAFs are delivered almost on a random schedule. Thus, there is seldom a set of $g_{ik \cdot t}$ defined for the same t after $t = 0$ (i.e., at baseline). Thus, the trajectory of changes must be generated from the partially (and randomly) observed processes in the experimental and control conditions.

The patterns of the λ_{kjl}^{++} estimated from the combined data sets were compared to the λ_{kjl}^{++} patterns derived from the HSF only. Most changes were a result of the higher prevalence of health problems generated on the CAFs. Mortality data were derived from Medicare administrative records as were data on S/HMO enrollment. There are approximately 10,800 enrollees in S/HMOs and 16,700 persons in the four FFS samples. About 23,000 HSFs were delivered and, for 3,234 persons, a total of 8,506 CAF interviews were conducted in 3 years of follow-up. The time between CAFs varied by individual as did the number of CAFs (up to 10 CAFs were delivered). CAFs were triggered by a re-HSF telephone contact in the community or at the initiative of a S/HMO.

The information from the CAF, when available, may cause the g_{ik} for an individual to *change*. The CAFs did not occur at fixed times. Consequently, the g_{ik} vary over nonfixed length episodes ($g_{ik \cdot e}$) with the CAF being one of the transitions defining both the end and the beginning of episodes. Other episodes defining events were death, end of study (EOS), or change between service venues (e.g., moving from FFS to HMO). To maintain a constant definition of health status, and one where case-mix scores can be compared between the FFS and S/HMO populations, the λ_{kjl}^{++} (i.e., the coefficients determining the multivariate definition of the case-mix types from J health measures) generated from the HSF and CAF data were fixed over time and service venue. In addition, we estimated, for each episode (separately for S/HMOs and FFS populations) external transition variables $\lambda_{kj(l_1 \cdot l_2)}$, where l_1 is the duration of time spent in the episode (coded into l_1 intervals) and l_2 represents the mode of termination (i.e., the "event") where one mode of termination is administration of a CAF interview (i.e., a discrete"jump" in information). Thus, the CAF both changes the $g_{ik \cdot e}$ and defines episodes through the $\lambda_{kj(l_1 \cdot l_2)}$.

To calculate ALE the unequal periods between HSF and CAF assessments are adjusted by taking the 3-year follow-up period and dividing it into 36 monthly intervals. If, between two months, no assessment (e.g., CAF) occurs, the $g_{ik \cdot t}$ at time 1 are applied at time 2. If there is an intervening CAF, then the $g_{ik \cdot t}$ at $t = 2$ are altered to reflect the new information. This defined $36 \times 27{,}500 \sim 1.0$ million monthly intervals during which a change in $g_{ik \cdot t}$ or death could occur. The λ_{kjl}^{++} for the six classes derived from 30 health and functional measures common to the CAF and HSF instruments are presented in Table 8.7 (Manton et al., 1994c) along with estimates of the λ_{kjl}^{++} made using only the HSFs from the baseline assessment (Manton et al., 1994b).

A comparison of the marginal frequencies shows that the HSF/CAF data have a higher frequency of most functional impairments and medical conditions; e.g., on the HSF only 1.4% of the population was bedfast

Table 8.7. λ_{kjl} Values for 30 Health and Functioning Measures for S/HMO Demonstration Projects, 1984–1989, for HSF* and HSF/CAF**

	Frequency %		Class											
			1		2		3		4		5		6	
	HSF	HSF/CAF	HSF	HSF/CAF	HSF	HSF/CAF	HSF	HSF/CAF	HSF	HSF/CAF	HSF	HSF/CAF	HSF	HSF/CAF
Needs Help With:														
1. Preparing meals	7.6	27.1	0.0	0.0	0.0	0.0	100.0	100.0	0.0	100.0	0.0	0.0	100.0	100.0
2. Laundry	9.8	19.9	0.0	0.0	0.0	0.0	100.0	100.0	0.0	0.0	0.0	0.0	100.0	100.0
3. Light housework	9.4	16.3	0.0	0.0	0.0	0.0	100.0	100.0	0.0	0.0	0.0	0.0	100.0	100.0
4. Grocery shopping	13.3	15.8	0.0	0.0	0.0	0.0	100.0	100.0	0.0	0.0	0.0	0.0	100.0	100.0
5. Managing money	6.8	21.2	0.0	0.0	0.0	0.0	100.0	100.0	0.0	42.1	0.0	0.0	98.9	100.0
6. Taking medicine	4.9	14.8	0.0	0.0	0.0	0.0	100.0	100.0	0.0	0.0	0.0	0.0	71.1	100.0
7. Making phone calls	4.4	11.4	0.0	0.0	0.0	0.0	96.1	86.5	0.0	0.0	0.0	0.0	41.7	100.0
8. Eating	0.9	7.9	0.0	0.0	0.0	0.0	0.0	0.0	0.0	0.0	0.0	0.0	40.9	100.0
9. Getting in/out of chairs or bed	2.3	22.2	0.0	0.0	0.0	0.0	0.0	0.0	0.0	100.0	0.0	0.0	100.0	100.0
10. Walking around inside	4.0	15.7	0.0	0.0	0.0	0.0	0.0	0.0	0.0	0.0	0.0	0.0	100.0	100.0
11. Driving/using public transportation	14.3	32.3	0.0	0.0	0.0	0.0	100.0	100.0	0.0	100.0	0.0	0.0	100.0	89.2
12. Toileting	1.7	20.3	0.0	0.0	0.0	0.0	0.0	0.0	0.0	100.0	0.0	0.0	100.0	75.9
13. Dressing	2.6	16.2	0.0	0.0	0.0	0.0	0.0	0.0	0.0	100.0	0.0	0.0	100.0	100.0
14. Bathing	4.2	20.9	0.0	0.0	0.0	0.0	0.0	0.0	0.0	100.0	0.0	0.0	100.0	100.0
15. Uses a wheelchair or walker	6.8	6.7	0.0	0.0	0.0	0.0	0.0	38.5	0.0	0.0	0.0	0.0	100.0	46.2
16. Uses a cane	11.3	18.0	0.0	0.0	0.0	100.0	48.5	0.0	87.8	0.0	0.0	0.0	68.8	0.0
17. Is bedfast	1.4	13.2	0.0	0.0	0.0	0.0	0.0	0.0	0.0	42.2	0.0	0.0	47.4	100.0

Medical Conditions														
18. Diabetes mellitus	8.9	17.6	0.0	100.0	0.0	100.0	0.0	100.0	100.0	0.0	0.0	100.0	10.5	0.0
19. Hypertension	37.1	31.7	0.0	27.6	100.0	100.0	0.0	29.7	100.0	0.0	0.0	100.0	0.0	0.0
20. Heart trouble	21.1	18.6	0.0	0.0	100.0	0.0	0.0	0.0	0.0	0.0	0.0	100.0	34.4	0.0
21. Neurological problems	7.0	11.2	0.0	0.0	0.0	100.0	73.8	0.0	0.0	0.0	0.0	0.0	37.8	40.8
22. Stroke	4.2	17.0	0.0	0.0	0.0	100.0	0.0	0.0	30.0	0.0	0.0	0.0	65.5	47.4
23. Lung or breathing problems	13.0	18.9	0.0	0.0	100.0	0.0	0.0	0.0	0.0	80.8	0.0	100.0	14.6	0.0
24. Chronic cough	6.2	6.8	0.0	0.0	100.0	0.0	0.0	0.0	0.0	0.0	0.0	73.5	3.4	0.0
25. Cancer	4.0	16.3	0.0	0.0	56.0	100.0	0.0	0.0	0.0	49.8	0.0	0.0	7.5	10.0
26. Hardening of the arteries	16.7	14.6	0.0	0.0	0.0	0.0	0.0	0.0	100.0	0.0	0.0	100.0	74.1	0.0
27. Stomach/bowel problems	17.4	20.9	0.0	0.0	100.0	100.0	0.0	0.0	0.0	0.0	0.0	100.0	65.8	38.6
28. Bladder problems	11.8	15.5	0.0	0.0	100.0	0.0	0.0	0.0	0.0	44.4	0.0	100.0	67.2	0.0
29. Rheumatism or arthritis	52.1	56.7	0.0	41.0	0.0	100.0	0.0	0.0	100.0	53.5	100.0	100.0	100.0	63.9
30. Other health problems	21.0	25.2	0.0	11.6	0.0	96.5	27.3	26.4	0.0	30.8	100.0	71.6	50.1	36.5
Weighted prevalence			44.7	52.2	11.3	7.0	9.1	9.8	11.9	11.7	18.9	11.5	4.1	7.8

*HSF data are based on 30,503 cases including Health Maintenance Organizations.

**HSF/CAF data are based on 34,000 episodes derived from 27,000 S/HMO enrollees and FFS sample members.

compared to 13.2% of the monthly experience in the CAF/HSF data. For bathing the prevalence of problems increased from 4.2 to 20.9% of the monthly experience. Overall, in the baseline HSF data an average of 1.06 functional impairments (the sum of the marginal proportions of persons with a specific attribute) was observed. In the combined CAF/HSF data this average is 2.99 impairments per person, i.e., nearly three times as many as were reported on the baseline HSF alone.

Differences between the two data sets in terms of the prevalence of medical conditions were less. On the baseline HSF data there were an average of 2.21 medical conditions reported per person. In the combined HSF/CAF data, an average of 2.71 conditions were reported, i.e., 22.6% more. The higher rate of medical conditions and functional impairment reporting is not surprising since the CAF is used to capture changes in health among the elderly when most such changes are due to the accrual of conditions and impairments with age. For example, Class 1 in both analyses had no functional impairments. Class 1 differs in the *combined* CAF/HSF data set because there are a few (though not directly life threatening) medical conditions with nonzero λ_{kjl}. Thus, this class may still be characterized as "relatively healthy." Class 2 in the baseline HSF analysis had *no* functional impairments, but had significant risks of cancer (56.0%), pulmonary problems, heart trouble, and hypertension. Thus, it appeared to be an acutely ill group. This was confirmed in a separate analysis when it manifested high levels of service needs. In the combined HSF/CAF analysis there was one functional impairment (uses a cane), but a number of medical conditions are characteristic of the type. The risk of cancer increased from 56 to 100%. Heart trouble, however, was no longer involved in Class 2 in the combined HSF/CAF data. Diabetes was found in Class 2 in both analyses. In the combined HSF/CAF analyses there was a risk of stroke or neurological problems in Class 2, which was not found in the Class 2 generated from the HSF only. This suggests that the Class 2 from the HSF/CAF data, though still medically "acute," may be somewhat older than the class formed from only the HSF data.

Class 3 in the baseline analysis had IADL impairments and neurological problems suggesting early dementia. In the combined HSF/CAF analysis there is a similar pattern of IADL impairments with a few risk factors. Thus, the new Class 3 is similar to that from the prior analysis. Therefore, we expect it to be close to the same age and to have high service needs. Class 4 in the HSF-only analysis had little impairment, but several medical conditions (e.g., diabetes, hypertension, stroke (30%), atherosclerosis, rheumatism). In the combined HSF/CAF there are more functional limitations, including ADL impairments and a significant probability of being bedfast. The primary medical conditions for this class shifted from circulatory disease and stroke to pulmonary problems (80.8%) and cancer (49.8%). Thus, Class 4 in the HSF/CAF analysis has been partially redefined to have a high degree of functional impairment, pulmonary problems and cancer.

Class 5 in the HSF-only analysis had no functional impairments and only one medical condition (rheumatism) though the class was old relative to Class 1. In the HSF/CAF analysis, Class 5 still does not have functional impairment, but has more medical problems—especially cardiopulmonary conditions (but no stroke) and atherosclerosis. The higher frequency of medical conditions in the HSF/CAF data shifted cardiopulmonary problems from the old Class 4 to the new Class 5. This accommodated the higher frequency of cancer in the new Class 4 generated by increases in cancer's marginal prevalence (i.e., from 4.0 to 16.3%). Thus, clients in this class should be older than clients in Class 5 in the HSF-only analyses.

Class 6 in the HSF analysis is highly impaired and frail. This is mirrored in the new analysis with even more impairments (e.g., at the baseline 47.4% were bedfast) in the new analysis, Class 6 is 100% bedfast. Though functional impairment is greater in the combined analysis there are fewer medical problems. Stroke and neurological problems were still associated with the class, as was a small amount of cancer (10%) and stomach and bowel problems.

In separate analyses we found Class 1 was still relatively young with a mean age of 71.7 (compared to 71.0 in the baseline analysis). Class 2 had a higher mean age (77.5) than in the baseline analysis (72.2). Class 3 has a lower mean age of 75.8 (compared to the prior Class 3 mean age of 78.8), as did Class 4 (80.5 years versus the prior analysis of 81.1 years). In Class 5 the mean age was 80.1 years (76.2). Class 6, as before, was still the oldest group, 83.2 years—though significantly younger than the mean of 89 years for Class 6 in the baseline analysis. Thus, Classes 5 and 2 are significantly older than in the baseline study with the age differentiation between classes less than before. This is consistent with the fact that the CAF *updates* health *over* time so that health is less strongly related to age than in the analysis of the baseline HSF only. The overall sample is older by a year due to the fact that CAF assessments are necessarily made after the baseline HSF.

In a secondary life table analysis we studied the length of time between CAFs for case-mix groups in both the S/HMO and FFS groups. The CAF was given more rapidly in the FFS population (e.g., 227 days compared to 270 days in the S/HMOs) and was seldom administered to Classes 1 and 2. It was frequently administered to Classes 3, 4, and 6. The proportion of persons who had a CAF was higher in the S/HMO than in FFS.

We created life tables from the differential equations in Chapter 7 where the effects of the $g_{ik \cdot t}$ are included (i.e., the CAF is used to update health over the 3 years of follow-up). In those analyses we compared FFS versus S/HMO for males and females separately and for the total population. In Table 8.8 we examine the quadratic hazard functions for the total population.

The diagonal elements are mortality risks for a person of the Kth class where the Gompertz term which multiplies the coefficient is evaluated at age 75 (i.e., a coefficient is $b^*e^{\theta \cdot 75}$). The upper-diagonal terms represent the

Table 8.8. Hazard Functions for FFS and S/HMO for the Total Population

	1	2	3	4	5	6
1 FFS	0.023	0.039	0.040	0.046	0.050	0.065
S/HMO	0.021	0.034	0.044	0.041	0.034	0.093
2 FF		0.067	0.068	0.077	0.085	0.111
S/HMO		0.053	0.070	0.065	0.055	0.147
3 FFS			0.069	0.078	0.086	0.112
S/HMO			0.092	0.086	0.072	0.194
4 FFS				0.089	0.099	0.128
S/HMO				0.080	0.067	0.181
5 FFS					0.109	0.142
S/HMO					0.056	0.151
6 FFS						0.184
S/HMO						0.408

contribution to mortality of a person with "partial" scores on multiple classes. Mortality is most different for Classes 5 (with FFS higher) and 6 (with S/HMO mortality higher). Overall the mortality in the S/HMO is better predicted with a relative risk of 19.4 to 1 for Class 6 versus Class 1. This may reflect closer monitoring of the S/HMO population. The χ^2 for both models were highly significant. The θ estimated for the hazard functions are small (i.e., about 3%). In standard hazard analyses of populations, the θ value is 8 to 9% (Manton et al., 1986a). This suggests that the g_{ik} strongly predict mortality and that age-related unobserved variables have small residual effects. The standard errors for all coefficients were significant beyond $p = .0001$; i.e., coefficients were estimated with high precision.

In Table 8.9 we present the separate hazard function for males and females. We find the same general patterns for the coefficients as in the total populations with higher risks in the S/HMO for class 6 and for males, Class 3. The sex-specific hazard functions are highly significant, both overall and for each coefficient, and the relative risks show a higher relative risk for both the male and S/HMO populations. The θ again are quite low, i.e., 2.7% and 3.8% (NLTCS; Manton et al., 1994c).

For females, we find large differences between the FFS and S/HMO population for Classes 5 and 6. The hazard functions are again significant both overall and for individual coefficients. The relative risks between Classes 6 and 1 are larger for females (FFS = 8.94; S/HMO = 20.01) with the θ values being 3.4 and 3.0%. The level of mortality is lower for females.

These tables are used with estimates of disability transitions to generate life tables where survival is independent on functional change. These life tables are illustrated for the total population and males and females in Table 8.10. Life expectancy (at selected ages) is higher, and remains higher, to at least age 95, for the FFS population. Though life expectancy is higher for

Table 8.9. Hazard Function for FFS and H/HMO for Males and Females

	1		2		3		4		5		6	
	Male	Female	Male	Female	Male	Female	Male	Female	Male	Female	Male	Female
1 FFS	0.032	0.017	0.057	0.028	0.050	0.032	0.063	0.035	0.066	0.040	0.086	0.051
S/HMO	0.029	0.015	0.043	0.027	0.065	0.032	0.063	0.031	0.050	0.025	0.129	0.072
2 FFS			0.099	0.047	0.088	0.054	0.059	0.115	0.066	0.150	0.085	
S/HMO			0.062	0.046	0.096	0054	0.092	0.054	0.074	0.042	0.189	0.124
3 FFS					0.079	0.061	0.098	0.067	0.103	0.075	0.133	0.097
S/HMO					0.147	0.064	0.142	0.064	0.114	0.050	0.290	0.146
4 FFS							0.122	0.073	0.128	0.082	0.166	0.106
S/HMO						0.137	0.063	0.109	0.049	0.280	0.144	
5 FFS									0.134	0.092	0.174	0.118
S/HMO									0.87	0.039	0.223	0.113
6 FFS											0.226	0.152
S/HMO											0.571	0.333

Males		Females	
FFS χ^2	= 587.6	FFS χ^2	= 1,334.3
S/HMO χ^2	= 643.9	S/HMO χ^2	= 736.8
FFS Dead	= 1,060	FFS Dead	= 1,573
S/HMO Dead	= 664	S/HMO Dead	= 665
FFS θ	= 0.0271	FFS θ	= 0.0343
S/HMO θ	= 0.0383	S/HMO θ	= 0.0302
FFS RR	= 7.06	FFS RR	= 8.94
S/HMO RR	= 19.69	S/HMO RR	= 20.81

Table 8.10. Life Expectancy and Active Life Expectancy by Age and FFS, S/HMO for the Total and Male and Female Populations

Age	e_x	Active e_x	Class ($\bar{g}_{k \cdot t}$) 1	2	3	4	5	6
			Total					
65 FFS	18.60	14.82	0.797	0.020	0.015	0.049	0.106	0.014
S/HMO	16.82	14.78	0.879	0.026	0.014	0.006	0.068	0.007
75 FFS	13.60	11.04	0.812	0.024	0.030	0.045	0.070	0.019
S/HMO	10.94	8.72	0.797	0.026	0.020	0.059	0.073	0.024
85 FFS	8.44	5.67	0.672	0.023	0.070	0.152	0.047	0.035
S/HMO	6.09	3.80	0.624	0.035	0.038	0.161	0.076	0.067
95 FFS	4.37	1.64	0.377	0.067	0.054	0.181	0.102	0.219
S/HMO	3.02	1.25	0.413	0.074	0.052	0.212	0.064	0.185
			Males					
65 FFS	15.18	10.74	0.708	0.016	0.008	0.094	0.152	0.022
S/HMO	14.88	13.20	0.888	0.026	0.010	0.010	0.060	0.007
75 FFS	11.34	8.53	0.753	0.032	0.040	0.050	0.110	0.010
S/HMO	9.59	7.98	0.832	0.028	0.015	0.040	0.060	0.025
85 FFS	7.07	4.62	0.654	0.029	0.077	0.143	0.063	0.035
S/HMO	5.59	4.10	0.734	0.040	0.023	0.095	0.055	0.054
95 FFS	3.59	1.45	0.403	0.086	0.056	0.182	0.102	0.171
S.HMO	2.74	1.75	0.639	0.073	0.021	0.043	0.066	0.158
			Females					
65 FFS	21.39	18.65	0.872	0.023	0.020	0.011	0.066	0.007
S/HMO	18.38	16.06	0.874	0.028	0.016	0.003	0.073	0.006
75 FFS	15.13	13.04	0.862	0.016	0.022	0.040	0.042	0.019
S/HMO	11.89	9.22	0.776	0.023	0.021	0.075	0.083	0.022
85 FFS	9.16	6.33	0.691	0.019	0.061	0.159	0.036	0.035
S/HMO	6.49	3.63	0.560	0.031	0.048	0.198	0.087	0.076
95 FFS	4.62	1.69	0.365	0.060	0.052	0.191	0.085	0.247
S/HMO	3.44	1.13	0.328	0.074	0.069	0.267	0.050	0.204

FFS, ALE is the same in the two groups at age 65. At age 75 and beyond ALE is absolutely higher in the FFS.

Males in the two groups have similar life expectancies though FFS is slightly (0.3 years) higher. ALE is higher for the S/HMO population at age 65. Interestingly, changes in the $\bar{g}_{k \cdot t}$ are small between ages 65 to 95. This is reflected in higher $\bar{g}_{k \cdot t}$ values for most classes.

Females have a higher life expectancy in the FFS population. There is nearly a 3.0-year advantage at age 65 for those in the FFS. This value (21.4) is similar to the life expectancy value obtained in analyses of the 1982–1984–

1989 NLTCS (e.g., female life expectancy was 20.9 years; Manton et al., 1994d) through it is higher than in standard U.S. period or cross-sectional life tables. This is because the institutional population is excluded from the FFS samples and because more extremely elderly persons are recorded in Medicare data. In contrast to males, ALE is higher for females in the FFS at all ages because the difference in Class 1 is less for males.

To assess change over age the proportional changes are presented in Table 8.11. In the total population, both expectancy and ALE decrease more rapidly in the S/HMO with more rapid changes at advanced ages. This pattern holds for both sexes. The rate of decline is more rapid at later ages. Likewise, declines in total and ALE are greater for the S/HMO.

We conducted two simulations with these life tables to isolate the effects of differences in mortality from differences in disability change. In the first, all persons start in a given class. Then the classes disperse over the other classes according to the estimated transition matrices and diffusion. Second, we started each person in a given class and forced them to remain in that class. This is done by changing parameters in the state dynamic equations and the diffusion matrices. These calculations tells us about the effect of diffusion and regression to the mean independent of mortality. The second simulation represents the effect of mortality only; i.e., disability transitions are eliminated.

In Table 8.12 we provide a S/HMO/FFS comparison for males females

Table 8.11. Percent Changes in Life Expectancy and Active Life Expectancy for FFS and S/HMO, by Sex

Age	FFS	S/HMO	FFS	S/HMO
		Total		
65–74	73.1	65.0	74.5	59.0
75–84	62.1	55.7	51.4	43.6
85–94	51.8	49.6	28.9	32.9
65–94	23.5	18.0	11.1	8.5
		Males		
65–74	74.7	64.5	79.4	60.5
75–84	62.3	58.3	54.2	51.4
85–94	50.8	49.0	31.4	42.7
65–94	23.7	18.4	13.5	13.3
		Females		
65–74	70.7	64.7	69.9	57.4
75–84	60.5	54.6	48.5	39.4
85–94	50.4	53.1	26.7	31.1
65–94	21.6	18.7	9.1	7.0

Table 8.12. Comparison of FFS and S/HMO Values for Classes 1, 2, and 6, for Males and Females

Age	e_x		l_x		Class 1		Class 2		Class 3		Class 4		Class 5		Class 6	
	Males	Females	Males	Females	Males	Females	Males	Females	Males	Females	Males	Females	Males	Females	Males	Females
Class 1																
65 FFS	17.91	22.68	(11.0)	(1.0)	1.00	1.00	0	0	0	0	0	0	0	0	0	0
S/HMO	15.63	19.09	(1.0)	(1.0)	1.00	1.00	0	0	0	0	0	0	0	0	0	0
75 FFS	12.45	15.65	(73.9)	(85.4)	0.943	0.941	0.011	0.007	0.008	0.008	0.018	0.023	0.015	0.011	0.006	0.010
S/HMO	9.80	12.14	(71.0)	(81.7)	0.883	0.845	0.022	0.016	0.011	0.014	0.035	0.063	0.029	0.044	0.022	0.018
85 FFS	7.37	9.31	(44.2)	(62.9)	0.752	0.734	0.022	0.016	0.050	0.047	0.124	0.146	0.025	0.026	0.028	0.032
S/HMO	5.62	6.52	(32.6)	(49.2)	0.745	0.576	0.038	0.030	0.022	0.045	0.094	0.194	0.049	0.080	0.053	0.075
95 FFS	3.61	4.64	(12.8)	(27.3)	0.424	0.374	0.085	0.059	0.054	0.051	0.182	0.191	0.084	0.081	0.170	0.244
S/HMO	2.74	3.44	(5.1)	(10.5)	0.639	0.329	0.073	0.074	0.021	0.069	0.043	0.267	0.066	0.058	0.158	0.204
Class 2																
65 FFS	11.49	13.63	(1.0)	(1.0)	0	0	1.00	1.00	0	0	0	0	0	0	0	0
S/HMO	10.29	12.44	(1.0)	(1.0)	0	0	1.00	1.00	0	0	0	0	0	0	0	0
75 FFS	9.90	12.10	(46.2)	(52.5	0.473	0.386	0.067	0.084	0.149	0.101	0.099	0.161	0.177	0.187	0.035	0.081
S/HMO	8.29	10.13	(41.0)	(50.1)	0.569	0.363	0.148	0.158	0.038	0.061	0.096	0.178	0.094	0.182	0.056	0.059
85 FFS	6.73	8.34	(20.4)	(28.6)	0.537	0.465	0.036	0.035	0.122	0.136	0.165	0.232	0.096	0.081	0.044	0.051
S/HMO	5.48	6.32	(14.5)	(23.4)	0.698	0.476	0.048	0.039	0.026	0.062	0.102	0.224	0.069	0.116	0.058	0.084
95 FFS	3.56	4.52	(5.0)	(10.4)	0.380	0.320	0.087	0.064	0.057	0.058	0.182	0.193	0.119	0.104	0.174	0.261
S/HMO	2.73	3.43	(2.2)	(4.8)	0.637	0.325	0.073	0.075	0.021	0.070	0.043	0.267	0.067	0.060	0.159	0.204
Class 6																
65 FFS	8.43	13.75	(1.0)	(1.0)	0	0	0	0	0	0	0	0	0	0	1.00	1.00
S/HMO	4.77	7.49	(1.0)	(1.0)	0	0	0	0	0	0	0	0	0	0	1.00	1.00
75 FFS	9.29	12.47	(31.7)	(52.8)	0.373	0.438	0.081	0.087	0.116	0.104	0.144	0.162	0.241	0.127	0.046	0.083
S/HMO	8.43	10.57	(15.5)	(26.6)	0.579	0.476	0.085	0.066	0.033	0.060	0.088	0.155	0.164	0.181	0.051	0.062
85 FFS	6.60	8.43	(12.8)	(29.9)	0.495	0.489	0.039	0.034	0.125	0.131	0.177	0.225	0.119	0.071	0.046	0.050
S/HMO	5.48	6.37	(5.6)	(13.2)	0.691	0.498	0.046	0.036	0.026	0.054	0.101	0.216	0.078	0.111	0.058	0.081
95 FFS	3.35	4.53	(3.0)	(11.0)	0.370	0.325	0.087	0.064	0.058	0.058	0.182	0.193	0.128	0.101	0.175	0.259
S/HMO	2.73	3.43	(0.9)	(2.7)	0.636	0.326	0.073	0.074	0.021	0.070	0.043	0.267	0.068	0.060	0.159	0.204

of Classes 1, 2, and 6 (i.e., the healthy, the acutely ill, and the frail classes). For persons starting in the healthy class in the FFS population, there is a greater persistence (even though the S/HMO enrolls relatively more healthy persons at the start; Manton et al., 1994a). This leads to a greater life expectancy for those starting in Class 1 at all ages. For Class 2 life expectancy is lower than in Class 1 with the FFS population having a slight advantage. Interestingly, for those starting in Class 2, few remain in that class by age 75. Significant numbers move to the healthy Class 1. Indeed the prevalence for Class 2 (i.e., $\bar{g}_{k \cdot t}$) increases at advanced ages because the relative rate of "in" transitions surpasses "exits" at advanced ages.

For Class 6 the FFS population has a higher life expectancy at age 65. By age 75 the population originally in Class 6 has largely moved to other states. Thus, life expectancy for the S/HMO increases by age 75 because 47.6% have moved to Class 1. In addition, mortality selection has occurred with only 26.6% of the original Class 6 surviving to age 75. Female life expectancy in Table 8.12 is higher for Class 1 in the S/HMOs. Life expectancy is also higher in the S/HMOs for Classes 2, 4, and 5 (3 is not very different). Thus, for these classes the S/HMOs do a better job preserving life expectancy. Class 6 females, in contrast, have better survival in the FFS population. In general, S/HMOs do better in managing cases unless they are frail. The advantage that the FFS population has is that transition rates to morbid states are smaller with more transitions to healthy states.

FFS Class 1 males have a life expectancy advantage due, at least initially, to a higher retention rate in Class 1. In Class 2 there is a higher transition to Class 1 for the S/HMO population. The life expectancy for FFS remains higher because transitions to other favorable states (e.g., Class 5) are higher. Class 6 shows a similar pattern with more transitions to Class 1 but with high early mortality for Class 6 (15.5% survive to age 75 in S/HMO and 31.7% in FFS).

Table 8.13 presents class-specific mortality over age for males and females with no transitions permitted out of the class. The advantage for S/HMO female clients in Class 1 is manifest at all ages. For males there is a small advantage at age 65. There is a clear advantage for S/HMO clients in Classes 2 and 5 for both males and females. The FFS population is advantaged in Classes 3 (primarily for females), 4 (for males only), and 6. Thus, there are strong sex differentials in disability transitions and mortality between the S/HMO and FFS.

8.4. SUMMARY

We examined how data sets may be combined using fuzzy set statistical models. We illustrated procedures with examples from several surveys and LTC demonstrations projects. In those examples, we used a commonly

Table 8.13. Age-Specific Life Expectancy If Persons Remained in Classes, for Males and Females

			Class											
	e_x		1		2		3		4		5		6	
Age	Males	Females	Males	Females	Males	Females	Males	Females	Males	Females	Males	Females	Males	Females
65 FFS	15.18	21.39	23.58	33.43	10.28	17.69	12.28	14.76	8.65	12.95	7.99	10.92	5.09	7.35
S/HMO	14.88	18.38	23.73	37.09	14.49	18.19	7.62	14.33	8.08	14.59	11.43	20.61	2.35	3.66
75 FFS	11.37	15.18	19.67	27.85	8.23	13.94	9.90	11.49	6.88	10.00	6.34	8.35	3.99	5.51
S/HMO	9.63	11.89	18.80	31.70	11.01	14.68	5.57	11.40	5.92	11.63	8.55	16.76	1.65	2.77
85 FFS	7.11	9.20	16.32	23.02	6.53	10.82	7.91	8.81	5.44	7.61	4.99	6.30	3.11	4.09
S/HMO	5.59	6.49	14.63	27.02	8.22	11.72	4.00	8.98	4.27	9.17	6.28	13.50	1.14	2.09
95 FFS	3.61	4.65	13.53	19.03	5.16	8.30	6.29	6.68	4.27	5.72	3.91	4.70	2.41	3.00
S/HMO	2.75	3.47	11.24	23.16	6.03	9.29	2.84	7.01	3.04	7.16	4.54	10.82	0.79	1.57

defined state space (i.e., λ_{kjl}^{++}) and the fact that each person's position in that state space was uniquely determined to control problems in mixing data sets of different types and dealing with deficiencies in the sample space due to either nonresponse or to systematic choice of an experimental condition.

The principles are applicable to more than surveys and demonstration projects. One area of particular importance is the development of clinical trials to study the effects of therapies. The strategy used most often as the "gold standard" in evaluating clinical trials is whether or not a significant difference in response rates, cure rates, or survival time is demonstrated between groups where individual heterogeneity is dealt with by randomization. Though used for many clinical purposes, there are areas where the application of RCTs is difficult. One is the study of the efficacy of chemotherapy in pre- and postmenopausal women—especially those in a node-negative state. Three reviews of the use of chemotherapy and hormonal therapy (Fisher et al., 1989; Henderson et al., 1991; Early Breast Cancer Trialist, 1992a,b) ended up, even after reviewing 30 years worth of trials, with ambiguous conclusions. The problem is that there are so many parameters characterizing the presentation of disease and modalities of therapy that heterogeneity is manifest in the structure of the different studies. There are some parameters of treatment which evolve too rapidly for RCTs to be conducted fast enough to keep pace. In the Fisher et al. (1989) review there were so many changes in chemotherapy (e.g., early single agent trials, multiple drug therapy using methotrexate as a base, different lengths and durations of therapy) that overall conclusions about the effects of optimal therapy were not possible. Furthermore, what may be the current "gold standard" for therapy, an anthracycline-containing multidrug regime (e.g., CAF or FAC) was not a significant part of the review. As disease states become more complex in the elderly, the problems of RCTs become more complicated.

As a consequence, we think eventually RCTs will prove too cumbersome to deal with the complexity of chronic disease states and rapidly evolving therapies. It is in this context that the principle of an individual as his "own clinical trial" will have to be developed and formalized. This in essence says that even in therapies with high rates of success (e.g., 80%) individual factors will modify responses so that to optimize treatment for the patient his physiological course over time has to be evaluated. In effect, one goes from a static comparison of fixed-time treatments to an evaluation of how therapy effects multiple parameters of the patient's physiological homeostasis. We can cite two recent examples of such treatment. AIDS is a lethal disease which, as of yet, cannot be cured. However, by prophylactic management of comorbid states and the control of the virus by AZT (and now DDI and DDC), some patients have extended survival. Likewise, in high-intensity chemotherapy one must balance the effects of therapy (e.g., bone marrow suppression) with the effects of disease.

In this context, the fuzzy set model, with the individual being scored on very large number of measures, is relevant. We have used the procedure successfully, not only in psychiatric diagnosis, but, for example, in examining the efficiency among different classes of stage I melanoma patients receiving immunoprophylatic therapy (Manton et al., 1991c). In that study, we were able to identify the likely classes of patients to respond (e.g., response declined with age) from a detailed evaluation. Obviously, the utility of such fuzzy set models for clinical evaluations (i.e., either in formal diagnostic or in evaluating treatment outcomes over individuals) has only begun to be exploited. We feel that the need will increase as understanding of disease processes on a genetic and molecular basis proceeds and the number of physiological parameters relevant to describing the clinically relevant state of a person increases.

REFERENCES

Applebaum, R., Brown, R., and Kemper, P. (1986). Evaluation of the National Long Term Care Demonstration: An analysis of site specific results. Channeling Evaluation Supplementary Report 85-29, November 1985, Revised May 1986. DHHS Contract No. HHS-100-80-0157.

Berkman, B., Foster, L. W. S., and Campion, E. (1989). Failure to thrive: Paradigm for the frail elder. *The Gerontological Society of America* **29**: 654–659.

Brown, R. S. (1986). Methodological issues in the evaluation of the National Long Care Demonstration. Channeling Evaluation Technical Report No. TR-86B-01. Mathematica Policy Research Inc., Princeton, New Jersey.

Cassel, C.-M., Sarndal, C.-E., and Wretman, J. H. (1977). *Foundations of Inference in Survey Sampling*. Wiley, New York.

Chiang, C.L. (1968). *Introduction to Stochastic Processes in Biostatistics*. Wiley, New York.

Corder, L. S., and Manton, K. G. (1991). National surveys and the health and functioning of the elderly: The effects of design and content. *Journal of the American Statistical Association* **86**: 513–525.

Dempster, A. P., Laird, N. M., and Rubin, D. B. (1977). Maximum likelihood from incomplete data via the EM algorithm (with discussion). *Journal of the Royal Statistical Society 39 Series B* **1**: 1–38.

Early Breast Cancer Trialists' Collaboraive Group (1992a). Systematic treatment of early breast cancer by hormonal, cytotoxic, or immune therapy—Part I. *The Lancet* **339**: 1–15.

Early Breast Cancer Trialist's Collaborative Group (1992b). Systematic treatment of early breast cancer by hormonal, cytotoxic, or immune therapy—Part II. *The Lancet* **339**: 71–85.

Eaton, W. W., McCutcheon, A., Dryman, A., and Sorenson, A. (1989). Latent class analysis of anxiety and depression. *Sociological Methods & Research* **18**: 104–125.

Eisenbeis, R. A., and Avery, R. B. (1972). *Discriminant Analysis and Classification Procedures: Theory and Application*. D.C. Heath, Lexington, Massachusetts.

Fienberg, S. E. (1989). Modeling considerations: Discussion from a modeling perspective. In *Panel Surveys* (D. Kasprzyk, G. Duncan, G. Kalton, and M. P. Singh, Eds.). Wiley, New York, pp. 566–574.

Fisher, B., Costantino, J., Redmond, C., et al. (1989). Randomized clinical trial evaluating tamoxifen in the treatment of patients with node-negative breast cancer who have estrogen-positive tumors. *New England Journal of Medicine* **320**: 479–484.

Gibbins, F. J., Lee, M., Davidson, P. R., et al. (1984). Augmented home nursing as an alternative to hospital care for chronic elderly invalids. *British Medical Journal (Clinical Research Ed.)* **284**: 330–333.

Goldsmith, H. F., Jackson, D. J., Doenhoefer, S., Johnson, W., Tweed, D. L., Stiles, D., Barbano, J. P., and Warheit, G. (1984). *The Health Demographic Profile System's Inventory of Small Area Social Indicators*. National Institute on Mental Health, Series BN, No. 4. DHHS Pub. No. (ADM) 84-1354. USGPO, Washington, D.C.

Grannemann, T., Applebaum, R., Grossman, J. B., and Stephens, S. (1986). Report to Congress on identifying individuals at risk of institutionalization. Mathematica Policy Research, Inc., Princeton, New Jersey, December.

Greene, V. L., Lovely, M. M., Miller, M., and Ondrich, J. I. (1989). Reducing nursing home use through community based long term care: An optimization analysis. Final Report for Grant No. 87ASPE184A, Metropolitan Studies Program and the All-University Gerontological Center, Maxwell School of Citizenship and Public Affairs, Syracuse University, Syracuse, New York, October.

Headen, A. E. (1992). Time cost and informal social support, determinants of black/white input differences in long-term care production. *Inquiry* **29**, 440–450.

Heckman, J., Hotz, J., Walker, J. (1985). New evidence on the timing and spacing of births. *American Economic Review* **75**: 179–184.

Henderson, B. E., Paganini-Hill, A., and Ross, R. K. (1991). Decreased mortality in users of estrogen replacement therapy. *Archives of Internal Medicine* **151**: 75–78.

Hoem, J. (1985). Weighting, misclassification and other issues in the analysis of survey samples of life histories. In *Longitudinal Analysis of Labor Market Data* (J. Heckman, and B. Singer, Eds.). Cambridge University Press, Cambridge, Massachusetts, pp. 249–293.

Hoem, J. M. (1989) The issue of weights in panel surveys of individual behavior. In *Panel Surveys* (D. Kasprzyk, G. Duncan, G. Dalton, and M. P. Singh, Eds.). Wiley, New York, pp. 539–559.

Kalbfleisch, J. D., and Prentice, R. L. (1980). *The Statistical Analysis of Failure Time Data*. Wiley, New York.

Manton, K. G. (1987). The linkage of health status changes and work ability. *Comprehensive Gerontology* **1**: 16–24.

Manton, K. G. (1988). A longitudinal study of functional changes and mortality in the United States. *Journal of Gerontology* **43**: S153–S161.

Manton, K. G. (1989). Epidemiological, demographic, and social correlates of disability among the elderly. *Milbank Quarterly* **67** (Suppl. 1 on Disability Policy: Restoring Socioeconomic Independence): 13–58.

Manton, K. G., and Soldo, B. J. (1985c). Dynamics of health changes in the oldest old: New perspective and evidence. *Milbank Memorial Fund Quarterly* **63**: 206–285.

Manton, K. G., and Stallard, E. (1990). Changes in health functioning and mortality. In *The Legacy of Longevity: Health and Health Care in Later Life* (S. Stahl, Ed.). SAGE Publications, London, pp. 140–162.

Manton, K. G., and Stallard, E. (1991). Cross-sectional estimates of active life expectancy for the U.S. elderly and oldest-old populations. *Journal of Gerontology* **48**: S170–182.

Manton, K. G., and Suzman, R. M. (1992). Conceptual issues in the design and analysis of longitudinal surveys of the health and functioning of the oldest old. Chapter 5 in *The Oldest-Old* (R. M. Suzman, D. P. Willis, and K. G. Manton, Eds.). Oxford University Press, New York, pp. 89–122.

Manton, K. G., Stallard, E., Vaupel, J. W. (1986a). Alternative models for the heterogeneity of mortality risks among the aged. *Journal of the American Statistical Association* **81**: 635–644.

Manton, K. G., Stallard, E., Woodbury, M. A., and Yashin, A. I. (1986b). Applications of the Grade of Membership technique to event history analysis: Extensions to multivariate unobserved heterogeneity. *Mathematical Modelling* **7**: 1375–1391.

Manton, K. G., Stallard, E., Woodbury, M. A., and Yashin, A. I. (1987). Grade of Membership techniques for studying complex event history processes with unobserved covariates. In *Sociological Methodology, 1987* (C. Clogg, Ed.). Jossey-Bass, San Francisco, pp. 309–346.

Manton, K. G., Vertrees, J. C., and Woodbury, M. A. (1990). Functionally and medically defined subgroups of nursing home populations. *HCF Review* **12**: 47–62.

Manton, K. G., Stallard, E., and Woodbury, M. A. (1991a). A multivariate event history model based upon fuzzy states: Estimation from longitudinal surveys with informative nonresponse. *Journal of Official Statistics* (published by Statistics Sweden, Stockholm) **7**: 261–293.

Manton, K. G., Woodbury, M. A., and Stallard, E. (1991b). Statistical and measurement issues in assessing the welfare status of aged individuals and populations. *Journal of Econometrics* **50**: 151–181.

Manton, K. G., Woodbury, M. A., Wrigley, J. M., and Cohen, H. J. (1991c). Multivariate procedures to describe clinical staging of melanoma. *Methods of Information in Medicine* **30**: 111–116.

Manton, K. G., Corder, L. S., and Clark, R. (1993a). Estimates and projections of dementia related service expenditures. Chapter 9 in *Forecasting the Health of Elderly Populations* (K. G. Manton, B. H. Singer, and R. M. Suzman, Eds.). Springer-Verlag, New York.

Manton, K. G., Vertrees, J. C., Lowrimore, G. R., Newcomer, R., and Harrington, C. (1994a). Patterns of health and functioning in the enrollment of Social/Health Maintenance Organizations. In review at *Medical Care*.

Manton, K. G., Lowrimore, G. R., Vertrees, J. C., Newcomer, R., and Harrington,

C. (1994b). Disenrollment and mortality in Social/Health Maintenance Organizations. In review at *The Gerontologist.*

Manton, K. G., Newcomer, R., Lowrimore, G., Vertrees, J. C., and Harrington, C. (1994c). Differences in health outcomes over time between S/HMO and fee for service delivery systems. *Health Care Financing Review*, Special Issue in Managed Care, in press.

Manton, K. G., Vertrees, J. C., and Clark, R. F. (1993b) A multivariate analysis of disability and health groups, and their longitudinal change in the National Channeling Demonstration data. *The Gerontologist* **33**(5): 610–618.

Manton, K. G., Stallard, E., and Singer, B. H. (1994d). Projecting the future size and health status of the U.S. elderly population. In Studies of the Economics of Aging (D. Wise, Ed.). National Bureau of Economic Research, University of Chicago Press, forthcoming.

National Center for Health Statistics (1966). *Vital and Health Statistics Data Evaluation and Methods Research, Computer Simulation of Hospital Discharges.* Public Health Service Pub. No. 1000-Series 2, No. 13. USGPO, Washingon, D.C.

Orchard, R., and Woodbury, M. A. (1971). A missing information principle: Theory and application. In *Proceedings Sixth Berkeley Symposium on Mathematical Statistics and Probability* (L. M. LeCam, J. Neyman, and E. L. Scott, Eds.). University of California Press, Berkeley, California, pp. 697–715.

Potthoff, R. F., Manton, K. G., and Woodbury, M. A. (1993). Correcting for nonavailability bias in surveys by weighting based on number of callbacks. *Journal of the American Statistical Association* **89**: 1197–1207.

Potthoff, R. F., Woodbury, M. A., and Manton, K. G. (1992). Equivalent sample size and equivalent degrees of freedom refinements for inference using survey weights under superpopulation models. *Journal of the American Statistical Association: Theory and Methods* **87**: 383–396.

Tolley, H. D., and Manton, K. G. (1985) Assessing health care costs in the elderly. *Transactions of the Society of Actuaries* **36**: 579–603.

Tolley, H. D., and Manton, K. G. (1987). A Grade of Membership approach to event history data. *Proceedings of the 1987 Public Health Conference on Records and Statistics*. DHHS Pub. No. (PHS) 88-1214, USPGO, Hyattsville, Maryland, pp. 75–78.

U.S. Department of Health and Human Services (1988). *Health Care Financing: Program Statistics, Medicare and Medicaid Data Book, 1988.* HCFA Pub. No. 03270, Baltimore, Maryland.

Woodbury, M. A., Manton, K. G., and Corder, L. S. (1992). Representing changes over time in the Grade of Membership Model. Presented at the Symposium on Computer Simulation, Health Care Section, Society for Computer Simulation, Prague, Czechoslovakia, November 9–11.

Woodbury, M. A., Manton, K. G., and Tolley, H. D. (1994). A general model for statistical analysis using fuzzy sets: Sufficient conditions for identifiability and statistical properties. *Information Sciences*, in press.

CHAPTER 9

Areas of Further Statistical Research on Fuzzy Sets

INTRODUCTION: A REVIEW OF RESEARCH PRIORITIES ON THE STATISTICAL PROPERTIES OF FUZZY SETS

In this section we briefly review some of the properties of the GoM model and their differences from other multivariate statistical strategies. From that review, we can identify some areas of needed research on (a) the application of fuzzy set models to different study designs and types of data, (b) their operational performance in different types of data vis-a-vis other methodologies, and (c) their computational and statistical properties.

9.1. THE FUZZY SET CONCEPT: THE BASIS FOR THEORETICALLY RATHER THAN COMPUTATIONALLY LIMITED MODELS

Most standard statistical models were developed when computing power was limited and the use of assumptions to deal with distribution functions with good analytic properties was mandatory. Now the computational power available to statisticians and analysts has increased manyfold—and continues to increase. Yet there has not yet been a paradigmatic shift that fully utilizes this qualitative change in the quantitative tools and computer technology available to the statistician.

It is interesting that the change in computational power *has* been recognized by theoretical mathematicians who utilize this vastly increased computational power to either (a) directly enumerate cases (e.g., the resolution of the four-color map problem) or (b) aid in symbolic manipulation of increasingly complex series and other expressions that are too ponderous to do by hand (e.g., use of symbolic manipulation programs such as *Mathematica*).

Statisticians have utilized increased computational power by modifying methods of analysis to include large data set methods where "edges" of the data are of interest rather than simply the mean or middle of the data. To date, most of these techniques entail many observations of a low dimensional multivariate random variable with several concomitant or regressor variables. Primary techniques include the use of graphical and visual exploratory methods and density estimation methods.

Statisticians have not yet taken full advantage of the available computational power for high dimensional problems. To do so effectively would require examining and extending the foundations and traditions of current statistical theory. Indeed, the available computational power has tended more to restrict the theoretical development of statistical models as more applied statisticians become dependent on packaged software routines—even to the extent of no longer knowing the computational details of the statistical models implemented in the software package or the assumptions employed in any particular software implementation. The development of new models that take advantage of the available computational power will require skills in numerical analysis.

The fuzzy set logic provides the basis for a statistical paradigm to take better advantage of new computational resources and technologies. This paradigm eschews the parametric manipulation of specific distributional families in favor of direct manipulation of large numbers of data moments. The necessary computational power to make this type of model fully practical was not readily available even as recently as five years ago.

What is interesting is that GoM, which started as a heuristic algorithm to solve a formal diagnostic problem that was not well resolved by classical discrete Bayesian classification procedures (e.g., Woodbury, 1963), has led to a procedure whose efficacy has been demonstrated on several hundred data sets. Theoretical work on the principles of the fuzzy set model was initiated only after the utility of the analysis was extensively proven in a wide range of empirical applications. It appeared that a model that had performed so well in so many contexts under such wide-ranging conditions must be founded on sound theoretical principles.

Such a course of development is not unique to the GoM model. The original proportional hazard specification due to Cox (1972) was rationalized on a basis (i.e., that of partial likelihood) that had to be reconceptualized by subsequent investigators (e.g., Kalbfleisch and Prentice, 1973). The maximum likelihood factor analysis model has been shown to be applicable under much more general conditions than first considered (e.g., Anderson and Amemiya, 1988). If these procedures did not have broad empirical success, it seems unlikely that the considerable effort to identify and describe the theoretical bases of the models would have been expended.

In the fuzzy set models, the paradigmatic shift ironically involves a return to the direct use of "moment" representations of data structures and the manipulation of high dimensional moment spaces that makes the models semiparametric in form. This exchanges considerable computational effort in manipulating complex moment spaces to the Jth order for analytic assumptions to restrict solutions to tractable distribution functions.

This is not to suggest that detailed model development and specification does not have a place in data analysis. But its place seems to be more in forecasting and simulation than empirical and statistical analysis. Modeling based on biological or other substantively derived model structures helps deal with certain special problems that emerge in forecasting and simulation. For example, ancillary information used in model specification restricts the selection of parameter spaces to those that are more likely to reproduce the behavior of basic mechanisms even though an alternative mechanism might prove to better describe certain data (Manton and Stallard, 1991).

As with any new paradigm, considerable effort is required in its theoretical elaboration and development. Below we indicate areas of research, both applied and theoretical, that need to be undertaken to more fully understand the properties of fuzzy set models of different types.

9.2. ALGORITHMIC DEVELOPMENT

The GoM model with the estimation of $I \times K - 1$, g_{ik} parameters is computationally intensive. With the availability of new supermicrocomputers it is now possible to deal with very large problems. For example, we have analyzed data sets with up to 150,000 individuals and 40 variables. In the nursing home example in Chapter 6 there were 4,525 cases, 111 variables (407 answers), and 11 fuzzy classes. The computational burden increases linearly in cases and in answers and cubically in the number of classes.

With the computational power currently available the limits to the size of problems that can be handled practically are probably less a function of the number of cases and variables (for large vital statistic data files one can use the aggregate data Poisson GoM model described in Chapter 3) than the maintenance of precision in optimizing over such a large number of parameters (i.e., numerical analysis problems affecting precision may outweigh simple computational burden).

We have changed our basic computational algorithm to one derived from a quadratic programming perspective in pursuit of precision. We have recently examined, for the empirical Bayesian model, the issues in estimation as $\lambda_{kjl} \rightarrow 0$ by modification of the so-called "Amoeba" algorithm in order to generate estimates when a zero relevance factor implies a fold in the solution space. We have developed approaches for imposing a priori constraints on the g_{ik} and the λ_{kjl} in order to evaluate multivariate hypotheses.

Thus, there is considerable capacity and flexibility with the currently employed algorithms. Nonetheless, a number of algorithmic issues remain. For example, there is a wide range of ancillary statistics that can be generated both to aid diagnostically the analysis and to further interpret features of the complex solution. Identifying which of these is most informative is a complex task because there is variability across types of analysis and data sets. One very likely outcome is that given the semiparametric nature of the GoM structure, it will be unlikely to develop into a fully "canned" software procedure. There is a considerable amount of substantive knowledge and mathematical analysis that precedes the construction of a GoM analysis and its empirical application (e.g., Berkman et al., 1989; Singer, 1989) which cannot be simply "hard wired" into an algorithm. Computer implementations may require the use of certain "artificial intelligence" features to guide applications.

9.3. ASYMPTOTIC THEORY OF FUZZY SET MODELS

We have provided theorems about a number of asymptotic properties of the GoM parameters (e.g., identifiability, consistency, sufficiency, efficiency, properties of statistical tests). There is much more that remains to be done on a theoretical level.

For example, we have demonstrated consistency for the g_{ik} in two different ways, i.e., as I increases using moment constraints on the g_{ik} distribution of order J (i.e., $H(x)$) and as J increases. The behavior of the model across many different data sets with quite different features suggests that, ultimately, a unified theory of consistency will be required where both information increases in the case and variable spaces are taken into account simultaneously.

For example, in the nursing home example in Chapter 6, where there were 111 variables and $L_j = 406$ answers, the model exhibited behavior suggesting that the g_{ik} themselves were consistent (e.g., the ability to estimate 11 classes) because the moments-based arguments for consistency would suggest the difficulty (i.e., the large number of moments implied by 11 classes) in estimating that many classes if information increased as indicated only by the moments model. This leads us to these other statistical issues that require investigation.

One is precise determination of the degrees of freedom assumed in the model. In the tests conducted in this book, we took an exceedingly conservative perspective in counting all parameters estimated. One could have discounted the degrees of freedom involved in parameters on the boundary (e.g., weight them by half). Alternatively, one might use information-based statistics for model comparisons using derivatives of the AIC statistic (e.g., Anderson, 1985). A third approach would be to use the concept of "equivalent" degrees of freedom (Blackman and Tukey, 1958; Satterthwaite, 1946) and "equivalent" sample size (e.g., Potthoff et al., 1992) to provide more precise estimates of the information consumed in the estimation of parameters in a particular solution.

Though these are three possible approaches to assessing the degrees of freedom for specific hypotheses, they all require generalization to deal with the "duality" issue more directly (i.e., of the characteristics of the moments as they are mapped from case to variable spaces). An alternative fourth approach is simply to constrain the model to a simpler form (e.g., imposing a multivariate Dirichelet distribution on the g_{ik} distribution) to, in effect, apply a semimetric smoothing operator to parameter estimates. The GoM model seems to do this already in the determination of model order where the addition of one further class requires a factorial increase in information (in the moments model of consistency). In LCM and other discrete mixture problems, the number of classes increases too rapidly because the appropriate information penalty is not extracted. Some investigators have suggested the use of an ad hoc likelihood penalty/smoothness operator in these cases (Behrman et al., 1991).

A second issue, touched upon above, is the use of smoothing conditions to restrict the parameter space. This is an extension of some of the empirical Bayesian conditions and approaches discussed in Chapter 6. There we suggested that a priori moments constraint through an assumed distribution (e.g., the Dirichelet) could be used to represent hypotheses about restricted classes of models. A mixture of Dirichelet distributions is also a possibility for dealing with cases on the boundaries.

A third issue is the question of determining if a global solution has been produced for a given set of data and model specification. We considered this problem initially from a "quasi-Bayesian" perspective by viewing the initial trial partition as a substantively informed prior initial distribution and examining how that prior information influenced the final solution achieved when the sample was limited. In any realistically sized analysis this problem or question (i.e., a local maximum) can emerge.

9.4. FUZZY SET STOCHASTIC PROCESS

We showed, in Chapter 7, how a multivariate Gaussian stochastic process might be translated into a fuzzy set formulation. The likelihood can be modified in a number of ways to generate parameter estimates for a wide range of stochastic processes. Then we showed that those parameters could be embedded in a stochastic process by modifying the diffusion variables to represent a random walk within the bounded fuzzy set simplex.

Clearly more theoretical work needs to be done on the mathematical translation of "infinite" discrete state stochastic processes to fuzzy set processes. However, the initial work appears promising in the behavior of the model process and its forecasts.

An issue of theoretical concern is the evaluation of such forecasts using multiple criteria. That is, estimating identifiable parameters from data alone can lead (Manton and Stallard, 1991) to "poor" forecasts if one (a) accepts without question the solution with the maximum likelihood value and (b) ignores the likelihood value and (b) ignores the likelihood of certain biological invariances of the age dependence of the underlying morbidity, disability, and mortality processes.

The issue is whether formal tests that appropriately weight multiple sources of information can be developed. On a practical level we have found that verifying estimates of parameters that are "external" to the general failure process, generating outcomes from ancillary data, produces restricted classes of models within which the likelihood criteria selected an optimal set. Without restrictions of the likelihood model, however, solutions can be generated that are substantively unreasonable. This is a concern not restricted to the GoM model but for which the GoM model may provide techniques with useful properties. In particular, the specification of the J variables defines the $\mathfrak{M}$ space which restricts the range of parameter estimates to comply with Condition I and Condition II. Indeed, in the most general formulations of the GoM model it can be shown that the structure of the data can always be reproduced despite uncertainty in the structural or location parameter.

9.5. EXAMINATION OF THE GEOMETRY OF THE FUZZY SET AND DISCRETE SET DESCRIPTIONS OF DATA

In the comparison of LCM and fuzzy set models we generally compared models of the lowest-order acceptable class. It was possible to generate LCMs of order M where $M \gg K$ (for the fuzzy set model). Though possibly equivalent in a given set of data this equivalence need not hold stochastically, i.e., in data set replication or forecasting (see Chapter 7). Representation of information by M sets of p_{kjl} versus the K sets of g_{ik} and λ_{kjl} in GoM produces parameter sets with very different operational characteristics.

For example, while clinically meaningful λ_{kjl} parameters in GoM could be generated by using the g_{ik} as a statistical filter to data interactions expressible only at the individual level, the p_{kjl} in LCM would tend to become less distinct and each discrete class became a less stable partition of the overall state space. This could be a useful area of investigation using computer simulation methods.

9.6. FUZZY CLASS MODEL ROBUSTNESS TO SAMPLING DEVIATION AND ANOMALIES

The GoM model has the useful property that individual g_{ik} parameters are estimated and that, in high dimensional data, these g_{ik} may be consistent. We utilized this property in discussing sample weighting and postweighting to reproduce the population parameters. Another type of "sampling" (unintentional of course) is censoring (left or right). We discussed explicit analytic approaches that can be employed to deal with censoring in a "list" sample by employing auxiliary information. Moreover, we observed a "robustness" in the GoM solution applied to the ECA data from Baltimore and North Carolina under synthetic sampling plans which arose from the marginal constraints implemented in the LCM which were not operational in the GoM solution.

We have observed a high degree of "robustness" to sample selection, missing data, and censoring in a number of analyses. More formal investigations of the problem, and investigations by simulation studies, would be useful (based on the good operating characteristics of the model in multiple types of data sets).

9.7. COMBINING DATA AND META-ANALYTIC FORMULATIONS

In chapter 8 we illustrated the use of the GoM model as a tool to combine multiple data sources. Under the assumption that, conditional on the g_{ik}, the observations are realizations of independent random variables, we found that differences in sample design, representativeness, and question type could be adjusted for under certain situations. With the increasing flow of data into the cognitive system of scientists, researchers, and policy makers, it is becoming important to establish methods of bringing together multiple sources of data to produce information efficiently. Though illustrated as a potential method for these problems, the use of the GoM model in this regard requires further development and explication.

9.8. ADVANCES IN MODELS AND MORE GENERAL FORMULATIONS

As general as the GoM model might seem there are even more general mathematical model structures of which GoM is, itself, a special case (Woodbury et al., 1993). We are in the process of investigating these structures in that they seem to lead to model structures that may be even more "fault" tolerant, i.e., they may appropriately reparameterize the data even despite misspecification of some assumptions. Furthermore, the identification of the more general geometric and stochastic properties of the GoM model may lead to resolution of some of the statistical research issues identified above.

Already this research has led to the specification of a L_2 (quadratic) specification of GoM that may have useful properties for studies of complex, nonlinear physical systems. In this instance, the bilinear form for the observable cell probabilities is a quadratic form with the matrix of λ values being in the middle position, pre- and postmultiplied by matrices of the generalization of the g_{ik} parameters.

9.9. SUMMARY

In the previous chapters we described the work produced to date on fuzzy set statistical models and demonstrated the use of those models in a wide range of data sets. The models generally manifested good operating characteristics and consistent behavior even in this broad mix of statistical and multivariate problems. It also seemed to be the "model of choice" for specific types of data (e.g., longitudinal studies of elderly populations with high attrition rates, with fixed time assessments of large numbers of discrete response variables and randomly occurring episode events). However, the model also performed well in many other data sets with different characteristics (e.g., in the analysis of data on systemic lupus erythematosus where disease was indirectly manifested, in the assessment of changes in neurological reaction times among the elderly (Clive et al., 1983), in the study of data on allergic responses (Buckley and Woodbury, 1989), and in the analysis of data on the cross cultural manifestation of schizophrenia (Manton et al., 1993). We believe the excellent operating characteristics of the model in a wide range of data provides good evidence on the soundness of the basic model foundations. This will justify the extensive theoretical work on characterization of those principles that remains to be done—work which we have begun but which will require considerable further efforts by ourselves and other investigators (e.g., Singer, 1989).

REFERENCES

Anderson, Dale N. (1985). Density model identification with the Akaike Information Criterion. Master's Thesis, Brigham Young University, Provo, Utah.

Anderson, T. W., and Amemiya, Y. (1988). The asymptotic normal distribution of estimators in factor analysis under general conditions. *Annals of Statistics* **16**: 759–771.

Behrman, J. R., Sickles, R., Taubman, P. and Yazbeck, A. (1991). Black–white mortality inequalities. *Journal of Econometrics* **50**: 183–203.

Berkman, L., Singer, B., and Manton, K. G. (1989). Black–white differences in health status and mortality among the elderly. *Demography* **26**(4): 661–678.

Blackman, R. B., and Tukey, J. W. (1958). The measurement of power spectra from the point of view of communications engineering—Part I. *Bell System Technical Journal* **37**: 185–282.

Buckley, C. E., and Woodbury, M. A. (1983). Poster Presentation at Duke Medical Center Symposium, 1989.

Clive, J., Woodbury, M. A., and Siegler, I. C. (1983). Fuzzy and crisp set-theoretic-based classification of health and disease. *Journal of Medical Systems* **7**: 317–322.

Cox, D. R. (1972). Regression models and life tables. *Journal of the Royal Statistical Society*, *Series B* **34**: 187–202.

Kalbfleisch, J. D., and Prentice, R. L. (1973). Marginal likelihoods based under Cox's regression and life table model. *Biometrika* **39**: 267–278.

Manton, K. G., and Stallard, E. (1991). Cross-sectional estimates of active life expectancy for the U.S. elderly and oldest-old populations. *Journal of Gerontology* **48**: S170–182, 1991.

Manton, K. G., Korten, A., Woodbury, M. A., Anker, M., and Jablansky, A. (1993). Symptom profiles of psychiatric disorders based on graded disease classes: An illustration using data from the WHO International Study of Schizophrenia. *Psychological Medicine*, in press.

Potthoff, R. F., Woodbury, M. A., and Manton, K. G. (1992). Equivalent sample size and equivalent degrees of freedom refinements for inference using survey weights under superpopulation models. *Journal of the American Statistical Association, Theory and Methods* **87**: 383–396.

Satterthwaite, F. E.. (1946). An approximatre distribution of estimates of variance components. *Biometrics Bulletin* **2**: 110–114.

Singer, B. H. (1989). Grade of membership representations: Concepts and problems. In *Festschreift for Samuel Karlin* (T. W. Anderson, K. B. Athreya, and D. Iglehardt, Eds.). Orlando, Florida, Academic Press, pp. 317–334.

Woodbury, M. A. (1963). Inapplicability of Bayes' Theorem to diagnosis. *Proceedings Fifth International Conference on Medical Electronics*, sponsored by the International Institute for Medical Electronics and Biological Engineering, Liege, Belgium, July 16–22.

Woodbury, M. A., Manton, K. G., and Tolley, H. D. (1993). A general model for statistical analysis using fuzzy sets and statistical properties. *Information Sciences*, in press.

Index